Development Trends of Sustainable Mobility

Development Trends of Sustainable Mobility

Guest Editors

Leonardo Caggiani
Luigi Pio Prencipe

MDPI

Basel • Beijing • Wuhan • Barcelona • Belgrade • Novi Sad • Cluj • Manchester

Guest Editors

Leonardo Caggiani
Department of Civil,
Environmental, Land, and
Building Engineering and
Chemistry
Polytechnic University of Bari
Bari
Italy

Luigi Pio Prencipe
Department of Civil,
Environmental, Land, and
Building Engineering and
Chemistry
Polytechnic University of Bari
Bari
Italy

Editorial Office
MDPI AG
Grosspeteranlage 5
4052 Basel, Switzerland

This is a reprint of the Special Issue, published open access by the journal *Sustainability* (ISSN 2071-1050), freely accessible at: https://www.mdpi.com/journal/sustainability/special_issues/2WI07HGX70.

For citation purposes, cite each article independently as indicated on the article page online and as indicated below:

Lastname, A.A.; Lastname, B.B. Article Title. *Journal Name* **Year**, *Volume Number*, Page Range.

ISBN 978-3-7258-5405-9 (Hbk)
ISBN 978-3-7258-5406-6 (PDF)
https://doi.org/10.3390/books978-3-7258-5406-6

Contents

About the Editors

Leonardo Caggiani

Leonardo Caggiani is an Associate Professor at the Department of Civil, Environmental, Land, Building Engineering and Chemistry (DICATECh) of the Polytechnic University of Bari, Italy. He was awarded a Ph.D. in Road and Transportation Systems, Land and Technological Innovation at Polytechnic University of Bari, Italy, in 2009. He received the best Ph.D. thesis award from the Italian Society of Transportation Academics in 2012 and has been rewarded, together with co-authors, with the paper award at the mobil.TUM 2016 conference. He participated in an Italian research project (PRIN2007) on container ports competitiveness in the Mediterranean; he also took part in the Green Intermodal Freight Transport European project (GIFT) of the SEE program and in the National Sustainable Mobility Center, Italy (Spoke 8). His research interests concern quantitative methods for sustainable mobility, transportation network design, and equity issues in transportation. Among others, bike-sharing systems are the subject of the important studies he has conducted.

Luigi Pio Prencipe

Luigi Pio Prencipe is an Assistant Professor at the Department of Civil, Environmental, Land, Building Engineering and Chemistry (DICATECh) of the Polytechnic University of Bari, Italy. He earned his Ph.D. with full honors in March 2021. He was recognized with a "Mention of Merit" at the 1st Edition of the Abertis Chair Italy Award for the best doctoral thesis in transport infrastructure management and was a finalist for the IEEE ITSS Italian Chapter's Best Ph.D. Dissertation Award in 2021. He has actively participated in the "National Center for Sustainable Mobility (MOST)—Spoke 8" under the National Recovery and Resilience Plan (PNRR) and the "Smart City Integrated Air Mobility Evolution (SCIAME)" project. His research interests are primarily focused on optimization models for sustainable mobility systems. This includes the design and management of shared mobility systems as well as naval transport planning, railway network design models, and urban logistics.

sustainability

Editorial

Development Trends of Sustainable Mobility

Leonardo Caggiani and Luigi Pio Prencipe *

Department of Civil, Environmental, Land, Building Engineering and Chemistry (DICATECh), Polytechnic University of Bari, Via Edoardo Orabona, 4, 70125 Bari, Italy; leonardo.caggiani@poliba.it
* Correspondence: luigipio.prencipe@poliba.it

1. Introduction

The global effort toward sustainable development has underscored the urgent need to transform urban mobility and transportation systems. These networks are essential to the vitality and functionality of modern cities, but they also present significant challenges related to environmental degradation, resource consumption, social equity, and public safety [1]. As urban populations grow and cities expand, there is an increasing need to switch to more sustainable, efficient, and resilient transportation systems. This is a key part of the UN's 2030 Agenda initiative [2]. This highlights the need for a comprehensive strategy that supports technological advancements, reevaluates existing infrastructure, optimizes logistics, and promotes a deeper understanding of users' behavior and social aspirations. The compendium of research presented in this Special Issue offers a panoramic exploration of these interconnected themes, providing a critical perspective on current advancements and the upcoming challenges in the pursuit of sustainable transportation. Achieving this sustainable shift is a complex task that requires more than just the advent of new technologies. It calls for a paradigm shift in how we plan urban spaces, integrate various transport modes, and encourage behavioral change [3]. Scenarios and robust indicator frameworks are indispensable tools in this journey, aiding in the deconstruction of real-world intricacies and the formulation of evidence-based policies. Therefore, it is crucial for academics and practitioners to constantly and critically evaluate these instruments to ensure they remain effective and relevant in shaping our urban futures. This collection of papers bravely enters this arena, offering diverse perspectives and critical insights.

2. Synopsis of Contributions

The foundation of effective mobility planning often requires an understanding of the complex relationship between urban form and movement patterns. De Vincentis et al. (2022) (Contribution 1) highlighted this with a data-driven analysis of mobility and the built environment across three Italian cities: Brescia, Catania, and Salerno. They utilized floating car data (FCD) and census information to map accessibility metrics, illustrating the effects of different urban layouts and street network characteristics on traffic flow, travel efficiency, and the daily experiences of urban residents. These insights are crucial for developing integrated transportation and land use strategies.

Through a bibliometric analysis of 2000 articles from 2018 to 2023, İnce (2025) (contribution 2) systematically mapped the emerging trends and innovative solutions in sustainable urban mobility. This comprehensive review identified dominant themes, such as smart mobility and technology integration, and highlighted emerging research advances, including the crucial links between transportation, public health, and urban walkability. The importance of sustainable modes of transportation, such as cycling, cannot be overstated. Ahmed et al. (2024) (Contribution 3) conducted a systematic review of bicycle infrastructure design principles within urban bikeability indices. They critically assessed whether current indices adequately incorporate the five core principles of safety, comfort, attractiveness, directness, and coherence. They found that, although bicycle infrastructure is a common evaluative component, the principle of "coherence" is often the least considered. This points to a vital area for improvement in planning truly bicycle-friendly cities.

The evolution of cities is coupled with the progress of the technological landscape of transportation, particularly with the rise of electromobility. The transition to electric vehicles (EVs) raises questions about infrastructure and the economy. Alrubaie et al. (2023) (Contribution 4) offered a comprehensive review of EV charging stations integrated with solar photovoltaic (PV) systems. They considered market dynamics, technical specifications, grid implications, and future challenges. They concluded that PV-grid charging holds promise for profitability and reduced carbon footprints. However, its cost-effectiveness can be constrained by PV and battery capacities, as well as the inherent intermittency of solar power. In addition, Bozhi et al. (2023) (Contribution 5) explored the new realm of wireless pavement systems based on inductive power transfer (IPT) for electric vehicles (EVs). The authors investigated the key parameters affecting pavement inductivity and conductivity, including IPT pad materials and coil dimensions. Their work highlights the significance of balancing power transfer efficiency with pavement durability through optimized geometry and material science.

The financial viability and public acceptance of such advanced charging infrastructure were further explored by Kim et al. (2023) (Contribution 6). They investigated South Korean households' willingness to pay for interactive charging stations facilitating vehicle-to-grid (V2G) systems. Their survey of 1000 respondents revealed that most households were willing to financially support V2G infrastructure, an initiative that holds promises in improving grid stability by using EVs as distributed energy storage, which is essential for integrating various renewable energy sources and promoting electric vehicle-sharing solutions [4].

Beyond electrification, the rise of autonomous and shared mobility services is poised to transform urban transit. However, their success depends significantly on user perception and acceptance. Kashani et al. (2023) (Contribution 7) investigated parents' opinions in Kerman, Iran, considering shared autonomous vehicles (SAVs) for their children's school travel. Their findings revealed a complex interplay of socioeconomic factors, past travel experiences, and psychological factors such as perceived usefulness, environmental concerns, and safety assurances, including the ability to monitor their children. Viejo et al. (2025) (Contribution 8) focused on how novel transportation systems will address various needs. In their study, 147 citizens from five European countries engaged in a co-creation dialogue, and the results revealed a collective vision for the future of autonomous mobility: electric or hydrogen-powered vehicles that significantly rely on shared services, such as autonomous buses and e-hailing pods, rather than private vehicle ownership.

The implications of transportation trends need to extend to regional development and global logistics networks. Letnik et al. (2024) (Contribution 9) offered a forward-looking perspective from the business sector on the impact of transportation trends on sustainability in the Western Balkans. Their survey of transportation and logistics experts suggests that, although trends such as alternative fuels and smart city initiatives will likely reduce emissions and accidents, these trends will also pose economic challenges due to increased operational costs and the need for significant infrastructure investments. On a broader scale, Madanat et al. (2024) (Contribution 10) examined the drivers of port competitiveness in low-, upper-, and high-income countries in the Middle East and North Africa (MENA) region. Their structural equation modeling revealed that, while national competitiveness and ease of doing business generally bolster port competitiveness, the dynamics can differ significantly with income levels. For instance, increased national competitiveness in low-income countries might divert resources from port development, thereby diminishing their maritime competitiveness if not managed strategically.

3. Sustainable Mobility Challenges: Future Trends

This collection of diverse contributions depicts a sector that is facing a profound transition. It is a field that is technologically innovative yet faces significant socioeconomic and environmental challenges. As we navigate this evolving landscape, the critical assessments, innovative proposals, and forward-thinking perspectives offered by these researchers are key to guiding us toward sustainable, resilient, equitable, inclusive, and life-enhancing transportation systems for all urban communities. Recent comprehensive reviews confirm this collective shift, highlighting critical areas ranging from modeling post-pandemic travel behavior to the realization of "X-minute" cities [5]. Nevertheless, a well-defined and definitive plan is yet to materialize.

The forthcoming era of mobility will be characterized by several interconnected and challenging frontiers. A primary challenge remains: the persistent disparity between planning and implementation. Although strategic plans such as sustainable urban mobility plans (SUMPs) exist, they consistently encounter political, financial, and institutional hurdles in the real world. Even leading examples of best practice, e.g., Copenhagen, illustrate that translating strategy into practice is an ongoing process of negotiation and adaptation. This emphasizes that even the most brilliant plan depends on its execution [6]. This implementation gap is intrinsically linked to a frequently underestimated factor: the human workforce. The transition to a digitized, sustainable transportation sector requires a parallel evolution in skills and competencies. Demographic, cultural, and socioeconomic shifts are fundamentally reshaping transportation jobs. This requires a proactive strategy for workforce development to ensure the necessary human capital for future mobility systems [7]. The digital revolution is reshaping not only how people move but also the flow of goods within urban areas. The disruptive technologies transforming passenger transport are also evident in the freight sector, where digitalization and automation ensure a redefinition of logistics, from last-mile delivery to urban consolidation centers. Therefore, a truly sustainable urban system must address the symbiotic association between personal mobility and freight transportation as an integrated whole [8]. Navigating the immense complexity of policy execution, human capital, and the convergence of passenger and freight logistics requires a new approach. The increasing dependence on simulation, optimization, and machine learning cannot be considered simply an academic pursuit; it is essential for designing, testing, and managing resilient and adaptive transportation systems in a world filled with uncertainty [9]. However, one of the most defining challenges is the inherent uncertainty about the future. The evolution of transformative concepts such as mobility as a service (MaaS) is not predetermined but contested, with multiple plausible futures ranging from closed "mobility walled gardens" dominated by a few large platforms to more open and collaborative ecosystems. Thus, exploring these potential futures through co-design and scenario-planning methodologies becomes a vital strategic imperative. These methodologies empower stakeholders to anticipate challenges and collaboratively shape desirable outcomes rather than merely reacting to change [10].

Collectively, the contributions presented in this Special Issue serve as milestone reference points, underscoring the fact that the objective is not solely the efficient and clean transportation of people and goods. The real task is to facilitate a complex, systemic transition that is technologically innovative, politically viable, and, above all, socially just. The objective is to shape urban environments that are not only smarter but also wiser, more equitable, and more profoundly human.

Author Contributions: L.C. and L.P.P. contributed to every part of the research described in this paper. All authors have read and agreed to the published version of the manuscript.

Acknowledgments: We would like to thank all the authors who contributed to this Special Issue for their professional work. We are grateful to the reviewers, whose critical support and valuable comments have significantly improved the quality of the collection. Special thanks are granted to the Editorial Assistance Office of MDPI for their support throughout the review and publication process of this Special Issue.

Conflicts of Interest: The authors declare no conflicts of interest.

List of Contributions:

1. De Vincentis, R.; Karagulian, F.; Liberto, C.; Nigro, M.; Rosati, V.; Valenti, G. A Data-Driven Approach to Analyze Mobility Patterns and the Built Environment: Evidence from Brescia, Catania, and Salerno (Italy). *Sustainability* **2022**, *14*, 14378.
2. İnce, E.C. Mapping the Path to Sustainable Urban Mobility: A Bibliometric Analysis of Global Trends and Innovations in Transportation Research. *Sustainability* **2025**, *17*, 1480.
3. Ahmed, T.; Pirdavani, A.; Wets, G.; Janssens, D. Bicycle Infrastructure Design Principles in Urban Bikeability Indices: A Systematic Review. *Sustainability* **2024**, *16*, 2545.
4. Alrubaie, A.J.; Salem, M.; Yahya, K.; Mohamed, M.; Kamarol, M. A Comprehensive Review of Electric Vehicle Charging Stations with Solar Photovoltaic System Considering Market, Technical Requirements, Network Implications, and Future Challenges. *Sustainability* **2023**, *15*, 8122.
5. Bozhi; Mohamed, M.; Gilani, V.N.M.; Amjad, A.; Majid, M.S.; Yahya, K.; Salem, M. A Review of Wireless Pavement System Based on the Inductive Power Transfer in Electric Vehicles. *Sustainability* **2023**, *15*, 14893.
6. Kim, J.-H.; Hyun, M.-K.; Yoo, S.-H. Households' Willingness to Pay for Interactive Charging Stations for Vehicle to Grid System in South Korea. *Sustainability* **2023**, *15*, 11563.
7. Kashani, M.A.; Kamyab, S.; Mamdoohi, A.R.; Sierpiński, G. What Makes Parents Consider Shared Autonomous Vehicles as a School Travel Mode? *Sustainability* **2023**, *15*, 16180.
8. Viejo, J.; Quijano, A.; Farrow, L.; Papanelopoulou, S. What Type of Self-Driving Vehicle Do Citizens Imagine? Results of a Co-Creation Dialogue Across Five European Countries. *Sustainability* **2025**, *17*, 3561.
9. Letnik, T.; Hanžič, K.; Mencinger, M.; Sever, D. Impact of Transport Trends on Sustainability in the Western Balkans: A Future-Oriented Business Sector Perspective. *Sustainability* **2024**, *16*, 272.
10. Mdanat, M.F.; Al Hur, M.; Bwaliez, O.M.; Samawi, G.A.; Khasawneh, R. Drivers of Port Competitiveness Among Low-, Upper-, and High-Income Countries. *Sustainability* **2024**, *16*, 11198.

References

1. European Commission. *Communication from the Commission to the European Parliament, the Council, the European Economic and Social Committee and the Committee of the Regions: A European Green Deal*; COM (2019) 640 Final; European Commission: Brussels, Belgium, 2019. Available online: https://eur-lex.europa.eu/legal-content/EN/TXT/?uri=CELEX:52019DC0640 (accessed on 10 June 2025).
2. IPCC. Climate Change 2022: Mitigation of Climate Change. In *Contribution of Working Group III to the Sixth Assessment Report of the Intergovernmental Panel on Climate Change*; Shukla, P.R., Skea, J., Slade, R., Al Khourdajie, A., van Diemen, R., McCollum, D., Pathak, M., Some, S., Vyas, P., Fradera, R., et al., Eds.; Chapter 10: Transport; Cambridge University Press: Cambridge, UK; New York, NY, USA, 2022. Available online: https://www.ipcc.ch/report/ar6/wg3/ (accessed on 10 June 2025).

3. Creutzig, F.; Niamir, L.; Bai, X.; Callaghan, M.; Cullen, J.; Díaz-José, J.; Figueroa, M.; Grubler, A.; Lamb, W.F.; Leip, A.; et al. Demand-side solutions to climate change mitigation consistent with high levels of well-being. *Nat. Clim. Chang.* **2022**, *12*, 36–46. [CrossRef]
4. Prencipe, L.P.; van Essen, J.T.; Caggiani, L.; Ottomanelli, M.; de Almeida Correia, G.H. A Mathematical Programming Model for Optimal Fleet Management of Electric Car-Sharing Systems with Vehicle-to-Grid Operations. *J. Clean. Prod.* **2022**, *368*, 133147. [CrossRef]
5. Chen, Y.; Li, C.; Wang, W.; Zhang, Y.; Chen, X.M. The Landscape, Trends, Challenges, and Opportunities of Sustainable Mobility and Transport. *Npj Sustain. Mobil. Transp.* **2025**, *2*, 8. [CrossRef]
6. Peters, N.V.; Reisch, L.A. Drivers of and Barriers to Sustainable Urban Mobility Plan Implementation: Lessons from Copenhagen. *Case Stud. Transp. Policy* **2025**, *20*, 101417. [CrossRef]
7. Pomoni, M.; Laiou, A.; Plati, C.; Yannis, G.; Loukea, M.; Bekiaris, E. Future Trends in Transport Workforce Based on Demographic, Behavioural, Cultural and Socioeconomic Factors. *Transp. Res. Procedia* **2020**, *48*, 2811–2820. [CrossRef]
8. Dong, C.; Akram, A.; Andersson, D.; Arnäs, P.O.; Stefansson, G. The Impact of Emerging and Disruptive Technologies on Freight Transportation in the Digital Era: Current State and Future Trends. *Int. J. Logist. Manag.* **2021**, *32*, 386–412. [CrossRef]
9. de la Torre, R.; Corlu, C.G.; Faulin, J.; Onggo, B.S.; Juan, A.A. Simulation, Optimization, and Machine Learning in Sustainable Transportation Systems: Models and Applications. *Sustainability* **2021**, *13*, 1551. [CrossRef]
10. Caiati, V.; Rasouli, S.; López, H. Exploring the Future of Mobility as a Service (MaaS): A Co-Design Approach to Scenario Planning in European Cities. *J. Urban Mobil.* **2025**, *8*, 100133. [CrossRef]

sustainability

MDPI

Article

What Type of Self-Driving Vehicle Do Citizens Imagine? Results of a Co-Creation Dialogue Across Five European Countries

Jonatan Viejo [1,*], Ana Quijano [1,*], Lucy Farrow [2] and Selini Papanelopoulou [2]

[1] CARTIF Technology Centre, Parque Tecnológico Boecillo, 205, 47151 Boecillo, Spain
[2] Thinks Insight and Strategy, Somerset House, Strand, London WC2R 1LA, UK; lfarrow@thinksinsight.com (L.F.); spapanelopoulou@thinksinsight.com (S.P.)
* Correspondence: jonvie@cartif.es (J.V.); anaqui@cartif.es (A.Q.)

Abstract: It is believed that the deployment of autonomous vehicles in Europe has the potential to achieve safer, more sustainable and more equitable mobility. This study provides new insight into the hopes and expectations of citizens across five European countries for autonomous transport systems via a qualitative co-creation activity. A diverse and inclusive sample of 147 citizens was invited to generate their own ideas about how self-driving transport could be usefully deployed. Participants were asked to describe, in detail, what the vehicle would be like, what kind of mobility service it would provide, and who the target users would be. Structured qualitative analysis of the answers identified 337 distinct scenarios for autonomous vehicles across seven archetypes. In this paper, we describe a number of features of autonomous services that citizens expect and analyze the differences between demographic groups. We found that citizens across all five countries share a vision of autonomous mobility as electric- or hydrogen-powered, with a much greater use of shared mobility than is currently the case. This understanding of the expectations of a diverse group of citizens offers valuable insights for industry and policymaker actors to implement regarding future technology and transport investment and policy and service development.

Keywords: autonomous vehicles; sustainable mobility; mobility needs; technology

1. Introduction

With transport contributing around 5% to the Gross Domestic Product (GDP) and employing more than 10 million people, this sector is crucial for Europe's economy [1]. Transport is also responsible for a quarter of European greenhouse gas emissions, making it critical to the goal of a 90% reduction needed to achieve the commitment of Europe to becoming climate-neutral by 2050 [2].

Multiple strategies have been adopted to ensure the competitiveness of the European transport industry at the same time that it is reshaping towards more sustainable, safe, and efficient mobility [3,4]. In this context, autonomous vehicles (AVs) have emerged as a new trend that, it is argued, can improve quality of life, increase the safety of roads, and provide more sustainable and inclusive mobility services [5].

Technology is evolving rapidly, and AVs are starting to transition from demonstration pilots to commercial deployments. This complex socio-technological transformation of the transport system must be attentive to questions regarding both social and technical systems to secure equitable public benefits [6,7]. In this regard, expert stakeholders have been involved in key projects funded by the European Commission to design future transport services enabled by automatization and vehicle functionalities through workshops aimed at reaching consensus among participants [8–10]. In terms of public engagement, a wide number of studies in recent years have explored the awareness, attitude, concerns, willingness, and expectations of citizens toward this type of vehicle, as well as the most desirable characteristics of driverless mobility. Studies include diverse methods (i.e., surveys, interviews, citizen dialogues, focus groups) applied to different types of samples in different continents [11–14]. There are also a number of studies that synthesize insights from user preferences and public acceptance of AVs [15–17] and factors that influence travel mode choice and travel behavior [18,19]. These studies typically take the form of presenting options or choices to individuals, describing scenarios generated by the researchers, and frequently reaching different conclusions about how individuals will choose to travel when self-driving vehicles become available.

Beyond these studies of individual preferences and responses to expert-generated scenarios, there is a gap in understanding of what diverse and inclusive groups of citizens imagine and expect from mobility services provided by autonomous vehicles when the scenarios themselves are generated by individuals. Co-creation is an increasingly widely used tool in transport development and research, as evidenced by the interest of European actors like Civitas [20] and their increasing prevalence in transport research, including on the topic of autonomous vehicles [21–24]. Pappers et al. [25] defined co-creation in transport research as a form of public participation involving creative methods and innovation, with high levels of citizen empowerment or control. As operationalized here, co-creation can provide insights into how individuals envisage the future, using open questions and prompts to allow research participants to lead the discussion.

The study reported here is part of the Horizon project Move2CCAM [26], which uses multiple methods of inquiry to understand and model the impacts of AVs. This paper reports on data collected from a diverse and inclusive sample of 147 citizens from five European countries through the deployment of a co-creation activity [27] inspired partly by [28], which demonstrated its potential to be replicable. It reports on the types of autonomous vehicles and services that citizens imagine could meet their transport needs in the short and medium term. It investigates the expectations of citizens within a range of demographic criteria and across five geographically distinct locations. In line with socio-technological transition framing, we propose that an in-depth understanding of the hopes and expectations of citizens should be the basis of the action taken by industry transport operators and policymakers to develop and deploy automated mobility solutions that meet genuine public needs so that they are widely accepted, and therefore, deliver on the potential benefits of autonomous vehicles.

2. Materials and Methods

The Move2CCAM project invites citizens across Europe to participate in the co-creation of a shared understanding of the role and impact of AVs in future transport systems. This concept was operationalized through the recruitment of a diverse cohort of citizens and organizations who took part in multiple co-creative activities over a period of 2 years. As co-creation methods are less standardized than other qualitative approaches [28], we provide a detailed account of our procedures for reference.

This paper reports on the findings of the first co-creation activity of the Move2CCAM project, which aimed to understand citizens' hopes and ideas for how AVs could be deployed via a semi-structured online discussion. This format was chosen because it allows for the formulation of novel ideas while providing significant flexibility to pursue themes that emerge during the course of discussions. A tailored research protocol was defined for the creation of use cases by the Move2CCAM partners, which comprised a social research company with experience in co-creation processes (Think Insights), the research entity in charge of the analysis of the insights compiled (CARTIF), and the technical coordinator of the project (UCL). These partners collaborated to identify a sampling strategy for participation and design the co-creation activity and the analytical approach. A detailed manual for the co-creation event with instructions and recommendations to guide the moderators was used to support collaborative discussions among participants and consistency between the different locations.

2.1. Sample Characteristics

Participants were recruited from five European countries with different features, which were Germany (DE), Spain (SP), France (FR), Cyprus (CY), and the United Kingdom (UK). Stratified sampling was used to recruit public participants and achieve a sample that was inclusive of people with a range of characteristics relevant to mobility needs, in particular those known to be under-represented in existing research on autonomous transport. The recruitment criteria were as follows:

- Age: To address different factors that contribute to shaping each generation's travel behavior, a range of participants was recruited, with 25% of participants aged 18 to 34, 25% over 65, and the remainder of adults between 35 and 64.
- Gender: An even split of men and women was recruited, noting that existing evidence shows variation in attitudes towards autonomous vehicles.
- Health Conditions: Greater accessibility for those with conditions affecting mobility has been claimed as one of the potential benefits of autonomous vehicles; this study, therefore, included 20% of participants reporting a disability or long-term health condition that limited their mobility, alongside any carers.
- Place of Residence: The mobility needs of urban and rural residents are quite different, and, to date, most autonomous services have been trialed in urban areas. Participants were, therefore, recruited living in rural, suburban, and urban settings.
- Driving Capacity: Increased mobility for those who are unable to drive is another stated benefit of autonomous vehicles; therefore, this study included at least 10% of participants who do not have a license or are prevented from driving by a health condition.
- Income: Information was recorded about income level (relative to national averages) for analysis purposes but not as a recruitment criterion.

Table 1 provides the demographic features of 147 participants from the study.

Table 1. Demographic features of participants involved in the co-creation activity.

Features	Categories	# Participants
Age	18–34	32
	35–64	68
	65+	31
	Unknown	16
Gender	Male	65
	Female	66
	Unknown	16
Health conditions	Yes, additional needs needed	20
	Yes, carer for someone with additional needs	9
	No additional needs	82
	Unknown	36
Place of residence [1]	City center	17
	City periphery	36
	Town	28
	Village < 2000 inhabitants	17
	Unknown	9
Driving capacity	Have a license and can drive	105
	Have a license but cannot drive	11
	No license	15
	Unknown	16
Income [2]	High	43
	Medium	37
	Low	33
	Unknown	34
	Total	147

[1] Participants from Germany were classified into different subgroups: urban areas, towns, and villages. [2] Participants from Cyprus did not provide income data. #: Number of.

Table 2 reflects the income ranges in each participant's country.

Table 2. Income ranges for citizens involved in the co-creation activity.

	DE, FR (EUR)	SP (EUR)	UK (EUR)
Low	<24,999	<27,999	<25,999
Medium	25,000–49,999	28,000–41,999	26,000–49,999
High	>50,000	>42,000	>50,000

2.2. Co-Creation Activity

Participation was via an online engagement platform called Recollective, which allowed for asynchronous participation and a mix of individual and group activities (e.g., during the reflection phase). All activities were delivered in the local language spoken by the majority, and participants were supported by a member of the research team who provided technical support and moderated the online discussion, encouraging the completion of responses and seeking clarification where responses were ambiguous. Data were collected over the course of five days during March 2023 for a total of one hour and a half of engagement per participant.

The co-creation activity involved the following three phases:

- Setting the scene aimed to establish a common understanding of the basic concepts under discussion and to introduce the co-creation process. The activity started by asking participants about their lived experiences of transport by exploring the mode that they used for daily journeys, the challenges faced, the type of people most likely to suffer from transport constraints in the area where they live, and the improvements needed. Then, since there was no requirement for participants to have any pre-existing interest in or knowledge of AVs, the concept of autonomous vehicles was introduced with a short video on how an autonomous car operates without a human and a brief overview of different vehicle types under development. Finally, to support the objective of hearing from people with a broad range of experiences and characteristics, we introduced ten proto-persons that presented different mobility needs, desires, travel behavior, and socio-economic features.
- Idea generation: the core of the activity required participants to define potential mobility services that could be provided by autonomous vehicles in the area where they lived. Participants were able to generate multiple ideas and detailed descriptions of vehicle design and operation, as well as the kind of mobility service that should be provided. Participants were given a series of open-text questions to respond to as prompts, for example, what would this/these vehicle (s) be? What kind of mobility service (s) will it/they provide? What will these services allow you to do that you cannot do now? What kind of a person might need or use this service?
- Reflection: the final phase asked participants to review and comment on the ideas generated by their peers, for example, to expand their descriptions or provide some initial feedback on the perceived usefulness of the concepts generated.

2.3. Data Management and Analysis

Robust data security measures were defined to ensure that the compilation of personal data from the recruitment process and during the co-creation event complied with the General Data Protection Regulation (GDPR) as well as to ensure the integrity of the data. These practices consisted of an informed consent form collected from all participants ahead of the co-creation activity, the use of a password to access raw data compiled in the recruitment questionnaire, the application of pseudonymization techniques to replace contact details, and the storage of pseudonymized data in the Move2CCAM consortium SharePoint and in a data warehouse from the project.

Once the data collection process concluded, the dataset acquired from the recruitment questionnaire and the online platform was unified in an Excel file through the ID number. An iterative analysis process was used to identify a set of variables that defined the different mobility services suggested for autonomous vehicles (e.g., the type of vehicle, trip purpose, frequency, target user, etc.). Data were coded independently by two researchers to reduce the risk of bias, with researchers then comparing notes to ensure a common understanding of the ideas provided by the participants. The researchers ensured each participant had a full set of responses to ensure that all contextual information provided was considered (e.g., where an idea was suggested in phase 2 and developed in phase 3); however, a small number of ideas were insufficiently clear, and were not included in the analysis.

To enable a comparison of co-created ideas between demographic groups, we further analyzed the coded data to group similar concepts into archetypes, by which we mean broad categories of vehicle and service types. A comparison of the ideas generated across demographic groups was performed at the European level and not at the country level to maintain sufficient sample sizes for comparison. More details on the Materials and Methods used can be found in [27].

3. Results

3.1. Analysis of Co-Created Ideas

Through the data analysis process described above, 337 scenarios for autonomous vehicles were identified. These scenarios could imply widely used services (e.g., a public vehicle covering a generic route) or specific services (e.g., a vehicle adapted to a particular type of traveler and serving a specific location or for a specific purpose).

The key differential variables considered for the definition of scenarios were the following:

- Vehicle typology;
- Vehicle size;
- Energy source;
- Target users;
- Trip purpose;
- Locations served;
- Distances traveled;
- Frequency;
- Ownership;
- Safety measures;
- Accessibility features;
- Methods of payment.

By grouping the most similar ideas together, seven archetypes were identified, which encompassed 80% of the ideas generated, with the remaining 20% describing a diversity of use cases, which were suggested by fewer than eight participants each.

The archetypes identified are described below.

1. Passenger transport use cases
    - An autonomous e-hailing shared pod refers to an autonomous vehicle with a capacity for between two and eight passengers that covers short and medium distances (i.e., up to 30 km) and provides a door-to-door service shared between different users. Participants typically described this as an on-demand mobility service hired by users through an app operated by a private organization. This case is similar to current car-sharing services but without a driver.
    - An autonomous private car refers to a small vehicle whose owner can use it according to their mobility needs. This is just like a current private car but without the need for a driver.
    - An autonomous bus consists of a vehicle with a capacity for 10–60 passengers that provides a service with a fixed or flexible timetable and routes. Suggestions provided by participants included a variety of distances traveled, destinations served, and trip purposes and encompassed public and private operators.
2. Freight transport use cases
    - Delivery robots are small-sized autonomous surface vehicles that transport freight over short distances, for example, packages delivered to homes. Participants often suggested vehicles operating outside of peak hours and specific examples like delivering medicines or deliveries within large facilities like a factory.

- Delivery by vans includes medium-sized autonomous vehicles to transport small, medium, and sporadically large freight to medium distances (i.e., up to 15 km) working on demand and operated by a private organization.
- Delivery drones consist of small aerial autonomous vehicles delivering small freight over short distances and operating on-demand. The type of freight to transport includes food, medicines, and small packages, and the area of coverage could be large or part of a facility, such as factories.
- Autonomous long-distance trucks refer to large-sized autonomous vehicles that transport medium and large goods over long distances through optimized supply chains to ensure timely and reliable freight transportation. Examples provided by participants include all kinds of trucks delivering freight by road, including platooning trucks.

3. Other.
    - This group involves types of vehicles or services suggested by only a small number of participants (less than eight). Many of the ideas collected here involve autonomous vehicles for specific functions such as emergencies, farming, construction, military use, garbage trucks, and platooning pods. This category includes both passenger and freight vehicles.

3.2. The Future That Citizens Imagine Is Dominated by Shared, Electric, Passenger Vehicles

Exploring the number of times a particular archetype was suggested by participants in the co-creation process, as per Figure 1, reveals how widespread expectations are for these use cases among participants.

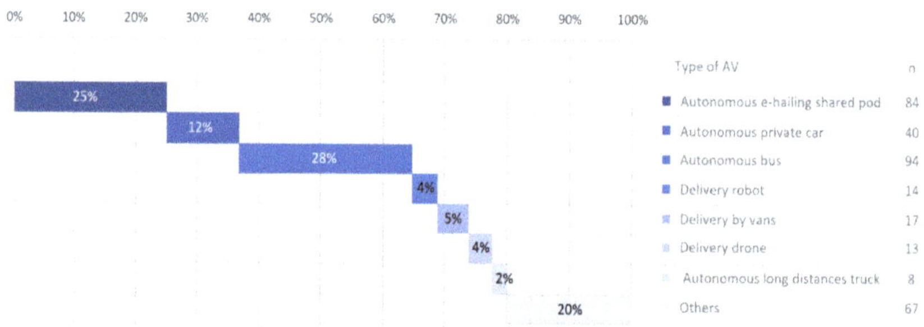

Figure 1. AV solutions co-created.

- Passenger transport was the most co-created solution. Of the 337 ideas generated, 218 were for passenger services (over two-thirds) compared with 52 for freight transport. There was a high degree of consistency in the use cases proposed for passenger transport, with autonomous buses and e-hailing shared pods being by far the most common.
- Freight solutions were much less frequently suggested. When participants mentioned freight, suggestions were much more variable, covering drones, robots, vans, and long-distance vehicles, among others.
- Shared mobility was the norm as, of the 218 passenger services proposed, the large majority were shared (buses or pods) rather than private vehicles, which were suggested just 40 times.

- AVs were assumed to be electric, with the exception of long-distance freight transport, which was considered powered by hydrogen, and participants consistently reported that vehicles would be electric.
- The archetypes were largely differentiated by vehicle type and size, the distance traveled, and ownership. Other variables like target users, methods of payment, and frequency were highly variable within each use case.

Appendix A includes more information about the number of AV co-created by each type of participant.

3.3. *Indicative Analysis of Co-Created Ideas by Demographic Group*

To analyze the relationship between demographic characteristics and perceptions of AVs, the demographic profiles of those suggesting passenger and freight vehicles were compared with the sample as a whole. Six demographic characteristics were included in this analysis, which correspond with the following:

- Socio-demographic aspects that cover age, gender, and income.
- Mobility needs and behavior associated with health conditions, driving capacity, and place of residence.

Given the small sample size of the co-creation study, these findings are indicative; however, they do offer novel insights into how different demographic groups may imagine the future. Charts have been provided to visualize the reported differences; however, these do not represent statistical analysis.

3.3.1. Who Is Imagining Autonomous Passenger Vehicles?

Looking across the three main categories of passenger vehicles (e-hailing shared pods, private cars, and buses), ideas were suggested by participants with a range of demographic criteria, largely consistent with the sample as a whole, as shown in Figure 2. These use cases were suggested frequently by participants in our study, indicating they are widely known and are seen to have broad applicability across age and gender boundaries in a range of urban and rural settings in all countries included in the sample and for those with a range of mobility needs.

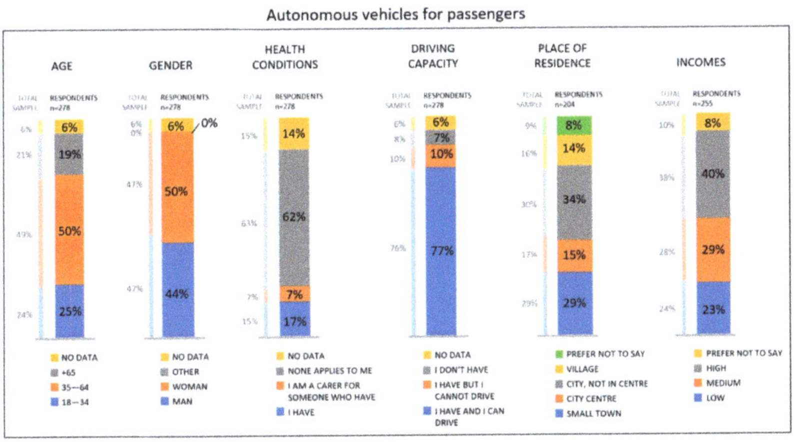

Figure 2. Demographics of participants suggesting any passenger vehicle.

Looking at specific use cases, minor differences occur in the patterns of users suggesting different approaches. For example, e-hailing services are slightly more frequently suggested by people in middle- or low-income categories; however, across other categories, including urban/rural locations, there are no strong trends. As with the broader category of passenger vehicles, the e-hailing shared pod is part of the future envisaged by a wide range of people in different demographic categories and mobility situations.

For private autonomous cars, there were slightly more suggestions from younger people (who were less likely to hold driver's licenses), those in higher income brackets (who are more likely to own a private car currently), and those in suburban and town settings (where public transport may be less available). These indicative findings suggest that the personal relevance of transport mode does have some influence on the futures people imagine, but not overwhelmingly so. Similar patterns were identified in the suggestions for autonomous buses, which are suggested to be used slightly more by women, those with health problems, those in small towns and suburbs, and all groups who may particularly benefit from improved bus services.

3.3.2. Who Is Imagining Autonomous Freight Vehicles?

As noted above, fewer citizens made suggestions related to freight transport uses, suggesting that they are a less well-established part of the shared imaginary of self-driving vehicles. Across the four main categories of freight use cases (delivery robots, drones, vans, and long-distance trucks), these were more frequently suggested by male participants, in contrast to the even split of men and women, suggesting passenger use cases (as shown in Figure 3). This finding may simply reflect the greater likelihood of male participants working in the male-dominated logistics industry, giving freight uses greater salience, rather than women who might not value freight services becoming autonomous.

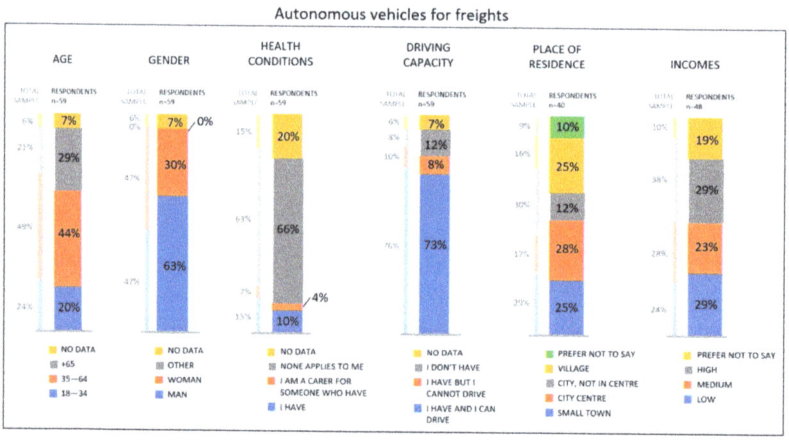

Figure 3. Demographics of participants suggesting any freight vehicle.

4. Discussion

The findings presented here represent a novel source of insight into public expectations, hopes, and aspirations for autonomous vehicles and services. It brings together data from five EU countries, a more diverse and heterogenous sample of citizens than is typical, and presents their spontaneous visions for the future of self-driving vehicles rather than their reactions to expert-derived scenarios. Here, the significance of these findings, particularly in relation to other data sources on public attitudes, is explored.

4.1. There Is a Consistent Shared Vision for Autonomous Passenger Transport Across Diverse European Countries

Analysis shows that in all five locations, and for a wide and diverse selection of citizens, automation is seen to offer useful transport solutions for passengers. Citizens generated over 300 different ideas for how automated vehicles and services could be deployed, of which two-thirds fell into three broad archetypes of passenger transport. Autonomous buses and e-hailing pods, in particular, are novel modes that appear to be part of a broadly shared understanding of the future of autonomous vehicles. This understanding seems to span beyond those who see themselves using these services immediately. In contrast with stated preference surveys, which find variation in their willingness to use autonomous vehicles among different demographic groups, our study found that both shared and autonomous services are envisaged by a wide range of individuals.

In contrast, there was little consensus among participants about the future of autonomous freight services, with many fewer and more varied suggestions made. The implications of this absence of consensus may be positive—leaving space for positive disruption—or negative, suggesting a number of individuals who are less accepting of new technological solutions that feel unfamiliar.

4.2. Citizens' Imagined Transport Systems with Automation Encompassing Electric and Shared Mobility

Looking across all demographic groups, co-creation data showed a shared vision of a transport system that is very different from today. Firstly, citizens imagined that all autonomous transport would use sustainable fuel—either electric or hydrogen—showing how widely accepted the move to sustainable fuels is across Europe. Secondly, there are many more suggestions for shared mobility services than for private vehicles, which is a reversal of current transport patterns where private cars account for the majority of trips across Europe. These findings suggest a greater willingness from citizens to adopt shared services than studies assessing the instrumental value of travel time or the affective and symbolic value of vehicle ownership (as per a recent literature review).

A useful future development would be to explore more explicitly how the expectation of participants, as expressed in the co-creation setting, may differ from their likely behavior—providing further explanation of the differences between studies such as ours and others that challenge the assumption of greater electric and shared mobility on the basis of stated preference studies.

4.3. How Citizens Imagine This Transport System Operating

Moving forward from the specification of the vehicle or service to the descriptions citizens used, the analysis of AV services provided by participants indicates how they see this new transport system operating. While participants in the co-creation activity did mention the value of reallocating time spent driving, they more frequently drew out other benefits over current transport systems. Participants commonly described new services as more frequent, more convenient, or more flexible than existing transport options, either because of extended schedules, door-to-door pickups, or demands for responsive scheduling. Many of the ideas co-created included reference to specific groups who would benefit from greater access to mobility, particularly those with disabilities, health conditions, or age-related issues that prevent them from driving. Participants often described autonomous services as more cost-effective than private transport, although perhaps more expensive than current public transport options, which is the price of greater convenience and flexibility.

These findings suggest that automated vehicles are largely valued for the benefits they can bring above and beyond the lack of a driver. This should inform the design of services and policy to support the autonomous vehicle rollout, which should seek to maximize wider benefits. For vehicle designers and manufacturers, these findings confirm the importance of accessibility if the desired goal of greater inclusion is to be met.

4.4. Who Is Imagining Which Future: The Risks of Replicating Inequalities

While our analysis of demographic differences is limited by the small sample size of the study, the high level of diversity and careful stratification relative to many larger convenience sample studies does allow us to draw out indicative findings. Looking across the use cases suggested by participants, it is notable that passenger transport use cases (e-hailing, private vehicles, and buses) are suggested by men and women, by people in a range of income brackets, with and without health conditions, and in a range of urban and rural settings across the five European countries of our study. This strongly suggests that these use cases are widely known, accepted, and seen as applicable in diverse settings.

However, there are differences that seem to correlate with current trends. For example, autonomous private cars were suggested more frequently by people in the high-income category and those without health problems, characteristics of which are both correlated with each other (e.g., people living with ill health and disabilities are disproportionately likely to live in poverty) and negatively correlated with car ownership. This effect is replicated with the autonomous bus service, which was suggested more often by women and people with health problems who are less likely to drive. This illustrates the potential for AV services to replicate existing patterns of transport usage, which in many cases exacerbates inequality—without active intervention to increase accessibility, there is a strong likelihood of replicating existing problems.

There are also some observable differences according to the place of residence—with both autonomous private cars, and autonomous buses suggested more often by those living in suburban and town settings, where public transport is typically less prevalent. This speaks to a perceived opportunity for autonomous services to fill existing transport gaps, adding value over and above the potential benefits of switching from existing modes.

4.5. Implications for Future Research

This study covers several regions of different European countries and includes the collaboration of a small but inclusive sample of citizenships. Working on this scale and using the open-ended and creative co-creation method allowed us to obtain a more detailed description of the mobility services imagined by citizens. It stimulated thinking both among research participants and researchers by bringing together diverse perspectives, leading to richer research findings. We encourage others to continue using co-creation activities to explore societal transformations like the move to autonomous vehicles for these reasons. They add a valuable and distinct perspective compared to more instrumentally focused studies such as stated preference surveys and other forms of opinion research.

However, future studies with a higher number of participants are required to enable a more robust characterization of the profile of citizens reporting each of the scenarios co-created. Future research on these directions could also explore more directly the difference between the desires and expectations of citizens in relation to autonomous vehicles.

5. Conclusions

With this pan-European co-creation activity, we set out to understand the kinds of futures citizens envisage for autonomous vehicles. In contrast with the existing literature on preferences and acceptability, it did not prescribe to participants what services were possible but invited them to describe their own visions of the future. We provide a detailed account of this methodology and encourage its wider use to further broaden our understanding of citizens' imagined futures.

This study found a high level of consistency across all five EU countries when it comes to the role of autonomous vehicles in passenger transport. Participants readily identified autonomous buses, e-hailing pods, and, to a lesser extent, private autonomous cars as key components of a future transport system despite their individual demographic characteristics. This suggests a base level of acceptance of these modes, even among those who are unlikely to be early adopters.

Going beyond vehicle type, we also found that autonomous vehicles were consistently imagined to be both electric and shared. While electrification is broadly seen as a fixed component of a future transport system in Europe, debates continue in research and policy about whether a shift to shared mobility is plausible. This finding suggests that the concept of shared mobility has entered the public consciousness and is strongly associated with autonomous vehicles, providing a basis from which modal shifts may emerge.

Finally, while we found a high degree of consistency in some elements of the way the public envisages autonomous vehicles, other aspects are less well defined—such as freight transport—which is likely to hinder public acceptance. We also identified indicative trends for individuals who envision futures that replicate current transport patterns, which in many cases are both unsustainable and inequitable. We encourage transport planners and autonomous vehicle developers to consider carefully how choices at this formative stage may reproduce or challenge these inequalities.

Author Contributions: Conceptualization, A.Q., J.V., L.F. and S.P.; methodology, A.Q. and S.P.; investigation, A.Q., L.F. and S.P.; formal analysis, J.V. and L.F., data curation and visualization, J.V.; writing—original draft preparation, A.Q., J.V. and L.F; writing—review and editing, L.F., A.Q. and J.V.; validation, A.Q. and L.F.; supervision, A.Q. All authors have read and agreed to the published version of the manuscript.

Funding: This research was funded by the EU-funded project Move2CCAM under the Horizon Europe program with grant agreement no. 101069852. Views and opinions expressed are however those of the authors only and do not necessarily reflect those of the European Union or CINEA. Neither the European Union nor CINEA can be held responsible for them.

Institutional Review Board Statement: According to the General Data Protection Regulation (GDPR), this study does not require additional ethical approval as it is a non-interventional study involving semi-structured online discussions. The project made an ethical self-statement that was positively appraised by the European Commission with no further specific approval necessary.

Informed Consent Statement: Informed consent was obtained from all subjects involved in this study.

Data Availability Statement: The data presented in this study are available upon request from the corresponding author.

Acknowledgments: The authors would like to thank EC and UKRI for funding and supporting the Move2CCAM project under GA# 101069852. Secondly, the authors would like to thank the rest of the consortium for their contributions to the definition of the methodology and for facilitating the data collection.

Conflicts of Interest: The authors declare no conflicts of interest.

Abbreviations

The following abbreviations are used in this manuscript:

AV	Autonomous Vehicle
CY	Cyprus
DE	Germany
FR	France
GDP	Gross Domestic Product
GDPR	General Data Protection Regulation
SP	Spain
UK	United Kingdom

Appendix A

This section includes the demographic features of participants involved in the co-creation activity and the number of autonomous vehicles co-created per each of the categories and the related percentages.

Table A1. AV co-created by each type of participant.

Features	Categories	# Participants	#AVs Co-Created	AVs Co-Created (%)
Age	18–34	32	35	24%
	35–64	68	72	49%
	65+	31	31	21%
	Unknown	16	9	6%
Gender	Male	65	158	47%
	Female	66	158	47%
	Unknown	16	21	6%
Health conditions	Yes, additional needs required	20	52	15%
	Yes, carer for someone with additional needs	9	22	7%
	No additional needs	82	211	63%
	Unknown	36	52	15%
Place of residence	City center	17	41	17%
	City periphery	36	74	30%
	Small town	28	70	29%
	Village < 2000 inhabitants	17	38	16%
	Unknown	9	21	9%
Driving capacity	Have a license and can drive	105	257	76%
	Have a license but cannot drive	11	33	10%
	No license	15	26	8%
	Unknown	16	21	6%
Income	High	43	115	38%
	Medium	37	85	28%
	Low	33	74	24%
	Unknown	34	29	10%
	Total	147	337	-

References

1. European Commission. Communication from the Commission to the European Parliament, the Council, the European Economic and Social Committee and the Committee of the Regions. *Sustainable and Smart Mobility Strategy-Putting European Transport on track for the Future. COM/2020/789.* Available online: https://eur-lex.europa.eu/legal-content/EN/TXT/?uri=CELEX:52020DC0789 (accessed on 20 December 2024).
2. European Commission. Communication from the Commission to the European Parliament, the European Council, the Council, the European Economic and Social Committee and the Committee of the Regions. *The European Green Deal. COM/2019/640.* Available online: https://eur-lex.europa.eu/legal-content/EN/TXT/?uri=COM:2019:640:FIN (accessed on 20 December 2024).
3. European Commission. Communication from the Commission to the European Parliament, the Council, the European Economic and Social Committee and the Committee of the Regions. A European Strategy on Cooperative Intelligent Transport Systems, a Milestone Towards Cooperative, Connected and Automated Mobility. *COM/2016/0766.* Available online: https://eur-lex.europa.eu/legal-content/EN/TXT/?uri=CELEX:52016DC0766 (accessed on 20 December 2024).
4. European Commission. Communication from the Commission to the European Parliament, the Council, the European Economic and Social Committee and the Committee of the Regions. On the Road to Automated Mobility: An EU Strategy for Mobility of the Future. *COM/2018/283.* Available online: https://eur-lex.europa.eu/legal-content/EN/TXT/?uri=CELEX:52018DC0283 (accessed on 20 December 2024).
5. European Commission. Commission Staff Working Document. *Status of Progress on Connected, Cooperative and Automated Mobility in Europe. SWD/2024/92.* Available online: https://research-and-innovation.ec.europa.eu/document/download/1720a5ef-01bf-498e-85f5-c61bb3a7bc31_en?filename=swd_2024_92.pdf (accessed on 20 December 2024).
6. Soh, E.; Martens, K. Public values in the socio-technical construction of autonomous vehicle futures. *Public Manag. Rev.* **2023**, *26*, 1322–1340. [CrossRef]
7. Roe, P.G. Qualitative research on intra-urban travel: An alternative approach. *J. Transp. Geogr.* **2000**, *8*, 99–106. [CrossRef]
8. Faul, N.; Dirnwöber, M.; di Pasquale, G.; Berrada, J.; Giro, C.; Elrofai, H. Systems and Service Thematic Areas: Challenges and Scenarios, Deliverable D3.4 of the H2020 Project ARCADE. 2018. Available online: https://ec.europa.eu/research/participants/documents/downloadPublic?documentIds=080166e5c9986a4c&appId=PPGMS (accessed on 20 December 2024).
9. Boghani, H.C.; Papazikou, E.; Zwart, R.d.; Roussou, J.; Hu, B.; Filtness, A.; Papadoulis, A. Defining the Future of Passenger Car Transport, Deliverable D6.1 of the H2020 Project LEVITATE. 2019. Available online: https://levitate-project.eu/wp-content/uploads/2019/11/D6.1-Defining-the-future-of-passenger-car-transport.pdf (accessed on 20 December 2024).
10. Hu, B.; Zwart, R.d.; Papazikou, E.; Boghani, H.C.; Filtness, A.; Roussou, J. Defining the Future of Freight Transport, Deliverable D7.1 of the H2020 Project LEVITATE. 2019. Available online: https://levitate-project.eu/wp-content/uploads/2019/10/D7.1-Defining-the-future-of-freight-transport.pdf (accessed on 20 December 2024).
11. European Commission. Expectations and Concerns of Connected and Automated Driving. *Special Eurobarometer 496.* 2020. Available online: https://europa.eu/eurobarometer/surveys/detail/2231 (accessed on 20 December 2024).
12. Rahman, M.M.; Thill, J.-C. What Drives People'sWillingness to Adopt Autonomous Vehicles? A Review of Internal and External Factors. *Sustainability* **2023**, *15*, 11541. [CrossRef]
13. Nordhoff, S.; de Winter, J.; Payre, W.; Van Arem, B.; Happee, R. What impressions do users have after a ride in an automated shuttle? An interview study. *Transp. Res. Part F Traffic Psychol. Behav.* **2019**, *63*, 252–269. [CrossRef]
14. Chng, S.; Kong, P.; Lim, P.Y.; Cornet, H.; Cheah, L. Engaging citizens in driverless mobility. Insights from a global dialogue for research, design and policy. *Transp. Res. Interdiscip. Perspect.* **2021**, *11*, 100443. [CrossRef]
15. Adrei, L.; Luca, O.; Gaman, F. Insigths from user preferences on automated vehicles: Influence of socio-demographic factors on value of time in Romania Case. *Sustainability* **2022**, *14*, 10828. [CrossRef]
16. Chng, S.; Plananska, J.; Cheah, L. Comparison of travel attitude study methods using online tools: The case of understanding public acceptance of autonomous vehicles. *Transp. Res. Interdiscip. Perspect.* **2023**, *20*, 100847. [CrossRef]
17. Milakis, D.; Seibert, D. The illusion of the shared electric automated mobility transition. *Transp. Res. Interdiscip. Perspect.* **2024**, *26*, 101171. [CrossRef]
18. Silm, S.; Tominga, A.; Saidla, K.; Poom, A.; Tammaru, T. Socio-economic and residential differences in urban modality styles based on a long-term smartphone experiment. *J. Transp. Geogr.* **2024**, *111*, 103810. [CrossRef]
19. Jamal, S.; Newbold, K.B. Factors associated with travel behaviour of millennials and older adults: A scoping review. *Sustainability* **2020**, *12*, 8236. [CrossRef]
20. Civitas Initiative. Available online: https://civitas.eu/thematic-areas/public-participation-co-creation (accessed on 20 December 2024).
21. Avila-Garzon, C.; Bacca-Acosta, J. Thirty Years of Research and Methodologies in Value Co-Creation and Co-Design. *Sustainability* **2024**, *16*, 2360. [CrossRef]

22. Flores, C.C.; Vanongeval, F.; Steenberghen, T. Identification of older adults' needs as future users of autonomous shuttles: A serious game co-creation approach for inclusiveness. *Trans. Transp. Sci.* **2023**, *14*, 14–23. [CrossRef]
23. Raats, K.; Lund, J.; Brodersen, M. Trust in autonomous vehicles: Insights from a Swedish suburb. *J. Responsible Innov.* **2024**, *11*, 2318825. [CrossRef]
24. Lampkin, S.R.; Barr, S.; Williamson, D.B.; Dawkins, L.C. Engaging publics in the transition to smart mobilities. *GeoJournal* **2023**, *88*, 4953–4970. [CrossRef]
25. Pappers, J.; Keserü, I.; Macharis, C. Co-creation or Public Participation 2.0? An Assessment of Co-creation in Transport and Mobility Research. In *Towards User-Centric Transport in Europe 2. Lecture Notes in Mobility*; Springer: Cham, Switzerland, 2020. [CrossRef]
26. Move2CCAM Project. Available online: https://move2ccam.eu/ (accessed on 20 December 2024).
27. Anciaes, P.; Chaniotakis, E.; Gkarzonikas, C.; Paraskeuas, N.; Skreki, I.; Mavromatis, K.; Economou, A.; Kamargianni, M.; Quijano, A.; Viejo, J.; et al. Primary and Secondary Data and the MOVE2CCAM Warehouse. Available online: https://ec.europa.eu/research/participants/documents/downloadPublic?documentIds=080166e506370bfd&appId=PPGMS (accessed on 20 December 2024).
28. Gebhardt, L. Understanding different car users as starting point for future mobility concepts—A co-creation approach. *Transp. Res. Interdiscip. Perspect.* **2021**, *12*, 100485. [CrossRef]

sustainability

MDPI

Article

Drivers of Port Competitiveness Among Low-, Upper-, and High-Income Countries

Metri F. Mdanat [1,*], Mohammad Al Hur [2], Omar M. Bwaliez [1], Ghazi A. Samawi [1] and Raed Khasawneh [1]

1 Business School, German Jordanian University, P.O. Box 35247, Amman 11180, Jordan; omar.bwaliez@gju.edu.jo (O.M.B.); ghazi.samawi@gju.edu.jo (G.A.S.); raed.khasawneh@gju.edu.jo (R.K.)
2 General Education Unit, Al Wasl University, Dubai P.O. Box 50106, United Arab Emirates; mohammadalhur@yahoo.com
* Correspondence: metri.mdanat@gju.edu.jo; Tel.: +962-799062815

Abstract: This study empirically investigated the drivers of port competitiveness among low-, upper-, and high-income countries in the Middle East and North Africa region. It explored the effects of country-level competitiveness, logistic performance, and ease of doing business on port competitiveness for 17 countries from the region using a 14-item scale and covering the years 2010 to 2022. Port competitiveness indicators were analyzed using confirmatory factor analysis to determine the research constructs' validity. Structural path analysis was deployed to verify hypotheses concerning effects between conceptualized variables. The findings demonstrate that in high-income countries, an increase in country competitiveness notably enhances port sustainability and competitiveness. Conversely, in low-income countries, higher country competitiveness appears to diminish port competitiveness. For countries with average income, the association is relatively neutral, exhibiting a slight positive trend. This study explains the specific drivers and interactions that improve port sustainability and competitiveness as countries move from low- to high-income levels of development.

Keywords: competitiveness; Middle East and North Africa; port management; economic growth; structural equation modeling; sustainable transportation

1. Introduction

Competitiveness encompasses a range of institutions, policies, and factors that contribute to an economic system's prosperity, directly influencing income levels and the overall health of the economy [1]. At the macroeconomic level, a key aspect of competitiveness is the quality of infrastructure, which includes vital components such as roads, railways, and ports. These infrastructures play a crucial role in facilitating trade and enabling economic activities. While national competitiveness covers a broader array of economic indicators and institutional frameworks, port competitiveness focuses explicitly on the operational efficiency and strategic advantages of individual ports within the global supply chain [2]. This distinction is significant: a country can markedly improve its overall economic performance by enhancing port competitiveness. Such enhancements reduce trade costs and improve market access, propelling national economic growth [3].

Global shipping service industry stakeholders use various key port hubs worldwide for global, regional, and national logistic purposes. The host countries and local areas of major ports achieve immense economic benefits and improved competitive potential for local, hinterland, and national economies. At the same time, a decline in the use of a particular port due to a lack of competitiveness relative to changing market conditions can have ruinous socioeconomic and commercial ramifications for ports and their dependent communities. Ports that are more flexible to the requirements of shipping lines and that can complement and add value will become preferred channels for shipping companies and be better positioned to achieve sustainable long-term economic growth.

While developing countries strive to enhance economic performance and move from one level of development to another, they face numerous challenges, and several key factors influence their progress towards sustainable development. However, countries that leverage their ports effectively as part of broader economic strategies are better positioned to achieve sustainable development goals and improve the well-being of their populations over the long term. Thus, the level of national income and availability of ports can be considered vital to competitiveness.

However, the sustainability of port competitiveness (SoPC) includes complex factors that affect ports' attractiveness and presence for shipping services. Sustainable port competitiveness means balancing the need for growth and efficiency with the responsibility to protect the environment and support social well-being. It refers to the ability of a port to maintain its competitive edge over time while also considering environmental, social, and economic factors [4]. For instance, many shipping lines choose ports that aim to strengthen the level of connectivity and integration with particular trade routes and regions. Other shipping lines choose ports based on their logistic chain solutions to minimize time and economic costs.

Ports are deemed pivotal to an efficient transport system and a well-organized supply chain [5], and the competitiveness of individual ports is formed from a web of infrastructural and service factors, external environmental and local resources, and connectivity with transport and communication networks. The complexity of these factors varies by location and service lines, and improving a port's competitiveness is a challenging task. While numerous studies have examined port competitiveness in the MENA region, limited consideration has been directed to these transition factors. Accordingly, this study fills this gap.

Similarly, geopolitical risks, such as conflicts and shifting trade policies, disrupt global supply chains. Therefore, the income level of a country plays a significant role in determining how well a port can respond to and recover from geopolitical disruptions. This is due to differences in infrastructure quality, financial resources, technological capabilities, and institutional stability. For instance, high-income countries are better positioned to manage these disruptions due to advanced infrastructure, technology, and financial resources, allowing them to quickly adapt and mitigate impacts. Conversely, low-income countries are more vulnerable, struggling with outdated infrastructure, limited technology, and weaker institutional stability, making adaptation to and recovery from geopolitical disruptions more challenging. Income levels thus shape each region's ability to respond to geopolitical risks, influencing port competitiveness and resilience.

Following the disruption of the COVID-19 pandemic and numerous ongoing conflicts, including in Ukraine and Gaza, new trade routes have developed, such as the India–Middle East–Europe (IME) corridor, which are affecting the dynamics of global trade, shifting trade routes, and influencing the performance and sustainability of many ports. The conflict in Gaza has severely affected the competitiveness and trade of ports in the Red Sea and eastern Mediterranean regions [6,7] and strained the capacity of regional ports, increasing risks for shippers and forcing them to seek alternative routes [8]. These events further complicate the operating environment for ports, requiring collaboration to remain competitive. In addition to disrupting trade, the conflict has highlighted the need for effective risk management strategies to maintain sustainable port competitiveness [9]. Furthermore, the establishment of new trade routes may also shift trade flows and impact port activities, highlighting the need for strategic adjustments in the post-conflict era.

Ports with advanced communication networks and associated technologies are essential to link the demand and supply sides of served markets [10]. The development in the importance of seaports as an enabler of international transportation and supply chains is profound. Continued growth and sustained operations of this magnitude are a major test of latent seaport capacity [11]. Whether legacy seaports are sufficient to accommodate the transport and logistic growth in the coming years is open to question [12,13]. To generate more business, there is intense competition among container ports to maximize their effectiveness and efficiency [14–16]. Such competition among ports is crucial for improving economic performance [17,18]. Cutting-edge progress in worldwide trade under globalization has resulted in a manifold increase in the importance of container transport [11].

This development in container transport is associated with numerous technological, logistic, and financial developments that are of increasing magnitude relative to traditional sea transportation. Seaports seem to be the most fundamental and distinguished part of the container transport system, commanding a pivotal nexus at the interface of land and ocean transportation [19]. To achieve a better edge in terms of competitiveness, ports need to improve their efficiency. Most ports worldwide have increasingly sought to incorporate modern technologies and IT systems and equipment within their operations. Though different ports have met with varying degrees of success, they have all generally achieved smoother container shipping processes in the multimodal transport system [18].

Several studies evaluated port competitiveness in terms of port operations, interests, and outcomes according to their needs using qualitative evaluation of experts' "experiences" and "perceptions," but few studies have undertaken a detailed examination of how related factors precisely affect the SoPC [20–22]. The current research seeks to add to the emerging research in assessing various effects arising from Logistic Performance Index (LPI), Country Competitiveness Index (GCI), and ease of doing business (EoDB) factors on the SoPC, as measured by the country's level of income and development.

Understanding the drivers of port competitiveness in the MENA region is essential in order to devise and enable strategic business models for countries with different levels of income and development. Consequently, this research aimed to discern the key drivers of port competitiveness affecting countries at different levels of development. The developed constructs applied in this research are novel in the regional context, and were designed in order to (a) examine which factors can enhance SoPC and (b) investigate the relationship of logistic performance with competitiveness and EoDB factors. Structural equation modeling (SEM) was deployed for port competitiveness impact assessment concerning logistic performance, competitiveness, and EoDB factors. Additionally, this study examines how country-level income acts as a moderating variable in these relationships.

Following this introduction, Section 2 reviews related literature, then Section 3 unpacks the developed model and tested hypotheses. Section 4 outlines the methodology deployed for data collection and statistical analysis. Section 5 presents and analyzes the empirical findings and conclusions from hypothesis testing, while Section 6 discusses these outcomes.

2. Literature Review

Ports are vital hubs in global cargo and logistics, contributing immensely to macroeconomic development by interfacing between overseas and domestic producers and consumers and facilitating the flow of capital, products, people, and data. Therefore, a comprehensive analysis of port sustainability, including warehousing and logistics in relation to emerging technologies, is necessary to understand their competitiveness and infrastructure requirements [23].

Port competitiveness is essentially defined as the ability of a port to acquire comparative advantages in terms of products, infrastructure, and services [24]. Over recent years, the competitiveness of the global port market has increased greatly, and various instrumental factors have been discerned, with new types of data being utilized to encompass a broader understanding of what port competitiveness can entail [25–29]. Port services and capabilities are affected by internal management and external factors such as national and international political, legislative, and economic considerations, and logistics (e.g., the length of maritime journey from one port to another, and the ports connectedness with inland transportation networks and infrastructure) [30,31]. Internal operational factors include tariffs levied by ports, processing and handling times, and onward and incoming terrestrial, maritime, and air transportation routes, etc. [15,32–37].

Other studies have offered various indices for the evaluation of ports' competitiveness, including a variety of pertinent considerations [38–43]. For example, Yeo et al. [36] advocated seven fundamental considerations, encompassing the services and regional centers of ports, interior conditions, logistic efficiency, and the connectivity and convenience of related infrastructure in relation to South Korean and Chinese ports. Ha and Yang [44] identified a hierarchy of six dimensions including 16 particular KPIs for ports' performance, with 60 total indices of port competitiveness. A systematic review by Parola et al. [43] encompassing major peer-reviewed international journals for the period 1983–2014 identified the main competitiveness drivers of ports as costs, connectivity to the interior, position relative to global markets, operational technology and infrastructure (and their efficiency), maritime and nautical connectivity and accessibility, quality of service, and specific site issues.

Based on the potential instrumentality of such diverse factors and approaches, various studies have proffered differing methodologies to assay global and local competitiveness among ports [44–49]. For example, Da Cruz and de Matos Ferreira [50] evaluated the efficiency of maritime ports in Spain and Portugal, finding that cargo amount was not an overwhelming determinant. Chen et al. [47] adopted global positioning system (GPS) traces and maritime open data to effectively monitor port performance. Ren et al. [49] developed a new "multi-attribute decision analysis method" to determine competitiveness at Shanghai, Hong Kong, and Singapore, and they found that the former could outperform the two other ports by leveraging effective market share ratio in the coming decade.

Other studies used the global maritime transport network (GMTN), including Peng et al. [23], who deployed it to analyze three normative transport networks for cargo vessels (bulk carriers, container ships, and oil tankers), leading to the conclusion that while some ports performed well in all studied categories, others varied greatly in terms of their network type. Ducruet [51] analyzed GMTN data from 1977 to 2008, finding that network growth and concentration in relation to key hubs rendered traffic distribution dependent on established ports' retrenched position.

Conversely, the competitiveness of ports is usually evaluated concerning both local socioeconomic profiles and prevailing geographical conditions, which converge to shape ports' operational status. For instance, individual wharves were the main focus of traditional port development, but combined ports' development and the nexus of advanced technologies offer a paradigm shift in flows of capital, technologies, and data, in addition to the more efficient distribution of goods internationally and regionally. Consequently, analyzing port status with market share ratio is important to assess net competitiveness [23]. Munim and Saeed [52] conducted a systematic literature review collected from the Web of Science database investigating articles on how port competitiveness evolved based on maritime literature from 1990 to 2015. The seven main factors identified in their study were port competition, efficiency, institutional transformation, pricing, embeddedness, choice, and cooperation.

Wan et al. [53] developed a model to evaluate the risk factors of maritime supply chains by investigating a real case of a world-leading container shipping company. The study found that the most critical risk factors are the transportation of dangerous products, fluctuations in fuel prices, hard competition, unattractive markets, and fluctuations in exchange rates. A recent analysis of Indonesian ports concluded that port competitiveness significantly relies on government and business support as well as operational performance and that the operational performance of national ports is particularly important in competitiveness [24].

Munim et al. [54] investigated the container market in Bangladesh, due to the uniqueness of its regional context and factors of competitiveness. The study developed a novel approach to evaluate the competitiveness of transshipment ports on seven major dimensions: connectivity, port facility, efficiency, cost factor, policy and management, information systems, and green port management. These various factors are fundamental in contemporary strategic visions and business models for port competitiveness [55] and in macroeconomic development for nations and the global economy in general [56].

The intensification of competition due to globalization, container isolation, the integration of markets, and global flows of labor and capital have fundamentally altered container port governance, operations, and competitiveness [55]. Many researchers have studied the competition of port systems [24,57–60]. Additionally, the performance of seaports is shaped by profit maximization imperatives in the face of fierce local and global competition [61]. Trade connectivity can be contextualized with a triad of interdependent dimensions: (1) maritime networks (i.e., shipping characteristics and performance at sea); (2) port efficiency per se; and (3) "hinterland connectivity," referring to the various stakeholders involved in supply chains and local and national economic activities intersecting with ports.

Policies that function effectively for one dimension of port management may be counterproductive in relation to others, and holistic approaches encompassing all dimensions are naturally preferred [62]. Trade connectivity dimensions' efficiency and growth drivers vary and may span multiple dimensions, but industrial strategies of shipping lines remain the core maritime network drivers. Serious global actors (e.g., Compagnie Générale Maritime, Maersk, and Mediterranean Shipping Company) have been consolidating operational strategies to converge on a "hub-and-spoke" system, oriented toward regions and feeding into secondary Mediterranean ports. Numerous studies were conducted to examine the port competitiveness in the region, highlighting the need for ports to invest in infrastructure, enhance operational efficiency, adopt new technologies, and navigate geopolitical and economic challenges to remain competitive in the global maritime industry [11,63,64].

3. Proposed Model and Hypotheses

This study constructs a conceptual framework to assess the effects of country-level logistic performance, competitiveness, and EoDB dimensions on the SoPC using country-level income as a moderator. Figure 1 displays the envisioned constructs and hypothesized relationships between them, as discussed below.

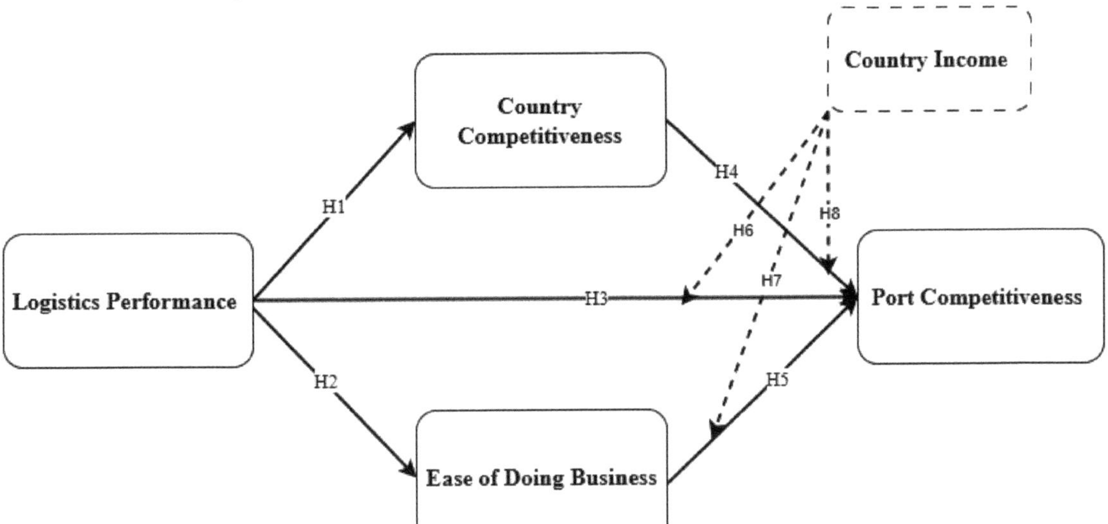

Figure 1. Proposed model.

3.1. Logistic Performance Dimensions

The association between logistic performance and port competitiveness is fundamental, whereby improving a country's logistic performance enhances the competitiveness of its ports. Logistic performance influences a port's efficiency, reliability, and ability to handle cargo effectively. Efficient logistics can reduce costs, improve service quality, and attract more shipping lines and cargo, thereby enhancing a port's overall competitiveness [65–68]. The logistic performance dimensions used in this study are explained below.

3.1.1. Customs

An extensive number of studies have found a positive relationship between customs procedures as the main components of logistic performance that affect port competitiveness [62,69]. Inefficient customs procedures increase costs and hinder trade and cargo shipments. Although tariffs for maritime containers have been significantly reduced due to the prevalence of larger container ships, the costs of onward (i.e., inland) transportation from ports remain relatively expensive [62]. Most studies examining the relationship between customs procedures and port competitiveness focused on regional contexts without looking into country-specific or other factors, like customs digitalization and integration with global trade networks, which can further improve port competitiveness. In addition, the majority of existing research focuses on high-income countries with already efficient customs systems, while few studies on how to improve customs efficiency have been conducted in low- and middle-income countries [70–73].

3.1.2. Timeliness

Timeliness refers to the ability to deliver goods or services within a specified period, ensuring that cargo handling, customs clearance, and overall transportation are completed without delays. Timeliness is critical for time-sensitive goods and is a key factor in determining the efficiency and reliability of a port or logistics system [70,74,75]. Therefore, ports that ensure timely handling, processing, and delivery of cargo attract more shippers and cargo flows, leading to higher efficiency, reduced costs, and better customer satisfaction. Delays, on the other hand, reduce a port's competitive advantage by increasing uncertainty and costs for users [76,77]. Thus, improving timeliness in logistic performance can enhance port competitiveness in different regional and economic contexts.

3.1.3. Tracking and Tracing

Tracking and tracing concern the ability to monitor and follow the progress and location of goods throughout the supply chain from origin to final destination. Studies show that ports that enhance tracking and tracing reduce delays and ensure timely deliveries, which in turn boosts competitiveness [78]. Tracking and tracing enable companies to monitor the movement of goods and ensure timely delivery, improving their competitiveness in the market [79]. Similarly, technologies such as electronic data interchange (EDI) and tracking and tracing systems employed in container terminals are expected to improve port competitiveness [80]. Based on the previous discussions on the links between port competitiveness and logistic performance dimensions, this study hypothesizes that:

H1: *Improving logistic performance dimensions enhances national competitiveness.*

H2: *Improving logistic performance dimensions affects a country's EoDB positively.*

H3: *Improving logistic performance dimensions enhances the SoPC.*

3.2. Competitiveness Dimensions

Competitiveness can be evaluated at various levels (e.g., countries, governments, organizations, and individuals). In the context of ports, economic performance, management efficiency, and infrastructure are widely recognized as core metrics of national competitiveness [1]. Many studies have historically tied port competitiveness to user perceptions and traditional metrics [81,82]. However, Parola et al. [43] expanded this understanding by incorporating additional factors such as cost, connectivity, location, services, logistics, inland transport, infrastructure, operational efficiency, and service quality.

This study extends the focus to include the broader structural and operational aspects of port competitiveness while highlighting the importance of professionalization of port officers [83]. Enhancing professional skills in areas like green port management can lead to gains in operational efficiency and sustainability, contributing to both short- and long-term competitiveness. The chosen competitiveness dimensions for this study reflect an integrated approach to the SoPC. Below, we provide a detailed explanation of these dimensions and their specific relevance to port competitiveness.

3.2.1. Institutions

Strong institutions provide the regulatory and governance framework necessary for stable and predictable business operations, which directly influence the SoPC. When governance is effective, it attracts foreign and domestic investment, enhances operational transparency, and ensures compliance with international standards. For example, a well-governed port with clear regulations and secure property rights facilitates smoother transactions, faster logistic processes, and better coordination among stakeholders [62]. Additionally, the institutional environment affects how other factors—like connectivity and location—are leveraged to improve port competitiveness. Effective institutions can therefore amplify the impact of infrastructural investments or mitigate risks associated with geographical location. Empirical evidence has developed and validated scales for the measurement of the quality of institutions [73,84]. However, other studies have examined the link between institutions and port competitiveness separately [85,86].

Notwithstanding the important developments offered by reviewed studies, more examination is needed concerning the impacts of institutions on port competitiveness in different regions and contexts. There is limited research on how government policies, institutional quality, and regulatory environments interact with other factors to influence port competitiveness.

3.2.2. Infrastructure

Infrastructure is a cornerstone of port competitiveness, influencing both operational efficiency and long-term strategic viability. The ongoing development of port facilities, technology, and connectivity affects port performance and competitiveness [4]. The quality of infrastructure affects transport costs, trade efficiency, and overall SoPC. Investments in inland terminals, logistic zones, and rail networks, as noted by [62], can expand a port's influence beyond its traditional hinterland. However, infrastructure is only one aspect; geographic location and connectivity also play crucial roles. Ports situated in strategic locations with strong inland and coastal links have a competitive advantage, underscoring the need to consider location as a moderating factor in infrastructure's contribution to the SoPC. High quality and functional infrastructure are essential for port competitiveness. However, such investment is never risk-free, and even the best ports in terms of infrastructure are vulnerable to market and political externalities [87]. Other factors affect port efficiency, which is closely linked to infrastructure quality that reduces transport costs and enhances trade [70].

3.2.3. Macroeconomic Environment

A country's macroeconomic environment is intricately linked to its ports' performance and competitiveness. Factors like inflation, exchange rates, and economic stability influence trade flows, investment, and thus the SoPC. Moreover, the macroeconomic environment interacts with social factors like education and income disparity. For example, a skilled labor force can enhance operational performance and support the adoption of new technologies, further strengthening the SoPC. Conversely, income disparity can limit access to skilled labor or create social challenges, potentially weakening the macroeconomic foundation for port competitiveness. Lirn et al. [88] demonstrated how a stable macroeconomic context can increase a port's attractiveness. However, the influence of these factors is moderated by elements like education levels, which affect the skill quality of the labor force in port regions, and income disparity, which can impact the overall socioeconomic stability of port operations. This study highlights the need to explore these connections in different economic contexts, particularly in developing regions where volatility may alter these dynamics [89,90].

3.2.4. Market Size and Efficiency

The relationship between market size and port competitiveness has been demonstrated in large markets, mainly in developing countries. Market size is closely linked to a port's geographical location and connectivity: ports situated near major trade routes or large consumer markets have a natural advantage in attracting cargo. Studies show that larger markets lead to higher accessibility to trade networks, container throughput, and better utilization of port infrastructure [91–93]. Arvis et al. [62] adumbrated seven indicators instrumental in ports' market share, the most significant of which were interconnectivity with the inland and destinations, road distance, and volume of throughput. Political stability, cultural considerations, and environmental policies also play a role in determining market size and efficiency [94–101]. Alongside such universal factors, a new trend of environmental protection has emerged, such as regulations on environmental issues and consumer and regulatory requests that logistics should be increasingly green. However, most related studies that provided criteria for port competitiveness did not systematically categorize them. For instance, Murphy et al. [102] and Wiegmans et al. [99] classified different groups of factors, but they did not prioritize, measure, or compare them. Based on the previous discussions on the link between port competitiveness and competitiveness dimensions, this study hypothesizes that:

H4: *Improving national competitiveness positively enhances the SoPC and overall economic performance.*

3.3. Ease of Doing Business Dimensions

EoDB refers to a set of indicators developed by the World Bank to measure how conducive a regulatory environment is for business operations in different countries [73]. EoDB dimensions such as trading across borders, enforcing contracts, protecting minority investors, and dealing with construction permits are directly linked to port competitiveness, because efficient regulatory and operational environments in a country make it easier for ports to facilitate trade. According to the World Bank [73], EoDB tends to be more efficient in handling cargo, leading to increased competitiveness in the global logistics network. The EoDB dimensions used in this study are explained below.

3.3.1. Protecting Minority Investors

Ports that ensure good corporate governance and protect minority investors are seen as more stable and trustworthy, which can attract global shipping companies [103], thus protecting minority investors to attract investment, fostering good corporate governance, and enhancing port competitiveness [73].

3.3.2. Trading Across Borders

The ease of trading across borders influences port competitiveness significantly, and ports with efficient logistics and streamlined trading procedures facilitate international trade to compete for global cargo flows [104]. On the other hand, enhanced customs procedures and reduced trade costs attract international shipping lines and traders, which enhances their global competitiveness.

3.3.3. Enforcing Contracts

Enforcing contracts ensures that disputes related to port services, contracts with shipping companies, or logistic operators are resolved promptly, maintaining port reliability by investors and shipping companies. Ports with efficient operations and high scores in enforcing contracts are better to handle larger cargo volumes, improve shipping schedules, and reduce costs for shippers [105].

3.3.4. Dealing with Construction Permits

The dimension of dealing with construction permits has direct impacts on port competitiveness by influencing the speed, cost, and quality of infrastructure development. This dimension is linked with regulatory ecosystems and shows how easy it is for investors to start and operate a business in different countries. Building on previous discussions, ports operating in countries with streamlined regulations, efficient customs procedures, and supportive business environments are more competitive globally. The ease with which businesses can operate and trade affects how efficiently ports handle cargo and serve international markets. Consequently, this study posits that:

H5: *EoDB enhances the SoPC and overall economic performance.*

3.4. National Income Level

The relationship between national income and port competitiveness has been explored through various theoretical lenses, demonstrating how income levels can influence LPI, GCI, and EoDB in relation to port efficiency. Previous studies show that the effects of income levels on port competitiveness vary significantly between low-income and high-income countries, influencing their investment capacity, workforce skills, technological adoption, and strategic focus, with high-income countries emphasizing advanced infrastructure and innovation, while low-income countries prioritize cost-effective, foundational improvements [106,107]. Generally, higher national income levels are associated with increased investment in infrastructure, enhancing port operations and services, and improving competitiveness [107,108]. Research indicates that as national income rises, ports are better positioned to optimize their logistic capabilities and operational efficiencies, contributing to an advantageous competitive stance in global trade [72].

This supports the notion that income levels play a positive role in moderating the connection between a nation's overall competitiveness and its port competitiveness, as wealthier countries can invest in advanced technologies and facilities that enhance port services while also lowering operational costs. On the other hand, some experts suggest that the advantages of rising national income may not necessarily lead to improved port competitiveness across the board [109]. In lower-income countries, for example, even when national income increases, challenges such as bureaucratic inefficiencies and insufficient regulatory frameworks can hinder the effective use of resources intended for port development [110,111]. Therefore, while higher national income has the potential to improve port infrastructure and services, the realization of these benefits is critically dependent on the wider economic environment and the effectiveness of governance structures. Hence, this study hypothesizes that:

H6: *Income level moderates the relationship between logistic performance and port competitiveness.*

H7: *Income level moderates the relationship between EoDB and port competitiveness.*

H8: *Income level positively moderates the relationship between a country's competitiveness and port competitiveness.*

3.5. Proposed Model

This study analyzes "port competitiveness" in multiple dimensions, arguing that the SoPC and the drivers thereof are significantly determined by notable developments in the maritime industry. Thus, building on previous literature, the current study posits that LPI, GCI, and EoDB are a function of interrelationships between the study constructs. Figure 1 shows the expected relationships between these factors.

4. Materials and Methods

4.1. Data Collection

This study utilized SEM with Smart PLS 4 to assess the effects of country-level competitiveness, logistic performance, and EoDB on the SoPC. The model was developed by integrating various factors of competitiveness, logistics, and EoDB factors to test their collective and individual influences on port competitiveness across 17 countries in the MENA region, classified according to their income level as per the World Bank classification: Algeria, Bahrain, Egypt, Iran, Israel, Jordan, Kuwait, Lebanon, Malta, Morocco, Oman, Qatar, Saudi Arabia, Tunisia, Turkey, United Arab Emirates, and Yemen. These countries were chosen based on their level of income according to the World Bank classification and their trade routes' connectivity, connections with international markets, and port availability. This comprehensive approach enables an in-depth assessment of how different income levels interact with critical logistic and economic factors, thus offering valuable insights into port competitiveness across varied economic landscapes. The study also included path analysis corresponding to H1–H8. These relationships suggest that increasing country competitiveness, logistic performance, and EoDB factors increase the SoPC. While SEM is considered a confirmatory technique, it also extends the possibility of relationships among the latent variables and encompasses two components: a measurement model (essentially CFA) and a structural model [112]. The applicability of port competitiveness factors to this study was investigated by applying CFA to test the hypotheses.

4.2. Measurement Items

A 14-item scale was developed by the researchers, derived from the World Economic Forum (WEF), GCI, LPI, and EoDB. The sustainability of the port competitiveness scale was measured using the weighted average of three different indices published by WEF, namely, the linear shipping connectivity index (with maximum value in 2004 = 100), quality of port infrastructure (with scores ranging from 1 = extremely underdeveloped to 7 = well developed and efficient by international standards), and international shipments. Country competitiveness uses a five-item scale derived from GCI. Logistic performance uses a five-item scale derived from LPI. EoDB uses a three-item scale derived from EoDB. All indicators were standardized on a five-point scale to be consistent with the study context. The most recently available data were used in this study, encompassing the years 2010, 2012, 2014, 2016, 2018, and 2022. The selected period ensured comprehensive and consistent data across all key factors influencing port competitiveness, including logistic performance, economic indicators, infrastructure development, and geopolitical impacts, encompassing a period marked by significant geopolitical events and conflicts in the MENA region.

Exploratory factor analysis (EFA) involved item-to-total correlations for 14 items, with all items showing satisfactory correlations. Subsequently, employing a principal component and varimax rotation resulted in the retention of 14 items and the identification of 4 factors suitable for EFA. The results for the Kaiser–Meyer–Olkin (KMO) measurement (0.708) and Bartlett's test of sphericity (reaching significance at the 0.001 level) (as shown in Table 1) confirmed that the data exhibited significant inherent correlations for EFA. The scree plot indicated a four-factor solution with 14 items as the optimal solution, accounting for 86.14% of the total variance in the factor pattern.

Table 1. KMO and Bartlett's test.

Kaiser–Meyer–Olkin (KMO)		0.708
Bartlett's test of sphericity	Approx. Chi-Squared	2437.651
	df	171
	Sig.	0.001

Table 2 shows that the reliability testing of the constructs utilizing Cronbach's alpha (CA) and composite reliability (CR) revealed that all constructs surpassed the 0.70 threshold recommended by Sarstedt et al. [113]. In addition, our assessment of convergent validity by determining AVE values revealed that all values exceeded the 0.50 threshold recommended by Henseler et al. [114].

Table 2. Factor loadings.

Indicators	Outer Loading	Composite Reliability	Cronbach's Alpha	AVE
Ease of doing business (EoDB)				
Protecting minority investors	0.932	0.953	0.956	0.878
Trading across borders	0.881			
Dealing with construction permits	0.986			
Enforcing contracts	0.945			
Country competitiveness				
Infrastructure	0.802	0.775	0.779	0.598
Macroeconomic environment	0.711			
Institutions	0.819			
Market efficiency	0.757			
Port competitiveness				
International shipments	0.976	0.955	0.956	0.917
Liner shipping connectivity	0.926			
Quality of port infrastructure	0.970			
Logistic performance				
Timeliness	0.944	0.917	0.919	0.859
Tracking and tracing	0.945			
Customs	0.890			

Additionally, we conducted Fornell–Larcker and heterotrait–monotrait (HTMT) ratio tests to assess discriminant validity. The HTMT ratio results showed acceptable values below the 0.80 threshold, as shown in Tables 3 and 4.

Table 3. Fornell–Larcker and heterotrait–monotrait ratio tests.

Indicators	Country Competitiveness	EoDB	Logistic Performance	Port Competitiveness
Country competitiveness	--------------			
EoDB	0.259	--------------		
Logistic performance	0.419	0.142	---------------	
Port competitiveness	0.454	0.51	0.399	----------------

Table 4. Discriminant validity regarding Fornell–Larcker criterion.

Indicators	Country Competitiveness	EoDB	Logistic Performance	Port Competitiveness
Country competitiveness	**0.773**			
EoDB	0.175	**0.937**		
Logistic performance	0.353	0.13	**0.927**	
Port competitiveness	0.392	0.488	0.375	**0.958**

Note: The square roots of the AVE are presented as bold values along the diagonal.

4.3. Hypotheses Testing

The structural model was calculated using the bootstrapping technique, and the hypothesized relationships were developed based on theoretical and empirical research. Table 5 shows the results of the hypothesis testing: it shows that the model fit is significant, with an SRMR value of 0.035. Additionally, the determination coefficient (R^2) value is greater than 0.25 [115]. These findings indicate that 50.4% of the variance in the SoPC was explained by country competitiveness, country logistic performance, and EoDB. Furthermore, the Q^2 value was higher than zero, demonstrating the significance and predictive relevance of the study model [116].

Table 5. Model fit.

Predictor(s)	F2	VIF	Outcome	R^2	Adj. R	Q^2 Predict	SRMR
Country competitiveness	0.093	1.327	Port competitiveness	0.504	0.474	0.134	0.035
Country logistic performance	0.084	2.494					
EoDB	0.117	3.068					

5. Results

The findings in Figure 2 show that country logistic performance has a significant and positive impact on the following factors: country competitiveness (β = 0.353, t = 4.485, $p \leq 0.000$), EoDB (β = 0.130, t = 2.780, $p \leq 0.000$), and SoPC (β = 0.317, t = 3.485, $p \leq 0.001$). Concerning H4, the results indicate that country competitiveness has a positive and significant impact on the SoPC (β = 0.239, t = 2.146, $p \leq 0.05$). EoDB also has a positive and significant impact on port competitiveness (β = 0.424, t = 4.042, and $p \leq 0.001$). Thus, we accepted the direct relationships of H1, H2, H3, H4, and H5 (Table 6).

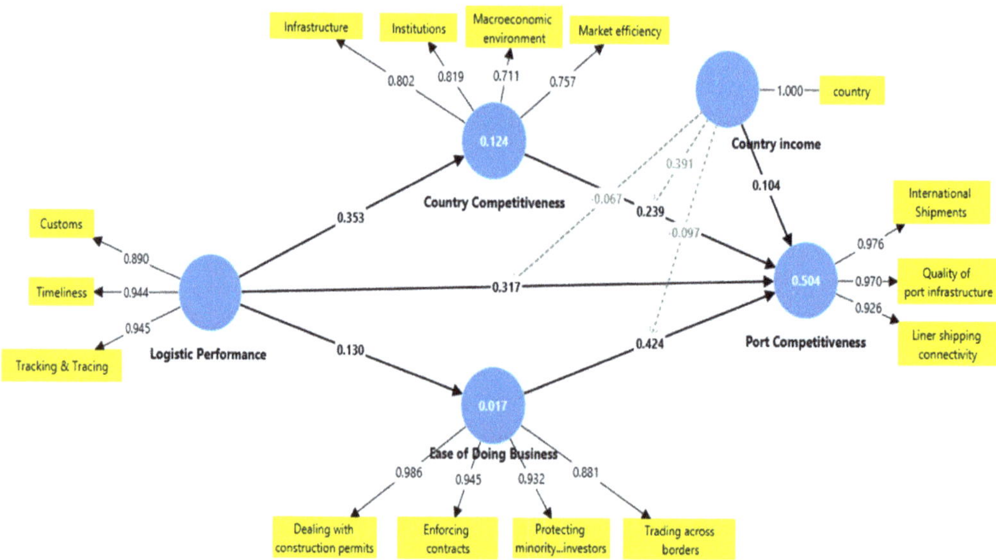

Figure 2. Proposed model.

Table 6. Results of the direct hypotheses.

Structural Path	Unstandardized Beta	SD	T Statistics	*p* Values	H
Logistic performance → Country competitiveness	0.349	0.078	4.55	***	H1
Logistic performance → EoDB	0.130	0.101	2.780	***	H2
Logistic performance → Port competitiveness	0.326	0.091	3.485	***	H3
Country competitiveness → Port competitiveness	0.243	0.111	2.146	0.032 *	H4
EoDB → Port competitiveness	0.417	0.105	4.045	***	H5

Note: SD = standard deviation, $p < 0.05$ *, $p < 0.001$ ***.

In addition, the study explored the influence of national income level on the relationship between independent variables (country logistic performance, country competitiveness, and EoDB) and the SoPC. The findings revealed that national income level moderates the relationship between country competitiveness and port competitiveness. It showed a positive and significant effect with a coefficient of 0.391, a t-value of 4.075, and a *p*-value of less than 0.05. However, the study found that the moderation role of national income level on the impact of the country's logistic performance and EoDB on port competitiveness was not significant, leading to the lack of support for hypotheses H6 and H8 (Table 7).

Table 7. Results of the moderation hypotheses.

Structural Path	Standardized Beta	SD	T Statistics	*p* Values
Country x logistic performance → Port competitiveness	0.069	0.095	0.700	0.484
Country x country → Competitiveness port competitiveness	0.384	0.096	4.075	***
Country x EoDB → Port competitiveness	0.082	0.084	1.148	0.251

Note: SD = standard deviation $p < 0.001$ ***.

Figure 3 demonstrates the interaction between country income level and country competitiveness and the relationship between country competitiveness and port competitiveness. The results show that in high-income countries (green line), an increase in country competitiveness notably improves port competitiveness. In contrast, in low-income countries (red line), higher country competitiveness appears to reduce port competitiveness. This could be caused by resource allocations toward sectors perceived to be more directly tied to economic growth, such as urban development or manufacturing, rather than toward improving port infrastructure. For countries with average income (blue line), the association is relatively neutral, with a small positive trend. These findings indicate that country competitiveness is influenced by national income in terms of the impact it exerts on the SoPC.

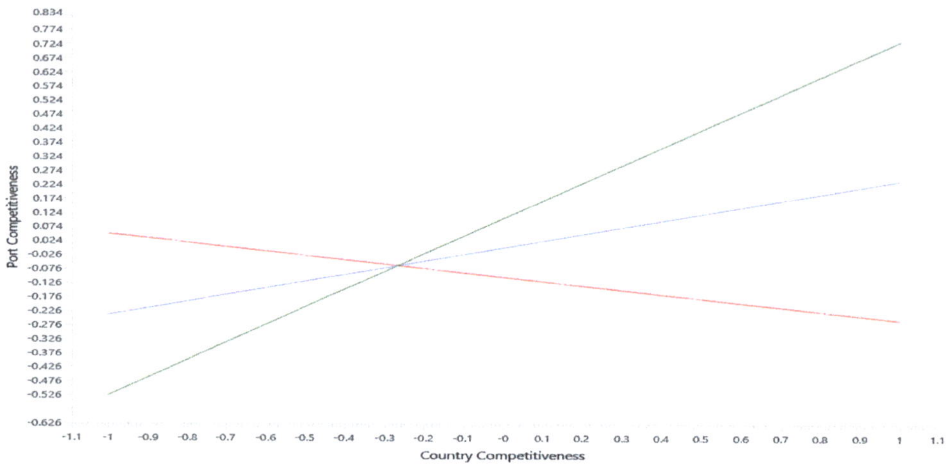

Figure 3. The moderating role of country income level.

6. Discussion

A country's income level plays a significant role in determining how its competitiveness affects the SoPC. It can also show how ports in different regions can respond and recover from geopolitical disruptions and risks. The ability of ports to respond to these risks is mainly determined by the level of port competitiveness, which is due to differences in infrastructure quality, financial resources, technological capabilities, and institutional stability. The capabilities of ports in the MENA region, with the ongoing challenges posed by the conflicts in Ukraine and Gaza along with the evolving dynamics of the IME corridor, underscore the need for ports to continuously upgrade their infrastructure, integrate advanced technologies, and develop tailored strategies to maintain their global competitiveness.

This study's findings confirm that the SoPC is closely linked with the country's logistic performance, competitiveness, and EoDB, but the associated effects vary significantly across countries by income. In low- and high-income countries, an increase in country competitiveness reduces or improves the SoPC (respectively), while in middle-income countries, the association is relatively neutral (with a small non-significant positive trend). Moreover, a country's logistic performance has a significant and positive impact on the EoDB and on the country's competitiveness. An important finding for low-income countries that affects port competitiveness is that resource allocation is typically skewed toward sectors perceived to be more directly tied to economic growth than ports. The findings suggest that low-income countries should adopt a balanced approach to economic development, ensuring that gains in national competitiveness are aligned with targeted investments in port infrastructure and capabilities for long-term economic development. This requires prioritizing port modernization, enhancing logistic networks, and providing training for port staff to avoid performance decline. It also highlights the need for supportive government policies that facilitate investments in maritime infrastructure, and the importance of international partnerships and capacity-building programs aimed specifically at ports in developing economies.

These findings must be interpreted in light of countries at all different income levels facing global competition, as well as unequal challenges related to infrastructure, technology, green practices, and financing [117]. In addition to disrupting trade, conflicts have highlighted the need for effective risk management strategies to maintain port competitiveness [9]. Furthermore, the establishment of new trade routes may also shift trade flows and impact port activities, highlighting the need for strategic adjustments in the post-war era. Low-income countries can benefit from international support and cost-efficiency-focused strategies. Likewise, middle-income countries must balance rapid growth with technological and environmental considerations, while high-income countries focus on innovation and maintaining advanced infrastructure.

By improving port infrastructure and operations, countries can enhance their logistic performance, overall competitiveness, and business environment. Tailored strategies that consider the economic context of low-, middle-, and high-income countries are essential for optimizing port performance and fostering sustainable economic growth. For low-income countries, enhancing port infrastructure and logistic efficiency can improve the SoPC. For middle-income countries, balancing growth with sustainable practices can enhance the SoPC. In addition, investment in modernizing ports and improving regulatory environments can lead to better performance in logistics, competitiveness, and business operations. For high-income countries, continuous innovation and infrastructure maintenance are key to sustaining port competitiveness. High-income countries must focus on integrating advanced technologies and maintaining regulatory efficiency to uphold their competitive edge. Therefore, high-income countries must continuously invest in maintaining and upgrading their advanced port infrastructure to stay competitive. This includes addressing aging facilities and integrating new technologies. As a result, port managers operating in countries with different income levels must prioritize strategic investments in infrastructure, professional training, and sustainable practices to improve efficiency and competitiveness. They can strengthen operational performance by adopting advanced technologies and green approaches and mitigate risks while fostering stronger partnerships.

By focusing on long-term growth, managers can align investments with industry trends, benchmark performance, and introduce innovative services to ensure continued success in a competitive environment. This study contributes to the literature by showing how ports in different regions and income levels respond to changes in logistic performance, competitiveness, and EoDB dimensions over time. The main limitations of this study are related to the availability of data over recent years and the inclusion of all dimensions in the model. Future studies should focus on comparative case studies from low-, middle-, and high-income countries to examine how these dimensions enhance the sustainability of ports' competitiveness. Furthermore, other dimensions such as insolvency, credit, and improved technology might be included in the model to increase reliability.

Author Contributions: Methodology, M.A.H.; validation, G.A.S.; formal analysis, O.M.B.; investigation, M.A.H.; data curation, O.M.B.; writing—original draft, M.F.M.; writing—review and editing, M.F.M.; review, R.K.; visualization, G.A.S.; supervision, M.F.M. All authors have read and agreed to the published version of the manuscript.

Funding: This research received no external funding.

Institutional Review Board Statement: Not applicable.

Data Availability Statement: World Economic Forum (WEF). https://www.weforum.org/ (accessed on 24 August 2024).

Conflicts of Interest: The authors declare no conflicts of interest.

References

1. Cho, H.; Lee, J. Does transportation size matter for competitiveness in the logistics industry? The cases of maritime and air transportation. *Asian J. Shipp. Logist.* **2020**, *36*, 214–223. [CrossRef]
2. Hussein, K.; Song, D.W. Port supply chain integration and sustainability: A resource-based view. *Int. J. Logist. Manag.* **2024**, *35*, 504–530. [CrossRef]
3. Irwin, D.A. Does Trade Reform Promote Economic Growth? A Review of Recent Evidence. World Bank Research Observer, Working Paper 25927. Available online: http://www.nber.org/papers/w25927 (accessed on 24 August 2024).
4. Notteboom, T.; Rodrigue, J.P. Port regionalization: A new conceptual framework. *Marit. Policy Manag.* **2005**, *32*, 281–294.
5. Meersman, H.; Van de Voorde, E. Port management, operation, and competition: A focus on North Europe. In *The Handbook of Maritime Economics and Business*, 2nd ed.; Grammenos, C., Ed.; Routledge: London, UK, 2010; pp. 921–936. ISBN 978-1-8431-1880-0.

6. El-Sayed, M.; Farouk, A. Geopolitical tensions and their impact on maritime trade in the Middle East: A case study of port resilience. *J. Glob. Trade Logist.* **2023**, *58*, 183–199.
7. Liu, X.; Lin, Y. The impact of geopolitical conflicts on port competitiveness and global trade routes: A focus on the Middle East. *J. Marit. Econ.* **2023**, *45*, 245–262.
8. Ahmed, A.; Saleh, M. The influence of regional conflicts on maritime logistics and trade competitiveness: An analysis of Middle Eastern ports. *Int. J. Shipp. Trade* **2023**, *12*, 321–338.
9. Sharma, R.; Kumar, V. The effects of geopolitical conflicts on global trade routes and port operations: A focus on South Asia and the Middle East. *J. Glob. Trade Marit. Stud.* **2023**, *27*, 200–217.
10. Schøyen, H.; Odeck, J. Comparing the productivity of Norwegian and some Nordic and UK container ports—An application of Malmquist productivity index. *Int. J. Shipp. Transp. Logist.* **2017**, *9*, 234–256. [CrossRef]
11. Mustafa, F.S.; Khan, R.U.; Mustafa, T. Technical efficiency comparison of container ports in Asian and Middle East region using DEA. *Asian J. Shipp. Logist.* **2021**, *37*, 12–19. [CrossRef]
12. Odeck, J.; Bråthen, S. A meta-analysis of DEA and SFA studies of the technical efficiency of seaports: A comparison of fixed and random-effects regression models. *Transp. Res. Part A Policy Pract.* **2012**, *46*, 1574–1585. [CrossRef]
13. International Transport Forum. *Transport Outlook 2023*; International Transport Forum: Paris, France, 2023; ISBN 978-9-2821-6726-7.
14. Luo, M.; Liu, L.; Gao, F. Post-entry container port capacity expansion. *Transp. Res. Part B Methodol.* **2012**, *46*, 120–138. [CrossRef]
15. Yuen, A.C.L.; Zhang, A.; Cheung, W. Port competitiveness from the users' perspective: An analysis of major container ports in China and its neighbouring countries. *Res. Transp. Econ.* **2012**, *35*, 34–40. [CrossRef]
16. Port Technology International. 2038: Future Visions. Available online: https://www.porttechnology.org/news/2038-future-visions/ (accessed on 24 August 2024).
17. Cullinane, K.; Wang, T.F.; Song, D.W.; Ji, P. The technical efficiency of container ports: Comparing data envelopment analysis and stochastic frontier analysis. *Transp. Res. Part A Policy Pract.* **2006**, *40*, 354–374. [CrossRef]
18. Feng, M.; Mangan, J.; Lalwani, C. Comparing port performance: Western European versus eastern Asian ports. *Int. J. Phys. Distrib. Logist. Manag.* **2012**, *42*, 490–512. [CrossRef]
19. Chao, S.L.; Yu, M.M.; Hsieh, W.F. Evaluating the efficiency of major container shipping companies: A framework of dynamic network DEA with shared inputs. *Transp. Res. Part A Policy Pract.* **2018**, *117*, 44–57. [CrossRef]
20. Rodrigue, J.P.; Notteboom, T.E. The Role of Ports in the Global Supply Chain. *Marit. Policy Manag.* **2013**, *40*, 420–431.
21. Wang, T.F.; Cullinane, K.P.B.; Song, D.W. Industrial concentration in container ports. In Proceedings of the International Association of Maritime Economists Annual Conference, Izmir, Turkey, 30 June–2 July 2004.
22. Yeo, G.T.; Roe, M.; Dinwoodie, J. Measuring the competitiveness of container ports: Logisticians' perspectives. *Eur. J. Mark.* **2011**, *45*, 455–470. [CrossRef]
23. Peng, P.; Yang, Y.; Lu, F.; Cheng, S.; Mou, N.; Yang, R. Modelling the competitiveness of the ports along the Maritime Silk Road with big data. *Transp. Res. Part A Policy Pract.* **2018**, *118*, 852–867. [CrossRef]
24. Wahyuni, S.; Taufik, A.A.; Hui, F.K.P. Exploring key variables of port competitiveness: Evidence from Indonesian ports. *Compet. Rev.* **2020**, *30*, 529–553. [CrossRef]
25. Yap, W.Y.; Lam, J.S.; Notteboom, T. Developments in container port competition in East Asia. *Transp. Rev.* **2006**, *26*, 167–188. [CrossRef]
26. Cheon, S.; Dowall, D.E.; Song, D.W. Evaluating impacts of institutional reforms on port efficiency changes: Ownership, corporate structure, and total factor productivity changes of world container ports. *Transp. Res. Part E Logist. Transp. Rev.* **2010**, *46*, 546–561. [CrossRef]
27. Bichou, K. An empirical study of the impacts of operating and market conditions on container-port efficiency and benchmarking. *Res. Transp. Econ.* **2013**, *42*, 28–37. [CrossRef]
28. Hales, D.N.; Chang, Y.T.; Lam, J.S.L.; Desplebin, O.; Dholakia, N.; Al-Wugayan, A. An empirical test of the balanced theory of port competitiveness. *Int. J. Logist. Manag.* **2017**, *28*, 363–378. [CrossRef]
29. Wan, Y.; Zhang, A.; Li, K.X. Port competition with accessibility and congestion: A theoretical framework and literature review on empirical studies. *Marit. Policy Manag.* **2018**, *45*, 239–259. [CrossRef]
30. Guridi, M.; Perea, J. The impact of global crises on port operations and trade efficiency: A case study of Middle Eastern ports. *J. Marit. Econ. Logist.* **2023**, *29*, 115–132.
31. Wang, L.; Chang, H. Port resilience and competitiveness in the face of global disruptions: Insights from the Middle East and Asia. *J. Marit. Trade Logist.* **2024**, *31*, 45–61.
32. Yuen, A.C.L.; Zhang, A.; Cheung, W. Foreign participation and competition: A way to improve the container port efficiency in China? *Transp. Res. Part A Policy Pract.* **2013**, *49*, 220–231. [CrossRef]
33. Suárez-Alemán, A.; Trujillo, L.; Cullinane, K.P. Time at ports in short sea shipping: When timing is crucial. *Marit. Econ. Logist.* **2014**, *16*, 399–417. [CrossRef]
34. De Oliveira, G.F.; Cariou, P. The impact of competition on container port (in) efficiency. *Transp. Res. Part A Policy Pract.* **2015**, *78*, 124–133. [CrossRef]
35. Tovar, B.; Hernández, R.; Rodríguez-Déniz, H. Container port competitiveness and connectivity: The Canary Islands main ports case. *Transp. Policy* **2015**, *38*, 40–51. [CrossRef]

36. Kang, D.J.; Woo, S.H. Liner shipping networks, port characteristics and the impact on port performance. *Marit. Econ. Logist.* **2017**, *19*, 274–295. [CrossRef]
37. Martínez Moya, J.; Feo Valero, M. Port choice in container market: A literature review. *Transp. Rev.* **2017**, *37*, 300–321. [CrossRef]
38. Yeo, G.T.; Roe, M.; Dinwoodie, J. Evaluating the competitiveness of container ports in Korea and China. *Transp. Res. Part A Policy Pract.* **2008**, *42*, 910–921. [CrossRef]
39. Salhieh, L.; Mdanat, M.; Al-Shboul, M.; Samawi, G. Transportation landed cost as a barrier to intra-regional trade. *J. Borderl. Stud.* **2019**, *34*, 89–103. [CrossRef]
40. Lee, P.T.; Hu, K.C. Evaluation of the service quality of container ports by importance-performance analysis. *Int. J. Shipp. Transp. Logist.* **2012**, *4*, 197–211. [CrossRef]
41. Yeo, G.T.; Ng, A.K.; Lee, P.T.W.; Yang, Z. Modelling port choice in an uncertain environment. *Marit. Policy Manag.* **2014**, *41*, 251–267. [CrossRef]
42. Kim, A.; Lu, J. A study on the effects of network centrality and efficiency on the throughput of Korean and Chinese container ports. In Proceedings of the 5th International Conference on Transportation Engineering, Dalian, China, 26–27 September 2015; American Society of Civil Engineers: Reston, VA, USA, 2015; pp. 760–769, ISBN 978-0-7844-7938-4. [CrossRef]
43. Parola, F.; Risitano, M.; Ferretti, M.; Panetti, E. The drivers of port competitiveness: A critical review. *Transp. Rev.* **2016**, *37*, 116–138. [CrossRef]
44. Ha, M.H.; Yang, Z. Comparative analysis of port performance indicators: Independency and interdependency. *Transp. Res. Part A Policy Pract.* **2017**, *103*, 264–278. [CrossRef]
45. Onut, S.; Tuzkaya, U.R.; Torun, E. Selecting container port via a fuzzy ANP-based approach: A case study in the Marmara Region, Turkey. *Transp. Policy* **2011**, *18*, 182–193. [CrossRef]
46. Zhen, L. Tactical berth allocation under uncertainty. *Eur. J. Oper. Res.* **2015**, *247*, 928–944. [CrossRef]
47. Chen, Z.; Chen, Y.; Li, T. Port cargo throughput forecasting based on combination model. In Proceedings of the 2016 Joint International Information Technology, Mechanical and Electronic Engineering Conference, Xi'an, China, 4–5 October 2016; Atlantis Press: New York, NY, USA, 2016; pp. 148–154, ISBN 978-94-6252-234-3. [CrossRef]
48. Song, M.G.; Yeo, G.T. Analysis of the air transport network characteristics of major airports. *Asian J. Shipp. Logist.* **2017**, *33*, 117–125. [CrossRef]
49. Ren, J.; Lützen, M.; Rasmussen, H.B. Identification of success factors for green shipping with measurement of greenness based on ANP and ISM. In *Multi-Criteria Decision Making in Maritime Studies and Logistics: Applications and Cases*; International Series in Operations Research & Management Science; Springer: Berlin/Heidelberg, Germany, 2018; Volume 260.
50. Da Cruz, M.R.P.; de Matos Ferreira, J.J. Evaluating Iberian seaport competitiveness using an alternative DEA approach. *Eur. Transp. Res. Rev.* **2016**, *8*, 1. [CrossRef]
51. Ducruet, C. Multilayer dynamics of complex spatial networks: The case of global maritime flows (1977–2008). *J. Transp. Geogr.* **2017**, *60*, 47–58. [CrossRef]
52. Munim, Z.H.; Saeed, N. Seaport competitiveness research: The past, present and future. *Int. J. Shipp. Transp. Logist.* **2019**, *11*, 533–557. [CrossRef]
53. Wan, C.; Yan, X.; Zhang, D.; Qu, Z.; Yang, Z. An advanced fuzzy Bayesian-based FMEA approach for assessing maritime supply chain risks. *Transp. Res. Part E Logist. Transp. Rev.* **2019**, *125*, 222–240. [CrossRef]
54. Munim, Z.H.; Duru, O.; Ng, A.K. Transhipment port's competitiveness forecasting using analytic network process modelling. *Transp. Policy* **2022**, *124*, 70–82. [CrossRef]
55. Dang, V.L.; Yeo, G.T. A competitive strategic position analysis of major container ports in Southeast Asia. *Asian J. Shipp. Logist.* **2017**, *33*, 19–25. [CrossRef]
56. Hu, Y.; Zhu, D. Empirical analysis of the worldwide maritime transportation network. *Phys. A* **2009**, *388*, 2061–2071. [CrossRef]
57. Han, C.H. Strategic positioning analysis on the Asian container ports. *J. Korean Assoc. Shipp. Stud.* **2002**, *34*, 99–115.
58. Lam, J.S.L.; Yap, W.Y. Competition for transhipment containers by major ports in Southeast Asia: Slot capacity analysis. *Marit. Policy Manag.* **2008**, *35*, 89–101. [CrossRef]
59. Notteboom, T.; Yap, W.Y. Port competition and competitiveness. In *The Blackwell Companion to Maritime Economics*; Talley, W.K., Ed.; Wiley-Blackwell: Oxford, UK, 2012; pp. 549–570. ISBN 978-1-444-33024-3.
60. Rimmer, P.J. *Asian-Pacific Rim Logistics: Global Context and Local Policies*; Edward Elgar Publishing: London, UK, 2014; ISBN 978-1-84720-628-2.
61. Da Cruz, M.R.P. Competitiveness and Strategic Positioning of Seaports: The Case of Iberian Seaports. Ph.D. Thesis, Universidade da Beira Interior, Covilhã, Portugal, 2012. Available online: https://hdl.handle.net/10400.6/2633 (accessed on 24 August 2024).
62. Arvis, J.F.; Vesin, V.; Carruthers, R.; Ducruet, C. *Maritime Networks, Port Efficiency, and Hinterland Connectivity in the Mediterranean*; World Bank Publications: Washington, DC, USA, 2018; ISBN 978-1-4648-1274-3. [CrossRef]
63. Almawsheki, E.S.; Shah, M.Z. Technical efficiency analysis of container terminals in the Middle Eastern region. *Asian J. Shipp. Logist.* **2015**, *31*, 477–486. [CrossRef]
64. Kutin, N.; Nguyen, T.T.; Vallée, T. Relative efficiencies of ASEAN container ports based on data envelopment analysis. *Asian J. Shipp. Logist.* **2017**, *33*, 67–77. [CrossRef]
65. Notteboom, T.; Rodrigue, J.P. The future of containerization: Perspectives from maritime and inland freight distribution. *GeoJournal* **2009**, *74*, 7–22. [CrossRef]

66. Wilmsmeier, G.; Sánchez, R.J. The relevance of international transport costs on food prices: Endogenous and exogenous effects. *Res. Transp. Econ.* **2009**, *25*, 56–66. [CrossRef]
67. Yap, W.Y.; Lam, J.S.L. 80 million-twenty-foot-equivalent-unit container port? Sustainability issues in port and hinterland development. *Transp. Res. Part A Policy Pract.* **2013**, *46*, 1617–1629.
68. Arvis, J.F.; Mustra, M.A.; Ojala, L.; Shepherd, B.; Saslavsky, D. Connecting to Compete 2010: Trade Logistics in the Global Economy. Available online: https://www.worldbank.org/en/news/feature/2016/06/28/connecting-to-compete-2016-trade-logistics-in-the-global-economy (accessed on 24 August 2024).
69. Hoffman, J.; Wilmsmeier, G.; Venus Lun, Y.H. Connecting the world through global shipping networks. *J. Shipp. Trade* **2017**, *2*, 2. [CrossRef]
70. Clark, X.; Dollar, D.; Micco, A. Port efficiency, maritime transport costs, and bilateral trade. *J. Dev. Econ.* **2004**, *75*, 417–450. [CrossRef]
71. Tongzon, J.L. Port choice and freight forwarders. *Transp. Res. Part E Logist. Transp. Rev.* **2009**, *45*, 186–195. [CrossRef]
72. Munim, Z.H.; Schramm, H.J. The impact of port infrastructure and logistics performance on economic growth: The mediating role of seaborne trade. *J. Shipp. Trade* **2018**, *3*, 1. [CrossRef]
73. World Bank. *Doing Business 2020: Comparing Business Regulation in 190 Economies*; World Bank Publications: Washington, DC, USA, 2020; ISBN 978-1-4648-1441-9. [CrossRef]
74. Martínez-Zarzoso, I.; Márquez-Ramos, L. The impact of trade facilitation on sectoral trade. *B.E. J. Econ. Anal. Policy* **2008**, *8*, 1–34.
75. Arvis, J.F.; Wiederer, C.; Raj, A.; Dairabayeva, K.; Kiiski, T. *Connecting to Compete 2018: Trade Logistics in the Global Economy*; World Bank Publications: Washington, DC, USA, 2018; Available online: https://hdl.handle.net/10986/29971 (accessed on 24 August 2024).
76. Merk, O.; Dang, T. Efficiency of World Ports in Container and Bulk Cargo (Oil, Coal, and Grain). In *OECD Report on Port Competitiveness*; OECD: Paris, France, 2013. [CrossRef]
77. Puertas, R.; Martí, L.; García, L. Logistics performance and export competitiveness: European experience. *Empirica* **2014**, *41*, 467–480. [CrossRef]
78. Harrison, A.; van Hoek, R. *Logistics Management and Strategy: Competing Through the Supply Chain*; Pearson Education: New York, NY, USA, 2011; ISBN 978-0-2737-1276-3.
79. Banomyong, R.; Supatn, N. Supply chain assessment tool development in Thailand: An SME perspective. *Int. J. Proc. Manag.* **2011**, *4*, 244–258. [CrossRef]
80. Satta, G.; Maugeri, S.; Panetti, E.; Ferretti, M. Port labour, competitiveness and drivers of change in the Mediterranean Sea: A conceptual framework. *Prof. Plan. Control.* **2019**, *30*, 1102–1117. [CrossRef]
81. Acosta, M.; Coronado, D.; Mar Cerban, M. Port competitiveness in container traffic from an internal point of view: The experience of the Port of Algeciras Bay. *Marit. Policy Manag.* **2007**, *34*, 501–520. [CrossRef]
82. Brooks, M.R.; Schellinck, T.; Pallis, A.A. A systematic approach for evaluating port effectiveness. *Marit. Policy Manag.* **2011**, *38*, 315–334. [CrossRef]
83. Mahmud, K.K.; Chowdhury, M.M.H.; Shaheen, M.M.A. Green port management practices for sustainable port operations: A multi method study of Asian ports. *Marit. Policy Manag.* **2023**, *50*, 254–280. [CrossRef]
84. Kaufmann, D.; Kraay, A.; Mastrorillo, M. The Worldwide Governance Indicators: Methodology and Analytical Issues. World Bank Policy Research Working Paper No. 5430. 2020. Available online: https://documents1.worldbank.org/curated/en/630421468336563314/pdf/WPS5430.pdf (accessed on 24 August 2024).
85. Hsu, C.C.; Hu, H.-C. The impact of institutional factors on port competitiveness: Evidence from Asian ports. *Transp. Res. Part A Policy Pract.* **2014**, *59*, 196–207.
86. Pallis, A.A.; Vergara, S. Port reform and the role of institutional frameworks in the port industry. *Transp. Rev.* **2017**, *37*, 1–20.
87. Nguyen, P.N.; Woo, S.H. Port connectivity and competition among container ports in Southeast Asia based on Social Network Analysis and TOPSIS. *Marit. Policy Manag.* **2022**, *49*, 779–796. [CrossRef]
88. Lirn, T.C.; Thanopoulou, H.A.; Beynon, M.J.; Beresford, A.K.C. An application of AHP on transshipment port selection: A global perspective. *Marit. Econ. Logist.* **2004**, *6*, 70–91. [CrossRef]
89. Blonigen, B.A.; Wilson, W.W. Port efficiency and trade flows. *Rev. Int. Econ.* **2008**, *16*, 21–36. [CrossRef]
90. Notteboom, T.E. Economic Analysis of the European Seaport System: Report Commissioned by the European Sea Ports Organisation (ESPO). Available online: https://www.espo.be/media/espopublications/ITMMAEconomicAnalysisoftheEuropeanPortSystem2009.pdf (accessed on 24 August 2024).
91. Sanchez, R.J.; Hoffmann, J.; Micco, A.; Pizzolitto, G.V.; Sgut, M.; Wilmsmeier, G. Port efficiency and international trade: Port efficiency as a determinant of maritime transport costs. *Marit. Econ. Logist.* **2003**, *5*, 199–218. [CrossRef]
92. Cullinane, K.; Song, D.W. Estimating the relative efficiency of European container ports: A stochastic frontier analysis. *Res. Transp. Econ.* **2006**, *16*, 85–115. [CrossRef]
93. Liu, Q.; Wang, K. Port competitiveness in a changing environment: A case of north Europe. *Transp. Res. Part A Policy Pract.* **2010**, *44*, 54–67.
94. Peters, H.J. Structural Changes in International Trade and Transport Markets: The Importance of Logistics. *Ports Harb.* **1990**, *35*, 9. Available online: https://api.semanticscholar.org/CorpusID:151205163 (accessed on 24 August 2024).

95. Lirn, T.C.; Thanopoulou, H.A.; Beresford, A.K. Transhipment port selection and decision-making behaviour: Analysing the Taiwanese case. *Int. J. Logist. Res. Appl.* **2003**, *6*, 229–244. [CrossRef]
96. Luo, M.J.; Zhou, Z.; Magni, K.; Christoforides, C.; Rappsilber, J.; Mann, M.; Reed, R. Pre-mRNA splicing and mRNA export linked by direct interactions between UAP56 and Aly. *Nature* **2001**, *413*, 644–647. [CrossRef]
97. Tongzon, J. Determinants of competitiveness in logistics: Implications for the ASEAN region. *Marit. Econ. Logist.* **2007**, *9*, 67–83. [CrossRef]
98. Department for Transport. National Policy Statement for Ports: Presented to Parliament Pursuant to Section 5(9) of the Planning Act 2008. Available online: https://www.gov.uk/government/speeches/national-policy-statement-for-ports--4 (accessed on 24 August 2024).
99. Wiegmans, B.W.; Hoest, A.V.D.; Notteboom, T.E. Port and terminal selection by deep-sea container operators. *Marit. Policy Manag.* **2008**, *35*, 517–534. [CrossRef]
100. Comtois, C.; Dong, J. Port competition in the Yangtze River delta. *Asia Pac. Viewp.* **2007**, *48*, 299–311. [CrossRef]
101. Rozenblat, C. *Les Villes Portuaires en Europe: Analyse Comparative*; Université de Lausanne: Montpellier, France; CNRS: Paris, France, 2004; Available online: https://www.mgm.fr/PUB/IRSIT.pdf (accessed on 24 August 2024).
102. Murphy, P.R.; Dalenberg, D.R.; Daley, J.M. Analyzing international water transportation: The perspectives of large US industrial corporations. *J. Bus. Logist.* **1991**, *12*, 169–190.
103. La Porta, R.; Lopez-de-Silanes, F.; Shleifer, A.; Vishny, R.W. Investor protection and corporate governance. *J. Financ. Econ.* **2000**, *58*, 3–27. [CrossRef]
104. Korinek, J.; Sourdin, P. To what extent are high-quality logistics services trade facilitating? In *OECD Trade Policy Working Papers*; No. 108; OECD: Paris, France, 2011. [CrossRef]
105. Merk, O. The Competitiveness of Global Port-Cities: Synthesis Report. Available online: https://www.oecd.org/en/about/directorates/centre-for-entrepreneurship-smes-regions-and-cities.html (accessed on 24 August 2024).
106. Lee, S.W.; Song, J.M.; Park, S.J.; Sohn, B.R. A study on the comparative analysis of port competitiveness using AHP. *KMI Int. J. Marit. Aff. Fish.* **2014**, *6*, 53–71. [CrossRef]
107. Jayathilaka, R.; Jayawardhana, C.; Embogama, N.; Jayasooriya, S.; Karunarathna, N.; Gamage, T.; Kuruppu, N. Gross domestic product and logistics performance index drive the world trade: A study based on all continents. *PLoS ONE* **2022**, *17*, e0264474. [CrossRef] [PubMed]
108. Dwarakish, G.S.; Salim, A.M. Review on the role of ports in the development of a nation. *Aquat. Procedia* **2015**, *4*, 295–301. [CrossRef]
109. Porter, M.E. *Competitive Advantage: Creating and Sustaining Superior Performance*, 2nd ed.; Free Press: New York, NY, USA, 1998; ISBN 978-06-8484-146-5.
110. Koeswayo, P.S.; Handoyo, S.; Abdul Hasyir, D. Investigating the relationship between public governance and the Corruption Perception Index. *Cogent Soc. Sci.* **2024**, *10*, 2342513. [CrossRef]
111. Wainer, L.; Daels, M. What makes local planning effective in a megacity? The overlapping agendas and scale inconsistencies in developing Buenos Aires' affordable land markets. *Plan. Theory Pract.* **2024**, *25*, 1–23. [CrossRef]
112. Schreiber, J.B.; Nora, A.; Stage, F.K.; Barlow, E.A.; King, J. Reporting structural equation modeling and confirmatory factor analysis results: A review. *J. Educ. Res.* **2006**, *99*, 323–338. [CrossRef]
113. Sarstedt, M.; Hair, J.F.; Ringle, C.M. PLS-SEM: Indeed a silver bullet—Retrospective observations and recent advances. *J. Mark. Theory Pract.* **2023**, *31*, 261–275. [CrossRef]
114. Henseler, J.; Ringle, C.M.; Sarstedt, M. A new criterion for assessing discriminant validity in variance-based structural equation modeling. *J. Acad. Mark. Sci.* **2015**, *43*, 115–135. [CrossRef]
115. Hair, J.F.; Hult, G.T.M.; Ringle, C.M.; Sarstedt, M.; Danks, N.P.; Ray, S. *Partial Least Squares Structural Equation Modeling (PLS-SEM) Using R: A Workbook*; Research Methods in Applied Linguistics (Vol. 1, Issue 3); Springer International Publishing: Cham, Switzerland, 2021; ISBN 978-3-030-80519-7. [CrossRef]
116. Wong, K. Mediation analysis, categorical moderation analysis, and higher-order constructs modeling in Partial Least Squares Structural Equation Modeling (PLS-SEM): A B2B Example using SmartPLS. *Mark. Bull.* **2016**, *26*, 1–22.
117. Saleh, E.; Mdanat, M.; Alsoud, A. Sustainability of Fiscal and Monetary Policies under Fixed Exchange Rate Regime in Jordan. *Sustainability* **2023**, *15*, 14625. [CrossRef]

sustainability

MDPI

Article

Impact of Transport Trends on Sustainability in the Western Balkans: A Future-Oriented Business Sector Perspective

Tomislav Letnik, Katja Hanžič, Matej Mencinger * and Drago Sever

Faculty of Civil Engineering, Transportation Engineering and Architecture, University of Maribor, Smetanova 17, 2000 Maribor, Slovenia; tomislav.letnik@um.si (T.L.); katja.hanzic@um.si (K.H.); drago.sever@um.si (D.S.)
* Correspondence: matej.mencinger@um.si

Abstract: This study examines the impact of emerging transport trends on the Western Balkans and their potential to improve the sustainability and efficiency of the transport sector in the region. In the context of global efforts to reduce environmental impacts and improve transport safety, the study aims to better understand the perspective of the business sector in order to align strategic planning for the introduction of new transport technologies and practices with the expectations of the business sector. Using a questionnaire-based survey on the Mentimeter platform, the study gathers insights from 49 transport and logistics experts. The analysis uses violin plots and heat maps to visualise the expected impact of the identified trends. The results show a consensus among the experts that the emerging trends are likely to lead to a reduction in accidents, congestion, and emissions, while at the same time increasing operating costs and investment in infrastructure. In particular, trends such as alternative fuels, electrification, and smart city initiatives are highlighted as important drivers for reducing emissions. The study concludes that while these trends are likely to lead to the achievement of environmental and safety goals, they also lead to economic challenges as operating costs increase and significant infrastructure investment is required. The study emphasises the need for strategic investment, policy adjustments, and capacity building to lead the Western Balkans towards a sustainable transport future.

Keywords: transport trends; congestion; accidents; infrastructure investment; operational costs; policy adaptation

1. Introduction

Transport is an essential part of modern society. It enables the smooth movement of people, goods, and services and plays a central role in global networking and economic growth [1,2]. However, its significant impact on the environment, characterised by pollution, accidents, and congestion, is a growing concern [3]. These negative impacts are exacerbated by the sector's vulnerability to unpredictable global events such as pandemics, economic recessions, geopolitical tensions, and fluctuating energy prices [4,5]. These dynamics emphasise the urgent need for resilient transport systems supported by diverse infrastructures that can adapt to changing circumstances.

In response to these challenges, the European Union (EU) has taken proactive steps to promote a resilient and sustainable transport system [6]. Recognising the integral role of transport in society and its impact on the environment, the EU has embarked on a path towards sustainability and anchored these goals in ambitious initiatives such as the Green Deal and the Fit for 55 programme [7,8]. These comprehensive strategies aim to radically reshape the continent's environmental landscape. They aim to achieve a significant reduction in greenhouse gas emissions and a gradual transition to a carbon-free society. EU policy focuses in particular on developing a transport system that takes into account not only the environmental but also economic, social, and safety aspects [6]. The EU's approach is to improve the resilience and adaptability of transport infrastructure to withstand various types of disruptions, from natural disasters to socio-economic shocks [9]. This includes investing in sustainable and versatile modes of transport, such as electric and hydrogen-powered vehicles, and improving public transport systems to reduce dependence on private cars.

Furthermore, the EU's vision is in line with the concept of Society 5.0, which envisages a harmonious fusion of digitalization and environmental awareness. This forward-looking approach is transforming the transport sector and emphasising the importance of integrating sustainable practices into its core activities [10]. The introduction of intelligent transport systems that utilise advanced technologies such as AI, IoT, and big data is central to this transformation [11,12]. These systems improve the efficiency, safety, and sustainability of transport while providing adaptable solutions for rapidly changing scenarios. In addition, accessibility and inclusivity are important aspects of ensuring that all members of society have equal access to efficient and sustainable transport options, regardless of their socio-economic background. This is crucial for social cohesion, economic equality, and the overall quality of life.

Despite the momentum of global initiatives to improve transport systems, their regional impact needs to be carefully analysed. As Europe sets out to put these transformative strategies into action, it becomes clear that the impact and feasibility of these initiatives can vary greatly from region to region, as each region has its own unique challenges and contexts [13]. This brings us to the specific focus of our study: the Western Balkans. Distinct in its geographical and socio-economic characteristics, the Western Balkans is an important area for analysing the implementation and effectiveness of EU transport policy. Our analysis aims to shed light on how the Western Balkans, with their strategic importance and specific needs, can adapt to and benefit from the EU's vision for a sustainable transport future.

The Western Balkans, located at the crossroads of important cultural and trade routes, have always been a nexus of various influences [14]. However, this strategic location has not been fully reflected in a robust transport infrastructure. The region faces significant infrastructure bottlenecks that hinder efficient mobility and economic growth [15,16]. Major challenges include an underdeveloped road and rail network, limited transport corridors, and inadequate maintenance of existing infrastructure. These inadequacies not only affect regional connectivity but also the prospects for international trade and tourism. One of the most pressing problems is the lack of comprehensive transport corridors. The existing network is often characterised by missing links and sections that do not meet modern standards, which affects the efficiency and safety of transport. This is particularly evident in cross-border areas, where different standards and systems in different countries further complicate the situation. The fragmentation of transport infrastructure hinders the seamless movement of goods and people, which is a crucial factor for regional integration and economic development [17].

In addition, the Western Balkans face low levels of transport and logistics services, which is reflected in the region's performance in various logistics indices, such as the Logistics Performance Index (LPI). This is attributed to several factors, including inadequate infrastructure, bureaucratic hurdles, inefficient customs procedures, and a lack of modern logistics and supply chain management procedures [18–20]. As a result, the region is struggling to attract significant foreign investment and effectively integrate into global supply chains. The development of the transport sector in the Western Balkans must be an urgent priority if integration into the EU is to be successful. This includes not only the modernisation and expansion of physical infrastructure but also the implementation of policy reforms to facilitate cross-border transport, the introduction of advanced logistics and transport management systems, and the promotion of regional cooperation to ensure a harmonised approach to transport challenges [21]. By addressing these issues, the Western Balkans can capitalise on their geographical position, strengthen their role in international trade, and pave the way for sustainable economic growth and regional integration.

As can be seen, the Western Balkans face many challenges and are therefore worth considering as a study case in the field of the transport sector. Many researchers have dealt with transport issues in this region, but none of them are from the perspective of emerging trends, especially from the perspective of the business sector. It has been shown that business stakeholders have an advanced understanding of emerging technologies and their likely impacts, as well as strategic practices, which could be of great benefit for policymaking [22]. Their opinion is often not taken into account in this very demanding process. For this study, a comprehensive but straightforward methodology was developed to gather stakeholders' opinions and assess the most promising trends that could drive the development of a sustainable transport system in the Western Balkans. In this way, we provide, for the first time, a comprehensive overview of the expected impact of emerging trends on emissions, congestion, accidents, operational costs, and infrastructure investments, which can inform decision-makers and make an important contribution to the discourse on sustainable transport from a regional perspective.

In the following sections, we provide an examination of the current transport landscape and a detailed analysis of key trends, including a literature review, and their potential impact on sustainability in the Western Balkans. Through this comprehensive approach, the study aims to provide valuable insights and practical recommendations for promoting a sustainable and resilient transport ecosystem in the region that meets both regional aspirations and global environmental goals.

2. Transportation System Trends—Literature Review

In order to understand the trends in the transport system, a comprehensive examination is required. For this study, we have analysed the academic literature to identify the most frequently cited trends in this area. By examining existing research and studies, we have gained valuable insights into the direction in which transport systems are evolving [23].

In this context, we analysed the identified trends in terms of their impact on two important fronts. On the one hand, we assess their impact on congestion, accidents, and emissions, as these are key indicators of the efficiency of a transport system and its environmental impact. On the other hand, we assess the trends in terms of costs related to infrastructure investments and operational expenses. This dual analysis is important to understand not only the benefits of these trends in improving the transport system but also their economic feasibility and their impact on future planning and development.

The identified global trends are presented in Table 1, which includes concise definitions of opportunities for the advancement of the transport system, thereby contextualising their significance. Furthermore, a curated list of the literature, reviewed for analysis, is also presented.

Table 1 shows the diverse range of trends shaping the future of transportation. Each trend represents a significant change in how transportation systems are designed, operated, and experienced. From the electrification of vehicles and the digitization of traffic systems to the development of intelligent transport systems and the adoption of new business models, these trends collectively aim to enhance the efficiency, safety, and sustainability of transport. While they offer substantial benefits, each also brings a set of challenges that require thoughtful planning and investment to fully realise their potential in shaping future transport systems.

The in-depth literature review focuses mainly on the expected impact on the transport sector. Table 2 outlines the expected impact of each identified trend on various aspects, such as congestion, accidents, emissions, operational costs, and infrastructure investment. When assessing the impact, not only the aspect of increase or decrease is considered, but also a scale ranging from 'potential' to 'likely', and 'expected' to 'significant'. In cases where divergent views are found in the literature, the term 'varies' is used to reflect these discrepancies. In addition, temporal aspects were also identified in certain cases, taking into account initial (meaning short-term) and long-term aspects.

Table 1. List of identified trends and corresponding literature.

Trend	Opportunities	Literature
Electrification	Shift towards electric vehicles (EVs) to reduce emissions and air pollution, including EV infrastructure development.	[24–31]
Digitization	Integration of digital technologies like IoT and AI into transportation for improved traffic management and logistics optimisation.	[11,12,26,27,32–34]
Automation	Incorporation of autonomous vehicles into the transport system to increase efficiency and reduce human error-related accidents.	[33,35–38]
Intelligent Transport Systems (ITS)	Use of advanced technologies in transportation infrastructure and vehicles for improved traffic management, safety, and efficiency.	[39–42]
Informatics Process	Application of IT and data analytics in transport systems for optimised operations and planning.	[43–50]
Change in Travel Habits	Shifts in preferences towards sustainable transport modes and teleworking.	[51–55]
Smart Cities and Communities	Development of urban areas integrating digital technologies, including transportation.	[10,43,44,56,57]
Increased Regionalisation	Creating shorter, localised supply chains to reduce transport route pressure and distances.	[58–63]
Alternative Fuels	Adoption of non-traditional fuels like electricity, hydrogen, and biofuels in transportation to reduce environmental impact.	[64–68]
Development of Transport Infrastructure	Expansion and modernisation of infrastructure to improve transportation efficiency and safety.	[69–73]
New Business and Logistics Models	The emergence of new models like e-commerce and shared mobility is changing transportation demand and usage.	[22,74–80]
Transport Techniques and Technology	Advances in vehicle design, traffic management systems, and routing algorithms.	[75,81–85]

Table 2. Assessment of trends and their influence on various transport aspects.

Trend	Congestion	Accidents	Emissions	Oper. Costs	Infr. Invest.
Electrification	Potential increase	Potential decrease	Significant decrease	Long-term decrease	High initial increase
Digitization	Significant decrease	Likely decrease	Decrease	Long-term decrease	Considerable initial increase
Automation	Likely decrease	Decrease	Likely decrease	Long-term decrease	High initial increase
Intelligent Transport S.	Decrease	Decrease	Decrease	Long-term decrease	Substantial initial increase
Informatics Process	Decrease	Likely decrease	Decrease	Long-term decrease	Significant initial increase
Change in Travel Habits	Likely decrease	Varies	Decrease	Varies	Initial increase (adoption)
Smart Cities and Comm.	Decrease	Decrease	Decrease	Long-term decrease	High initial increase

Table 2. *Cont.*

Trend	Congestion	Accidents	Emissions	Oper. Costs	Infr. Invest.
Increased Regionalisation	Decrease in major routes	Likely decrease	Decrease	Varies	Increase (adaptation)
Alternative Fuels	Varies	Varies	Significant decrease	Long-term decrease	High initial increase
Development of Transport Infrastructure	Varies	Likely decrease	Varies	Varies	Considerable increase
New Business and Logistics Models	Varies	Varies	Varies	Long-term decrease	Adaptation needed
Transport Techniques and Technology	Varies	Likely decrease	Decrease	Varies	Varies

As shown in Table 2, the trends identified through a comprehensive review of the relevant literature show multiple implications for the transport sector. These trends not only highlight the dynamic interplay between technological innovation, infrastructural change, and societal shifts but also emphasise the evolving nature of transportation challenges and opportunities.

The trend of electrification brings a significant reduction in emissions, a major benefit of switching to electric vehicles (EVs), as highlighted in references [24–26]. However, due to the increasing popularity of EVs, this trend could lead to more traffic congestion, particularly in urban areas, due to the growing popularity of EVs, as noted in [27]. The integration of advanced driver-assistance systems in EVs indicates a potential reduction in accidents [28], although the need for significant investment in charging infrastructure is a notable consideration [29]. Over time, a decrease in operational costs is expected due to the higher energy efficiency of EVs [30,31].

Digitization, which includes the implementation of IoT and AI in traffic management, is poised to reduce traffic congestion significantly [26]. Enhanced data analytics can lead to more efficient transport and logistics processes and reduce the time vehicles spend on the road, thereby also reducing emissions [11,32]. This trend is also likely to decrease accidents by integrating advanced safety features into vehicles, as discussed in [12]. While the initial investments in digital infrastructure are considerable [27], the long-term perspective points towards reduced operational costs due to improved efficiency in transport logistics [33,34].

Automation, particularly the integration of autonomous vehicles, is expected to ease congestion [35], decrease emissions and fuel consumption [36], and reduce accidents, as these vehicles are designed to minimise human error [37]. Significant investments in infrastructure are required to support this transition [38], but the promise of long-term savings in operational costs due to reduced maintenance and fuel expenses is a key factor [33].

Intelligent Transport Systems (ITS) are anticipated to decrease congestion and accidents through enhanced traffic efficiency and safety features [39]. The implementation of ITS contributes to a reduction in emissions by optimising traffic flow and vehicle operations [40]. In addition, Cooperative Intelligent Transport Systems (C-ITS) enhance these capabilities by enabling direct communication between vehicles and road infrastructure [41]. While the initial investment for ITS is substantial, the potential for long-term operational savings is significant [39,42].

The informatics process involves using IT and data analytics to optimise transport operations, expected to decrease congestion [43–45]. It also has the potential to improve safety and reduce accidents through data-driven decision-making [46–48]. The reduction in emissions is another anticipated benefit [46], though the initial investment in IT infrastructure and data systems is noteworthy [49,50].

Changes in travel habits reflect a societal shift towards more sustainable transport modes, likely to decrease congestion [51,52] and emissions [53]. The impact on accidents and operational costs can vary depending on how transport infrastructure adapts to these changing patterns [54]. This trend also indicates a need for investment in public transport systems, cycle paths, and pedestrian walkways [55].

The development of smart cities and communities is expected to decrease urban congestion through the application of digital technologies for traffic optimisation [44]. These technologies also contribute to reducing emissions and improving safety [10,43]. The initial investment for developing smart city infrastructure is substantial [56,57] but is offset by the potential for long-term operational savings [10].

Increased regionalisation suggests a decrease in congestion on major transport routes as businesses shift to shorter, localised supply chains [58,59]. This change is also expected to decrease emissions and may reduce the risk of accidents [60]. However, the impact on operational costs [61] and the need for investment in regional transport infrastructure [62,63] can vary.

The adoption of alternative fuels is likely to significantly reduce emissions [64], particularly when EVs are powered by renewable energy sources [65]. The impact of this trend on congestion and accidents varies depending on the adoption rate and the characteristics of alternative fuel vehicles [66]. The shift to alternative fuels requires considerable investment in infrastructure, such as charging and fuel stations [67]. However, these initial significant investments are offset by long-term operational savings due to the higher energy efficiency, lower maintenance requirements of alternative fuel vehicles, internalisation of external costs, and economies of scale [66,68].

The development of transport infrastructure can have varying impacts on congestion and accidents, depending on the nature of the infrastructure and the extent of its development [69]. This expansion is crucial for reducing congestion and improving road safety, but it may also lead to increased traffic volumes and emissions [70,71]. The associated operational costs [72] and the need for investment in infrastructure development are significant considerations [71,73].

New business and logistics models have the potential to change traffic patterns, impacting congestion and accidents [74,75]. The increase in delivery vehicles due to e-commerce, for example, may contribute to urban congestion [22,76]. However, innovations in logistics efficiency could mitigate this impact [77,78]. The need for adjustments in transport infrastructure to support these evolving models, along with their influence on operational costs, is noteworthy [79,80].

Lastly, transport techniques and technology encompass advances in vehicle design and traffic management systems [81–83]. While these advancements have the potential to decrease emissions and improve safety, their impact on congestion and operational costs varies [84,85]. The investment required in technology development and implementation is an important factor in realising the benefits of these techniques [75].

3. Impact Assessment on the Western Balkans—Survey Methodology and Results

The methodology focuses on analysing the potential impact of various emerging trends on different aspects of transport. The main objective is to understand how the trends identified and explained in the previous section will affect congestion, accidents, emissions, operational costs, and infrastructure investments from a business expert point of view.

3.1. Methods and Material

The study employs a questionnaire-based survey method, utilising the Mentimeter platform for data collection (see Appendix A for the graphical results of the Mentimeter survey). Each of the identified trends was orally presented before experts were asked to evaluate the potential impact of each trend on congestion, accidents, emissions, operational costs, and infrastructure investment for the period until 2030 on a scale of −3 to 3. A score of −3 indicated that a particular trend is expected to have a high influence on reducing considered aspects, while a score of 3 indicated a high increase.

The main aim of the research is to uncover whether there is any specific trend that will have a particularly high impact on a certain aspect. Conversely, the aim is also to find out whether there is an aspect that will be particularly high across all trends. In essence, this study provides a forward-looking examination of the complex interaction between emerging trends and their potential impacts on different aspects of transport and logistics. The findings are expected to provide valuable insights for researchers, policymakers, and practitioners to enable informed decision-making for a sustainable and efficient future of transportation.

The expert group was composed of 49 professionals, mainly from the business sector, representing 76% of the total. In terms of experience, the majority of respondents had more than 15 years of experience (42%), followed by those with 8–15 years (30%), 2–8 years (26%), and a minority with less than 2 years (2%). In terms of educational qualifications, most have a Master's degree (57%), followed by a Bachelor's degree (26%), and 8% each have a PhD or other educational qualifications. Professionally, the experts were split between Transportation and Traffic Engineering (47%), Economics and Law (28%), and a combination of Civil, Mechanical, and Electrical Engineering (6%), with the remainder (18%) coming from various other fields. Geographically, the vast majority were from Serbia (82%), with the remainder from Bosnia and Herzegovina (6%), Montenegro (4%), Macedonia (4%), and North Macedonia (2%).

The sample size for this study, comprising 49 experts mainly from the business sector in the Western Balkans, provides a statistically sound basis for the analysis conducted. While larger samples can often increase the generalizability of results, the sample size of nearly fifty professionals is adequate for complex multivariate techniques, especially when the respondents are subject matter experts. Given that these individuals bring a high level of domain-specific knowledge, their insights are likely to be particularly informative and nuanced, thereby increasing the relevance of the results. Furthermore, the homogeneity of the respondents in terms of their professional background and regional focus increases the contextual relevance of the study and makes it a valuable contribution to understanding the impact of transport and logistics trends in this specific economic and geographical milieu.

3.2. Survey Results

In this section, we first discuss the statistical relevance of the survey, followed by an analysis of the distribution of responses, as depicted in Figure 1, which shows a violin plot across five aspects. Figures 2 and 3 show the median and mean distributions of the responses, respectively. Finally, we present some partial results of the relationship maps.

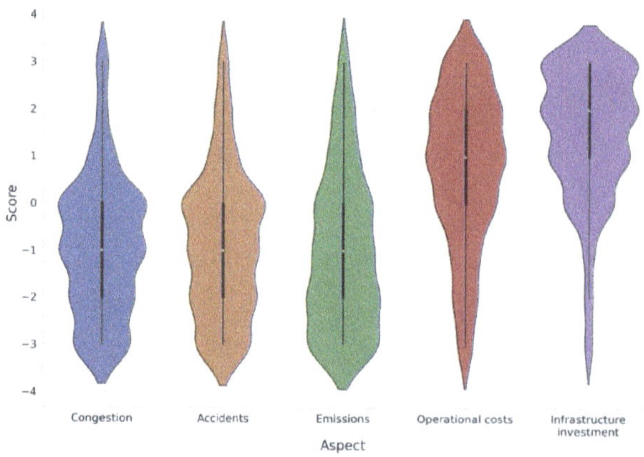

Figure 1. Violin plot distribution of scores for each aspect across all trends.

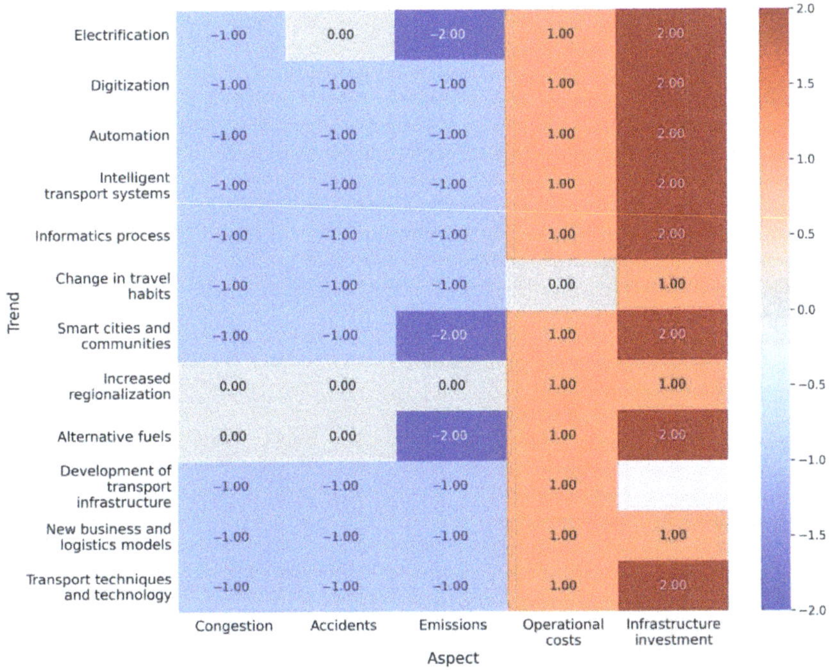

Figure 2. Median impact score heatmap of each trend on analysed aspects.

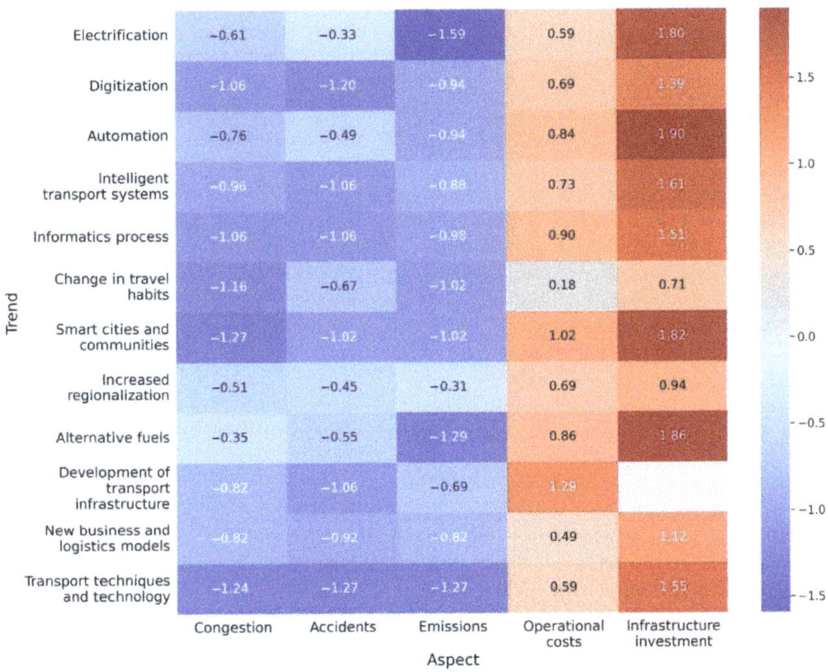

Figure 3. Mean impact score heatmap of each trend on analysed aspects.

3.2.1. Statistical Relevance of the Survey

To assess the statistical relevance of the survey and its results, several statistical tests were performed with SPSS (version 29.0, IBM, Chicago, IL, USA). For 12 trends and five aspects, we used 7-point Lickert Scale questions (including a neutral option) to collect the information on all 60 pairs that appeared in the survey. To test the reliability of all five variables along the 12 trends, we used Cronbach's alpha coefficient. A value of 0.912 was obtained in this study. Please note that the results are considered reliable if the Cronbach's alpha is above 0.6 [86,87].

The factor analysis indicates that the selected factors are relevant for explaining the variability of the assessed impacts. The commonalities across all 12 trends and the five aspects (congestion, accidents, emissions, operational costs, and infrastructure investments) are presented in Table 3.

Table 3. Communalities of aspects over the trends.

Trend	Congestion	Accidents	Emissions	Oper. Costs	Infr. Invest.
Electrification	0.067	0.371	0.734	0.799	0.766
Digitization	0.726	0.676	0.375	0.779	0.785
Automation	0.630	0.799	0.370	0.729	0.736
Intelligent transport systems	0.670	0.636	0.473	0.718	0.717
Informatics process	0.842	0.898	0.611	0.684	0.719
Change in travel habits	0.628	0.766	0.548	0.304	0.955
Smart cities and comm.	0.765	0.758	0.707	0.791	0.757
Increased regionalisation	0.828	0.804	0.752	0.802	0.815
Alternative fuels	0.658	0.725	0.467	0.814	0.832
Dev. of transport infr.	0.800	0.732	0.731	0.026	
New bus. and log. models	0.768	0.747	0.667	0.764	0.710
Transp. techn. and technol.	0.869	0.837	0.773	0.688	0.628

High commonalities, such as those observed for "Development of transport infrastructure" in congestion (0.800) and emissions (0.731), indicate a significant shared variance. On average, the factors account for approximately 71.7% of the variance across all measures, which is a robust indication of their relevance. Nevertheless, some trends, such as "Electrification" and its influence on congestion, show low communalities (0.067), indicating more complex relationships or external variables not captured by the analysis. Despite these outliers, the statistical evidence broadly supports the pertinence of the selected factors in transportation and logistics impact studies.

The Kruskal–Wallis test, a non-parametric statistical test, was used to compare the medians across different categories of transportation trends. The summary of the results is presented in Table 4.

Table 4. Kruskal–Wallis test.

	Hypothesis Test Summary			
	Null Hypothesis: The Medians of ASPECTS Are the Same across Categories of Trend	**Test**	**Significance (p)**	**Decision (Null Hypothesis)**
1	ASPECT: Congestion	Independent-Samples Median Test	<0.001	Reject
2	ASPECT: Accidents	Independent-Samples Median Test	0.002	Reject
3	ASPECT: Emissions	Independent-Samples Median Test	0.074	Retain
4	ASPECT: Operational costs	Independent-Samples Median Test	0.316	Retain
5	ASPECT: Infrastructure investment	Independent-Samples Median Test	0.005	Reject

For the Kruskal–Wallis test, the expected significance level is $p < 0.050$. The test results show significant differences in the medians for congestion ($p < 0.001$), accidents ($p = 0.002$), and infrastructure investment ($p = 0.005$), which leads us to reject the null hypothesis for these variables. These results suggest that these particular aspects of transportation are influenced by the trends in question. In contrast, the median for emissions ($p = 0.074$) approached the threshold of significance ($p = 0.05$), suggesting a potential impact of trends on emissions, whereas the median for operational costs ($p = 0.316$) showed no significant difference, meaning that we cannot reject the null hypothesis. The latter emphasises that operational costs are less relevant for selected trends.

3.2.2. Distribution of the Results

A violin plot presentation (Figure 1) is used to show the distribution of scores for each transport aspect across different trends. Each "violin" represents one aspect. The width of a violin at a given level represents the proportion of responses at that level, so a wider section of a violin indicates a higher density of responses. The violin plot and the heatmap (in continuation of the article) were created with Python (version 3.8), using the data visualisation functions of the Seaborn library.

As can be seen from the violin plots, most experts believe that the trends will lead to an overall reduction in accidents, congestion, and emissions while increasing operational costs and infrastructure investment.

In the case of "Congestion", "Accidents", and "Emissions", the median score is $\tilde{x} = -1$, indicating a consensus among the experts that these trends will have a moderately positive impact (i.e., a reduction) on these aspects. Specifically, the mean scores for "Congestion", "Accidents", and "Emissions" are $\overline{x} = -0.88$, -0.84, and -0.98, respectively. Furthermore, the 25th and 75th percentiles for each of these subcategories are -2 and 0, indicating that half of the scores lie within this range, demonstrating a moderate to strong belief in the positive impact.

Some important differences can be observed in the violin charts of "Operational costs" and "Infrastructure investment". The median and mean score for "Operational costs" are $\tilde{x} = 1$ and $\overline{x} = 0.74$, respectively, which indicates an expected increase in these costs. For "Infrastructure investment", the median score is $\tilde{x} = 2$ and the mean score is $\overline{x} = 1.47$, which implies a stronger belief in a significant increase in infrastructure investment. In addition, the 25th and 75th percentiles for "Operational costs" are in the range from 0 to 2, while for "Infrastructure investment", they are between 1 and 3. This indicates that for "Operational costs", experts' opinions vary from no change to a substantial increase, whereas for "Infrastructure investment", there is a stronger consensus towards a moderate to a large increase.

As can be seen, the median and mean scores vary between different aspects when all trends are considered together. For a more detailed overview, first the median and then the mean scores for each trend and each aspect are analysed as follows.

3.2.3. Median Impact Scores of Each Trend on Each Analysed Aspect

For the analysis of median impact, a heatmap is shown in Figure 2. The colour of each cell in the heatmap corresponds to the median impact score, with blue colours representing negative impact scores and red colours representing positive impact scores. The number in each cell is the actual median impact score.

The analysis shows that the experts largely agree on the positive effects of most trends on reducing traffic congestion, accidents, and emissions ($\tilde{x} = -1$). Exceptions are observed for: "Alternative fuels" with a neutral impact on congestion and accidents ($\tilde{x} = 0$) and a significant impact on reducing emissions ($\tilde{x} = -2$); "Electrification" with a neutral impact on accidents ($\tilde{x} = 0$) and a significant impact on emissions ($\tilde{x} = -2$), and "Increased regionalisation" with a neutral impact on congestion, accidents, and emissions ($\tilde{x} = 0$).

On the contrary, increased operational costs are expected for most trends ($\tilde{x} = 1$), except for "Change in travel habits", which is considered neutral ($\tilde{x} = 0$). Similarly, significant infrastructure investments ($\tilde{x} = 2$) are expected for most trends ($\tilde{x} = 2$). In this case, "Change in travel habit", "Increased regionalisation" and "New business and logistics models" are expected to have a neutral impact ($\widetilde{x} = 0$).

In summary, the heatmap analysis shows a consensus among the experts that the trends analysed are expected to have the greatest impact on reducing emissions and will have a major impact on infrastructure investment. The highest scores in relation to this observation are recorded for "Electrification", "Smart cities and communities" and "Alternative fuels".

3.2.4. Mean Impact Scores of Each Trend on Analysed Aspects

In addition to the median scores presented in the previous section, the mean impact scores were also analysed. This approach makes it possible not only to recognise the most likely impacts but also to understand potential extremes. This provides policymakers and stakeholders with a more complex and robust understanding of future dynamics in the sector to be considered for comprehensive strategic planning and risk assessment.

Also in this case, the heatmap (Figure 3) provides a graphical representation of the impact score of each trend for each aspect. The colour of each cell in the heatmap this time corresponds to the mean impact score, with blue colours representing negative impact scores and red colours representing positive impact scores. The number in each cell is the actual mean impact score.

As can be seen from Figure 3, the heatmap of the mean impact score represents a much more differentiated situation. The following conclusions can be drawn for a particular aspect:

- Congestion: "Smart cities and communities" ($\overline{x} = -1.27$), "Transport techniques and technology" ($\overline{x} = -1.24$), and "Digitization" ($\overline{x} = -1.06$) have the most negative mean impact scores, which indicates considerable potential for alleviating traffic congestion;
- Accidents: "Transport techniques and technology" ($\overline{x} = -1.27$), "Digitization" ($\overline{x} = -1.20$), and "Intelligent transport systems" ($\overline{x} = -1.06$) have the most negative mean impact scores, indicating a strong potential for reducing accidents;
- Emissions: "Electrification" ($\overline{x} = -1.59$), "Alternative fuels" ($\overline{x} = -1.29$), and "Transport techniques and technology" ($\overline{x} = -1.27$) have the most negative mean impact scores, indicating a strong potential for reducing emissions;
- Operational costs: "Development of transport infrastructure" ($\overline{x} = 1.29$), "Smart cities and communities' ($\overline{x} = 1.02$), and "Automation" ($\overline{x} = 0.84$) have the highest mean scores, implying that this trend will have the greatest impact on increasing operational costs;
- Infrastructure investments: "Automation" ($\overline{x} = 1.90$), "Alternative fuels" ($\overline{x} = 1.86$), and "Smart cities and communities" ($\overline{x} = 1.82$) have the most positive mean impact scores, indicating the highest expected impact on infrastructure investment.

Experts firmly believe in "Transport techniques and technology" as a trend that is expected to have a significant impact on all three negative aspects of transport: congestion, accidents, and emissions. On the other hand, "Smart cities and communities" are expected to have the greatest impact on operational and investment costs combined. Furthermore, "Electrification" is expected to have the largest impact on emissions; "Transport techniques and technology" is expected to have the predominant impact on accidents; and "Smart cities and communities" are expected to have the greatest impact on congestion. In relative terms, "Change in travel habits" is expected to have the most positive impact on congestion, accidents, and emissions, while the need for infrastructure investment is rather modest.

3.2.5. Interpretation of the Relationship Map

SPSS version 29 also allows the functionality to create relationship maps in the form of graphs, which enable the examination of connections between individual aspects for each trend. The strongest relationships illustrate the prevailing opinion of the survey cohort and consequently the most transformative changes in transport, according to the proposed survey.

The in-build algorithm counts the number of connections between nodes (in our case, −3, −2, −1, 0, 1, 2, and 3) of different colours (in our case, aspects: accidents, congestion, emissions, infrastructure investments, and operational costs) and returns a radius of a node that is proportional to the number of links (edges in the relationship map graph). In addition, the thickness of the links between the nodes is proportional to the number of identical answers. The relationship map graphs were created for all 12 trends using the following settings: the link size was 10 (the thickest link is 10 times wider than the thinnest), and the threshold for the presented link between two nodes was the existence of five connections between them. The node size is proportional to the number of links connected to a given node. According to the above, for each trend in the corresponding relationship map, the most significant relationships between different aspects correspond to the thickest links between two nodes (see the example below). Using these settings, the resulting graphs ensure clarity and readability for all 12 trends. For the sake of brevity, we present only a relationship map for the electrification trend (Figure 4); for other trends, we provide below just the main observations.

As seen from Figure 4 for the electrification trend, the strongest relationship is obtained between congestion and infrastructure investment. Since this is the link between −3 and +3, it indicates the inverse relation between emissions and infrastructure investment (thus, increasing electrification results in reducing emissions on the one hand and is conditioned by infrastructure investment on the other).

In examining the relationship maps between all emerging transportation trends and their impacts on key sectoral aspects, we can identify some groups of trends that encapsulate the key patterns between aspects observed in the dataset.

The first group encompasses "Electrification", "Automation", and "Intelligent Transport Systems". These trends are unified by a salient relationship between a pronounced decrease in emissions (−3) and an associated increase in infrastructure investments (+3). The trends within this group represent a paradigm shift towards environmentally sustainable transportation solutions that, while significantly reducing emissions, also require substantial infrastructure outlays. Notably, "Intelligent Transport Systems" differentiates itself within this grouping by additionally exerting a considerable negative impact on congestion (−3), highlighting its dual potential to enhance not only environmental sustainability but also traffic flow efficiency.

The second group is characterised by the trend of "Digitization" alone, which shows a uniform reduction in congestion, accidents, and emissions (−2 for each), illustrating the multifaceted efficiencies enabled by digital technologies, suggesting an ability to enhance transportation systems in terms of safety, environmental impact, and traffic management without precipitating a proportional increase in infrastructure expenditure. This indicates the potential for digitalization to yield a high return on investment in the form of broad-ranging sectoral improvements.

The third group, consisting of "Smart Cities and Communities", demonstrates a marked reduction in both congestion and emissions (−3 for each). This group suggests a focus on leveraging technology within urban planning to optimise existing infrastructure thereby presenting a potentially cost-effective approach to addressing urban transportation challenges. Unlike the first group, this one does not correlate these benefits with a significant rise in infrastructure investments, indicating an optimisation of current resources rather than an addition of new ones.

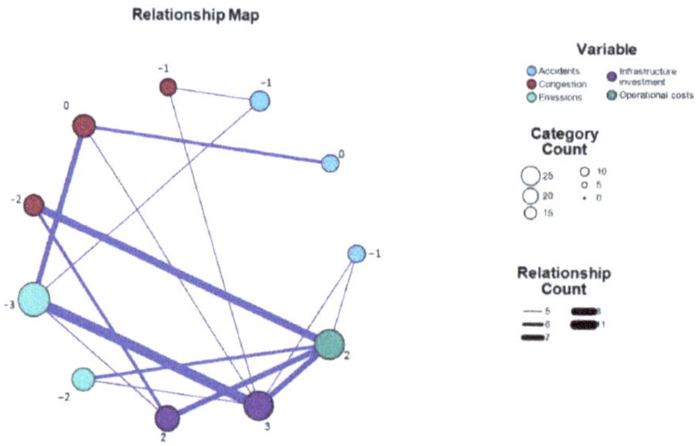

Figure 4. Relationship map for electrification trend.

The identification of these groups provides a reasonable framework for understanding how each trend interplays with different aspects of the transportation sector. This group analysis is instrumental for policymakers and urban planners as it identifies critical areas for targeted investments and research, guiding strategic decision-making towards the realisation of a more efficient, safe, and sustainable transportation future.

4. Discussion and Implications

The discussion is based on the main findings presented in the previous section and is consistent with the analysis of the relationships between the aspects analysed for the trends under consideration, highlighting the most uniform agreement among the experts regarding the developments to be expected for the Western Balkans in the future.

4.1. Discussion and Implications Based on the Expert Survey

The survey results show a clear expectation among experts that technology-driven trends such as "Transport techniques and technology", "Smart cities and communities", and "Digitalisation" will have the greatest positive impact when all three aspects (congestion, accidents, and emissions) are considered together. These trends are likely to be valued for their transformative potential, enabled by digital innovation and smarter infrastructures. On the other hand, these trends are expected to significantly increase investment in transport infrastructure at the same time. This conclusion is a clear signal to policymakers at the EU and Western Balkan levels that the greatest success in reducing the negative aspects of transport can only be expected if adequate funding and development of transport infrastructure are in place.

On the other hand, "Changing travel habits", "Increased regionalisation", and "New business and logistics models" are likely to have a positive impact on congestion, accidents, and emissions, while the need for infrastructure investment is relatively modest. These trends are so-called "soft" trends that can be implemented by changing the organisational aspects of transport processes. The introduction of these trends promises environmental and economic benefits without the need for large investments. It is suggested that policymakers in the EU and the Western Balkans give greater importance to these trends if they lack investment resources and want to promote sustainable transport.

Policymakers and industry leaders need to address the economic challenges associated with the introduction of green technologies by balancing environmental and safety improvements with financial realities. For the Western Balkans, this balance is crucial, as economic constraints can hinder rapid technological adaptation.

Finally, the experts have not identified any positive impact of the analysed trends on operating costs. One would expect new technologies to reduce operating costs. This finding is particularly true for regions such as the Western Balkans, where technological adaptation requires significant adjustments and business opportunities are expected in the long-term.

4.2. Comparison of Differences among the Literature Review and Expert Survey

When synthesising the comparative analysis between the literature review and the experts' responses, several key differences and similarities become apparent.

The literature review shows a broad spectrum of potential impacts across transportation trends, with a focus on technological and infrastructural developments. The scientific discourse points to a comprehensive shift towards environmentally friendly technologies, highlighting a spectrum of outcomes from "Electrification", "Digitization", and "Automation". The expert survey, on the other hand, places more emphasis on the practical implications and real-world feasibility of these trends, particularly focusing on the immediate and tangible impacts such as emissions reductions and the need for extensive infrastructure investments. A clear difference is that the experts emphasise the immediate need for infrastructure investment to support the adoption of advanced transportation technologies. In the literature, this tends to be discussed in the context of long-term strategic planning, while the experts emphasise it as an urgent, near-term concern, possibly reflecting their experience on the ground and their direct involvement with the current state of the sector.

Both the literature and the expert survey agree that advanced technologies play an important role in the transformation of the transportation sector. There is a consensus that trends such as "Intelligent Transport Systems", "Smart Cities and Communities", and "Digitization" have the potential to significantly reduce congestion and emissions. This consensus underlines the collective recognition of the value that these technologies have in creating more sustainable and efficient transportation systems. Another similarity is the recognition of the environmental benefits of the transition to "Alternative Fuels" and "Electrification". Both sources agree that these trends are capable of reducing emissions, which is critical to achieving sustainability goals in transportation.

4.3. Importance of Selected Trends for the Western Balkans

When looking at the 12 transportation trends discussed, their relevance for the Western Balkans hinges on the region's priorities, such as EU integration, economic development, and environmental sustainability. "Electrification" and "Intelligent Transport Systems" are particularly relevant, given the region's goals to align with EU environmental standards and to modernise its transportation infrastructure. "Electrification" addresses the urgent need for sustainable transport solutions, while "Intelligent Transport Systems" can significantly improve traffic management and safety, which is crucial for the region's growing transport networks.

Furthermore, the development of "Smart Cities and Communities" is highly relevant, considering the urbanisation trends and the need for efficient, tech-driven city planning in the region's expanding metropolitan areas. Emphasising smart solutions can lead to an improved quality of life and better integration into European urban networks.

Finally, "Increased Regionalization" reflects the Western Balkan countries' efforts to strengthen local economies and reduce dependence on distant markets by strengthening intra-regional connectivity and supply chains. This trend is in line with the region's strategic goal of creating a more self-sufficient and resilient economic structure.

These trends are characterised by their alignment with the Western Balkans' development path and the region's efforts to align with broader European standards while addressing its own transport and economic challenges.

4.4. Limitations, Applicability, and Recommendations for Future Research

The main limitation of the study is the fact that the survey was only conducted among a specific group of experts in a specific region. To improve the methodology, a Europe-wide survey could be conducted to allow comparisons based on different levels of infrastructure development. This survey should involve different stakeholders from the public and private sectors to understand the different perceptions that are crucial for the development of strategies and policies, practical solutions, and the implementation of new technologies and approaches.

Experts specialising in specific areas of transport (accidents, congestion, and emissions) could deepen the study and provide a more detailed understanding of the significance of individual trends. The results of this study can serve as a basis and guide for further, in-depth research.

Given the rapid development of technology and the fast pace at which new trends emerge, it is essential to constantly update the methodology to take into account new, emerging trends and phase out those that have become established.

5. Conclusions

This study provides an important roadmap for regions such as the Western Balkans and other regions with similar dynamics, guiding them through the complex transition towards more sustainable and efficient transport systems. Drawing on extensive expert opinions and data, the findings emphasise the need for strategic investment and policy adjustment to ensure that the transition to an advanced transportation framework is environmentally, economically, and socially sustainable.

The research emphasises the importance of a balanced approach to managing the benefits and challenges of emerging trends, particularly those related to green technologies and new infrastructure. These trends promise significant improvements in emissions reduction, safety, and congestion management, but they also entail considerations of higher operational costs and significant infrastructure investments.

For the Western Balkans, this transition represents both an opportunity and a challenge. The opportunity lies in potentially making the leap to an advanced transport system, positioning the region as an important player in sustainable transport solutions, enhancing global competitiveness, and contributing to climate change mitigation. However, the challenge lies in careful planning and resource allocation to manage the financial implications and ensure equitable access to these systems.

The crucial role of policymakers and stakeholders in this transition is evident. Policies that promote the adoption of environmentally sustainable technologies and support infrastructure development are essential. Innovative financing mechanisms, such as public–private partnerships and leveraging European Union funds, are especially relevant for Western Balkan countries in the context of EU integration. In addition, capacity building and workforce development are vital for the adoption of new technologies and systems. This includes not only technical training but also fostering a culture of innovation and adaptability within the transport sector.

In summary, this study provides a comprehensive guide for regions such as the Western Balkans to strategically shape the transition towards a sustainable transport system. By balancing environmental, economic, and social considerations and making informed policy decisions and investments, these regions can effectively harness the potential of emerging transport trends to build a more sustainable, efficient, and inclusive future.

Author Contributions: Conceptualisation, T.L. and D.S.; methodology, T.L. and K.H.; validation, T.L., M.M. and D.S.; formal analysis, T.L. and M.M.; investigation, D.S. and T.L.; data curation, K.H.; writing—original draft preparation, T.L. and K.H.; writing—review and editing, D.S.; visualisation, T.L.; supervision, D.S.; project administration, K.H.; funding acquisition, M.M. All authors have read and agreed to the published version of the manuscript.

Funding: This research was partially funded by ARIS (Slovenia), Research Programme No P1-0288.

Institutional Review Board Statement: Not applicable.

Informed Consent Statement: Informed consent was obtained from all subjects involved in the study.

Data Availability Statement: Generalized data are contained within the article. More detailed data is not publicly available for the privacy reasons.

Acknowledgments: We would like to express our sincere thanks to the many experts who responded to our survey and made this research possible.

Conflicts of Interest: The authors declare no conflicts of interest.

Appendix A

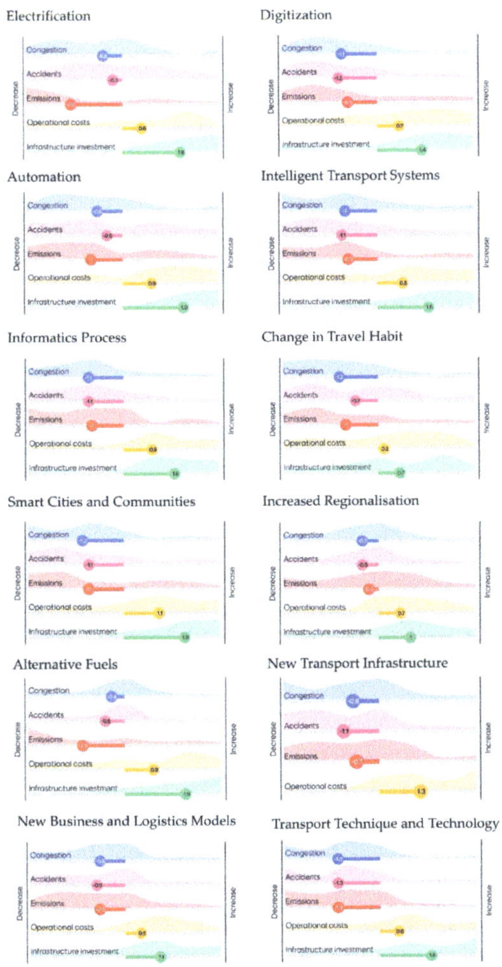

Figure A1. Graphical results of the Mentimeter survey.

References

1. Ketter, W.; Schroer, K.; Valogianni, K. *Information Systems Research for Smart Sustainable Mobility: A Framework and Call for Action*; INFORMS Institute for Operations Research and the Management Sciences: Catonsville, MD, USA, 2022; Volume 34.
2. Cigu, E.; Agheorghiesei, D.T.; Gavriluţă, A.F.; Toader, E. Transport Infrastructure Development, Public Performance and Long-Run Economic Growth: A Case Study for the Eu-28 Countries. *Sustainability* **2018**, *11*, 67. [CrossRef]
3. Jokanović, I.; Pavić, M.; Jokanović, I.; Pavić, M. Environmental Protection—Sustainable Development—Transport: Chronology of the Approach and Political-Strategic Framework. *AGG+* **2022**, *10*, 88–105. [CrossRef]
4. Khan, K.; Su, C.-W.; Tao, R.; Umar, M. How do geopolitical risks affect oil prices and freight rates? *Ocean Coast. Manag.* **2021**, *215*, 105955. [CrossRef]
5. Rothengatter, W.; Zhang, J.; Hayashi, Y.; Nosach, A.; Wang, K.; Oum, T.H. Pandemic waves and the time after COVID-19—Consequences for the transport sector. *Transp. Policy* **2021**, *110*, 225–237. [CrossRef] [PubMed]

6. European Commission. *Sustainable and Smart Mobility Strategy-Putting European Transport on Track for the Future;* COM(2020) 789 Final; European Commission: Brussels, Belgium, 2020.
7. European Commission. *The European Green Deal;* COM(2019) 640 Final; European Commission: Brussels, Belgium, 2019.
8. European Commission. *Fit for 55: Delivering the EU's 2030 Climate Target on the Way to Climate Neutrality;* COM(2021) 550 Final; European Commission: Brussels, Belgium, 2021.
9. Lim, Y.; Ninan, J.; Nooteboom, S.; Hertogh, M. Organizing resilient infrastructure initiatives: A study on conceptualization, motivation, and operation of ten initiatives in the Netherlands. *Resilient Cities Struct.* **2023**, *2*, 120–128. [CrossRef]
10. Roring, R.S.; How, B.C. Towards Society 5.0: A Pilot Study on Costless Smart Transportation Business Model. *Int. J. Bus. Soc.* **2022**, *23*, 73–87. [CrossRef]
11. Iliashenko, O.; Iliashenko, V.; Lukyanchenko, E. Big Data in Transport Modelling and Planning. *Transp. Res. Procedia* **2021**, *54*, 900–908. [CrossRef]
12. Baliyan, A.; Dhatterwal, J.S.; Kaswan, K.S.; Jain, V. Role of AI and IoT Techniques in Autonomous Transport Vehicles. In *AI Enabled IoT for Electrification and Connected Transportation. Transactions on Computer Systems and Networks;* Springer: Singapore, 2022; pp. 1–23. [CrossRef]
13. Ran, W.; Liu, R.; Liu, S. Area Differences in Regional Logistics Efficiency and the Law Governing Its Temporal and Spatial Evolution. *J. Adv. Transp.* **2022**, *2022*, 3596524. [CrossRef]
14. Chrzová, B.; Grabovac, A.; Hála, M.; Lalić, J.; Bashota, V.; Bjeloš, M.; Latal, S.; Naunov, M.; Semanić, H.; Wilson, A.; et al. *Western Balkans at the Crossroads: Assessing Influences of Non-Western External Actors;* Prague Security Studies Institute: Prague, Czech Republic, 2019.
15. Holzner, M.; Stehrer, R.; Vidovic, H. *Infrastructure Investment in the Western Balkans;* Vienna Institute for International Economic Studies: Vienna, Austira, 2015.
16. Roy, S.; Vulevic, A.; Hore, S.; Chaberek, G.; Mitra, S. Regional Classification of Serbian Railway Transport System through Efficient Synthetic Indicator. *Mechatronics Intell. Transp. Syst.* **2023**, *2*, 1–10. [CrossRef]
17. Charokopos, M. Networks as Political Tools in the Western Balkans: From the "Brotherhood and Unity" Highway to the Transport Community. *Eur. Asia. Stud.* **2022**, *74*, 1483–1506. [CrossRef]
18. Verlagsgesellschaft, N.; Affairs, S.; Europe, E. Logistical performance in the Balkans—Trend evidence. *J. Labour Soc. Aff. East. Eur.* **2016**, *19*, 223–238.
19. Karaduman, H.A.; Karaman-Akgül, A.; Çağlar, M.; Akbaş, H.E. The relationship between logistics performance and carbon emissions: An empirical investigation on Balkan countries. *Int. J. Clim. Chang. Strateg. Manag.* **2020**, *12*, 449–461. [CrossRef]
20. Aytekin, A.; Korucuk, S.; Karamaşa, Ç. Ranking Countries According to Logistics and International Trade Efficiencies via REF-III. *J. Intell. Manag. Decis.* **2023**, *2*, 74–84. [CrossRef]
21. Rezaei, J.; van Roekel, W.S.; Tavasszy, L. Measuring the relative importance of the logistics performance index indicators using Best Worst Method. *Transp. Policy* **2018**, *68*, 158–169. [CrossRef]
22. Letnik, T.; Hanžič, K.; Luppino, G.; Mencinger, M. Impact of Logistics Trends on Freight Transport Development in Urban Areas. *Sustainability* **2022**, *14*, 16551. [CrossRef]
23. Angelidou, M.; Politis, C.; Panori, A.; Barkratsas, T.; Fellnhofer, K. Emerging smart city, transport and energy trends in urban settings: Results of a pan-European foresight exercise with 120 experts. *Technol. Forecast. Soc. Change* **2022**, *183*, 121915. [CrossRef]
24. Bi, X.; Li, W.; Zhang, H. Transport Electrification: Opportunities and Future Challenges. *Highlights Sci. Eng. Technol.* **2023**, *46*, 14–18. [CrossRef]
25. Mohan, A.; Bruchon, M.; Michalek, J.; Vaishnav, P. Life Cycle Air Pollution, Greenhouse Gas, and Traffic Externality Benefits and Costs of Electrifying Uber and Lyft. *Environ. Sci. Technol.* **2023**, *57*, 8524–8535. [CrossRef]
26. Zhao, T.; Yan, H.; Liu, X.; Ding, Z. Congestion-Aware Dynamic Optimal Traffic Power Flow in Coupled Transportation Power Systems. *IEEE Trans. Ind. Inform.* **2023**, *19*, 1833–1843. [CrossRef]
27. Wang, Y.; Sarkis, J. Emerging digitalisation technologies in freight transport and logistics: Current trends and future directions. *Transp. Res. Part E Logist. Transp. Rev.* **2021**, *148*, 102291. [CrossRef]
28. Masello, L.; Castignani, G.; Sheehan, B.; Murphy, F.; McDonnell, K. On the road safety benefits of advanced driver assistance systems in different driving contexts. *Transp. Res. Interdiscip. Perspect.* **2022**, *15*, 100670. [CrossRef]
29. Waseem, M.; Fahad, S.; Alanazi, F. Electric Vehicles: Benefits, Challenges, and Potential Solutions for Widespread Adaptation. *Appl. Sci.* **2023**, *13*, 6016. [CrossRef]
30. Blonsky, M.; Nagarajan, A.; Ghosh, S.; McKenna, K.; Veda, S.; Kroposki, B. Potential Impacts of Transportation and Building Electrification on the Grid: A Review of Electrification Projections and Their Effects on Grid Infrastructure, Operation, and Planning. *Curr. Sustain. Energy Rep.* **2019**, *6*, 169–176. [CrossRef]
31. Liu, Z.; Song, J.; Kubal, J.; Susarla, N.; Knehr, K.W.; Islam, E.; Nelson, P.; Ahmed, S. Comparing total cost of ownership of battery electric vehicles and internal combustion engine vehicles. *Energy Policy* **2021**, *158*, 112564. [CrossRef]
32. Raja, R.; Venkatachalam, S. *Factors Influencing the Adoption of Digital Technology in Transportation Among Logistics Service Providers;* Taghipour, A., Ed.; Blockchain; IGI Global: Hershey, PA, USA, 2022.
33. Polydoropoulou, A.; Thanopoulou, H.; Karakikes, I.; Pronello, C.; Tyrinopoulos, Y. Adapting to the future: Examining the impact of transport automation and digitalization on the labor force through the perspectives of stakeholders in all transport sectors. *Front. Futur. Transp.* **2023**, *4*, 1173657. [CrossRef]

34. Easton, J.M. Digitalization and the Transport Industry—Are we Ready? In Proceedings of the 2022 IEEE International Conference on Big Data (Big Data), Osaka, Japan, 17–20 December 2022; pp. 3814–3820. [CrossRef]
35. Overtoom, I.; Correia, G.; Huang, Y.; Verbraeck, A. Assessing the impacts of shared autonomous vehicles on congestion and curb use: A traffic simulation study in The Hague, Netherlands. *Int. J. Transp. Sci. Technol.* **2020**, *9*, 195–206. [CrossRef]
36. Mersky, A.C.; Samaras, C. Fuel economy testing of autonomous vehicles. *Transp. Res. Part C Emerg. Technol.* **2016**, *65*, 31–48. [CrossRef]
37. Hopkins, D.; Schwanen, T. The expected speed and impacts of vehicle automation in passenger and freight transport: A Dissensus Delphi study among UK professionals. *Res. Transp. Bus. Manag.* **2023**, *50*, 100973. [CrossRef]
38. Grosso, M.; Cristinel Raileanu, L.; Krause, J.; Alonso Raposo, M.; Duboz, A.; Garus, A.; Mourtzouchou, A.; Ciuffo, B. How will vehicle automation and electrification affect the automotive maintenance, repair sector? *Transp. Res. Interdiscip. Perspect.* **2021**, *12*, 100495. [CrossRef]
39. Kazembe, M.D. Intelligent Transport Systems. *Int. J. Res. Appl. Sci. Eng. Technol.* **2022**, *10*, 45271. [CrossRef]
40. Mikhalevich, I.F. Intelligent Transport Systems Software as a Source of Transport Security Threats. In Proceedings of the 2023 Systems of Signals Generating and Processing in the Field of on Board Communications, Moscow, Russian, 14–16 March 2023. [CrossRef]
41. Lee, H.J.; Yoo, S.H.; Lim, S.; Huh, S.Y. External benefits of a road transportation system with vehicle-to-everything communications. *Transp. Policy* **2023**, *134*, 128–138. [CrossRef]
42. Stodola, J.; Stodola, P.; Furch, J. Intelligent Transport Systems. *Chall. Natl. Def. Contemp. Geopolit. Situat.* **2022**, *2022*, 41–49. [CrossRef]
43. Zubairi, J.A.; Idwan, S.; Haider, S.A.; Hurtgen, D. Smart City Traffic Management for Reducing Congestion. In Proceedings of the 2022 IEEE 19th International Conference on Smart Communities: Improving Quality of Life Using ICT, IoT and AI (HONET), Marietta, GA, USA, 19–21 December 2022; pp. 225–230. [CrossRef]
44. Jenifer, J.; Jemima Priyadarsini, R. Empirical Research on Machine Learning Models and Feature Selection for Traffic Congestion Prediction in Smart Cities. *Int. J. Recent Innov. Trends Comput. Commun.* **2023**, *11*, 269–275. [CrossRef]
45. Wang, C.; Atkison, T.; Park, H. Dynamic adaptive vehicle re-routing strategy for traffic congestion mitigation of grid network. *Int. J. Transp. Sci. Technol.* **2023**, *in press*. [CrossRef]
46. Aishwarya Gowri, E.S.; Bhoomika, M.S.; Nerella, K.; Roltsh, L.; Vineeth, N. Reduction of Traffic Congestion due to Accidents by Communicating Information using VANETs. In Proceedings of the 2021 12th International Conference on Computing Communication and Networking Technologies (ICCCNT), Kharagpur, India, 6–8 July 2021. [CrossRef]
47. Wu, J.; Wang, X.; Dang, Y.; Lv, Z. Digital twins and artificial intelligence in transportation infrastructure: Classification, application, and future research directions. *Comput. Electr. Eng.* **2022**, *101*, 107983. [CrossRef]
48. Eom, M.; Kim, B.I. The traffic signal control problem for intersections: A review. *Eur. Transp. Res. Rev.* **2020**, *12*, 1–20. [CrossRef]
49. Giannopoulos, A.G.; Moschovou, T.P. Estimating the Value of Information Technology in the Productivity of the Transport Sector. *Futur. Transp.* **2023**, *3*, 601–614. [CrossRef]
50. Cheng, Z.; Pang, M.S.; Pavlou, P.A. Mitigating Traffic Congestion: The Role of Intelligent Transportation Systems. *Inf. Syst. Res.* **2020**, *31*, 653–674. [CrossRef]
51. Ward, J.W.; Michalek, J.J.; Samaras, C. Air Pollution, Greenhouse Gas, and Traffic Externality Benefits and Costs of Shifting Private Vehicle Travel to Ridesourcing Services. *Environ. Sci. Technol.* **2021**, *55*, 13174–13185. [CrossRef]
52. Ruiz, T.; Arroyo, R.; Mars, L.; Casquero, D. Effects of a Travel Behaviour Change Program on Sustainable Travel. *Sustainability* **2018**, *10*, 4610. [CrossRef]
53. Pawluk De-Toledo, K.; O'Hern, S.; Koppel, S. Travel behaviour change research: A scientometric review and content analysis. *Travel Behav. Soc.* **2022**, *28*, 141–154. [CrossRef]
54. Li, R.; Chester, M.V.; Middel, A.; Vanos, J.K.; Hernandez-Cortes, D.; Buo, I.; Hondula, D.M. Effectiveness of travel behavior and infrastructure change to mitigate heat exposure. *Front. Sustain. Cities* **2023**, *5*, 1129388. [CrossRef]
55. Hamidi, Z.; Zhao, C. Shaping sustainable travel behaviour: Attitude, skills, and access all matter. *Transp. Res. Part D Transp. Environ.* **2020**, *88*, 102566. [CrossRef]
56. Biancardi, M.; Di Bari, A.; Villani, G. R&D investment decision on smart cities: Energy sustainability and opportunity. *Chaos Solitons Fractals* **2021**, *153*, 111554. [CrossRef]
57. Kumar, V.; Jain, V.; Sharma, B.; Chatterjee, J.M.; Shrestha, R. *Smart City Infrastructure: The Blockchain Perspective*; Wiley: Hoboken, NJ, USA, 2022; ISBN 9781119785569.
58. van Hassel, E.; Vanelslander, T.; Neyens, K.; Vandeborre, H.; Kindt, D.; Kellens, S. Reconsidering nearshoring to avoid global crisis impacts: Application and calculation of the total cost of ownership for specific scenarios. *Res. Transp. Econ.* **2022**, *93*, 101089. [CrossRef]
59. Kamakura, N. From globalising to regionalising to reshoring value chains? The case of Japan's semiconductor industry. *Camb. J. Reg. Econ. Soc.* **2022**, *15*, 261–277. [CrossRef]
60. Paciarotti, C.; Torregiani, F. The logistics of the short food supply chain: A literature review. *Sustain. Prod. Consum.* **2021**, *26*, 428–442. [CrossRef]

61. Dorneanu, B.; Masham, E.; Mechleri, E.; Arellano-Garcia, H. Centralised versus localised supply chain management using a flow configuration model. In *Computer Aided Chemical Engineering*; Elsevier: Amsterdam, The Netherlands, 2019; Volume 46, pp. 1381–1386.
62. Cedillo-Campos, M.G.; Piña-Barcenas, J.; Pérez-González, C.M.; Mora-Vargas, J. How to measure and monitor the transportation infrastructure contribution to logistics value of supply chains? *Transp. Policy* **2022**, *120*, 120–129. [CrossRef]
63. Tikoudis, I.; Sundberg, M.; Karlström, A. The effects of transport infrastructure on regional economic development: A simulated spatial overlapping generations model with heterogenous skill. *J. Transp. Land Use* **2012**, *5*, 77–101. [CrossRef]
64. Teixeira, A.C.R.; Machado, P.G.; Collaço, F.M.d.A.; Mouette, D. Alternative fuel technologies emissions for road heavy-duty trucks: A review. *Environ. Sci. Pollut. Res.* **2021**, *28*, 20954–20969. [CrossRef]
65. Ghandi, A.; Paltsev, S. Global CO_2 impacts of light-duty electric vehicles. *Transp. Res. Part D Transp. Environ.* **2020**, *87*, 102524. [CrossRef]
66. Salvi, B.L.; Subramanian, K.A.; Panwar, N.L. Alternative fuels for transportation vehicles: A technical review. *Renew. Sustain. Energy Rev.* **2013**, *25*, 404–419. [CrossRef]
67. Dimanchev, E.; Fleten, S.E.; MacKenzie, D.; Korpås, M. Accelerating electric vehicle charging investments: A real options approach to policy design. *Energy Policy* **2023**, *181*, 113703. [CrossRef]
68. Johansson, B. The economy of alternative fuels when including the cost of air pollution. *Transp. Res. Part D Transp. Environ.* **1999**, *4*, 91–108. [CrossRef]
69. Verhetsel, A. The impact of planning and infrastructure measures on rush hour congestion in Antwerp, Belgium. *J. Transp. Geogr.* **2001**, *9*, 111–123. [CrossRef]
70. Metz, D. Tackling urban traffic congestion: The experience of London, Stockholm and Singapore. *Case Stud. Transp. Policy* **2018**, *6*, 494–498. [CrossRef]
71. Chen, W.; Klaiber, H.A. Does road expansion induce traffic? An evaluation of Vehicle-Kilometers Traveled in China. *J. Environ. Econ. Manag.* **2020**, *104*, 102387. [CrossRef]
72. Holl, A. A Review of the Firm-Level Role of Transport Infrastructure with Implications for Transport Project Evaluation. *J. Plan. Lit.* **2006**, *21*, 3–14. [CrossRef]
73. Savchenko, L. Relationship between transport infrastructure expenditures and costs and transport indicators—An overview of European and Ukrainian situation. *Electron. Sci. J. Intellect. Logist. Supply Chain Manag.* **2020**, *2022*, 29–45. [CrossRef]
74. Dintén, R.; García, S.; Zorrilla, M. Fleet management systems in Logistics 4.0 era: A real time distributed and scalable architectural proposal. *Procedia Comput. Sci.* **2023**, *217*, 806–815. [CrossRef]
75. Nekrasov, A.G.; Sinitsyna, A.S. Digital transformation infrastructure and transportation logistics systems. *IOP Conf. Ser. Mater. Sci. Eng.* **2020**, *832*, 012052. [CrossRef]
76. Viu-Roig, M.; Alvarez-Palau, E.J. The Impact of E-Commerce-Related Last-Mile Logistics on Cities: A Systematic Literature Review. *Sustainability* **2020**, *12*, 6492. [CrossRef]
77. Grazia Speranza, M. Trends in transportation and logistics. *Eur. J. Oper. Res.* **2018**, *264*, 830–836. [CrossRef]
78. Ferrero, F.; Perboli, G.; Rosano, M.; Vesco, A. Car-sharing services: An annotated review. *Sustain. Cities Soc.* **2018**, *37*, 501–518. [CrossRef]
79. Jiang, J.; Zhang, D.; Meng, Q. Impact analysis of investment coordination mechanisms in regional low-carbon logistics network design. *Transp. Res. Part D Transp. Environ.* **2021**, *92*, 102735. [CrossRef]
80. Zhang, D.; Zhan, Q.; Chen, Y.; Li, S. Joint optimization of logistics infrastructure investments and subsidies in a regional logistics network with CO_2 emission reduction targets. *Transp. Res. Part D Transp. Environ.* **2018**, *60*, 174–190. [CrossRef]
81. Ait Ouallane, A.; Bakali, A.; Bahnasse, A.; Broumi, S.; Talea, M. Fusion of engineering insights and emerging trends: Intelligent urban traffic management system. *Inf. Fusion* **2022**, *88*, 218–248. [CrossRef]
82. Shahzad, K.; Iqbal Cheema, I. Low-carbon technologies in automotive industry and decarbonizing transport. *J. Power Sources* **2024**, *591*, 233888. [CrossRef]
83. Soto, I.; Calderon, M.; Amador, O.; Urueña, M. A survey on road safety and traffic efficiency vehicular applications based on C-V2X technologies. *Veh. Commun.* **2022**, *33*, 100428. [CrossRef]
84. Avetisyan, H.G.; Miller-Hooks, E.; Melanta, S.; Qi, B. Effects of vehicle technologies, traffic volume changes, incidents and work zones on greenhouse gas emissions production. *Transp. Res. Part D Transp. Environ.* **2014**, *26*, 10–19. [CrossRef]
85. Maheshwari, T.; Axhausen, K.W.; Amirgholy, M.; Nourinejad, M. How Will the Technological Shift in Transportation Impact Cities? A Review of Quantitative Studies on the Impacts of New Transportation Technologies. *Sustainability* **2021**, *13*, 3013. [CrossRef]
86. Cortina, J.M. What Is Coefficient Alpha? An Examination of Theory and Applications. *J. Appl. Psychol.* **1993**, *78*, 98–104. [CrossRef]
87. Cronbach, L.J. Coefficient alpha and the internal structure of tests. *Psychometrika* **1951**, *16*, 297–334. [CrossRef]

sustainability

MDPI

Article

What Makes Parents Consider Shared Autonomous Vehicles as a School Travel Mode?

Mahsa Aboutorabi Kashani [1], Salehe Kamyab [1], Amir Reza Mamdoohi [1,2,*] and Grzegorz Sierpiński [3]

[1] Faculty of Civil & Environmental Engineering, Transportation Planning Department, Tarbiat Modares University, Tehran P.O. Box 14115-111, Iran; mahsa.abootorabi@modares.ac.ir (M.A.K.); s.kamiyab@modares.ac.ir (S.K.)

[2] Department of Civil, Geological & Mining Engineering, Polytechnique Montréal, Montréal University, Montréal, QC H3C 3J7, Canada

[3] Department of Transport Systems, Traffic Engineering and Logistics, Faculty of Transport and Aviation Engineering, Silesian University of Technology, 44-100 Gliwice, Poland; grzegorz.sierpinski@polsl.pl

* Correspondence: armamdoohi@modares.ac.ir; Tel.: +98-21-8288-4925

Abstract: The integration of shared mobility and autonomous vehicles (AVs) could potentially change the way parents decide to transport their children to and from school. A better understanding of the factors influencing parents' intentions to use shared autonomous vehicles (SAVs) for school transportation is necessary to enhance their children's mobility. Unlike prior research, this paper significantly contributes to the literature by exploring the impacts of socioeconomic, travel-related, and psychological factors and their interactions. Using Google Forms for an online survey, the authors collected 1435 valid responses from parents in Kerman city schools in Iran. The estimation results of the generalized ordered logit model indicate the significant impact of parents' socioeconomic status (occupation, education, income), travel behavior (accident experience, crash severity, travel cost), and attitude (innovativeness, perceived usefulness, environmental concern, pro-driving, safety), and their children-related factors (gender, the most frequently used travel mode, the possibility of tracking the child). The findings show that an increase in parents' education, perceived usefulness, and environmental concern increase the likelihood of their intentions to use SAVs. Finally, based on the findings, several implications are suggested to increase parents' intentions to use SAVs for transporting their children and to make SAVs a safe, affordable, and sustainable transport solution.

Keywords: shared autonomous vehicle; rideshare services; school travel mode; generalized ordered logit; intention to use

1. Introduction

There has been a growing concern about the safety of school children, especially primary and secondary school students, whose ages usually range from 5 to 14 years as a subgroup of vulnerable individuals (e.g., those without a driving license, children, elderlies, and people with disabilities) [1]. Worldwide, it is estimated that approximately 186,300 children lose their lives in road traffic incidents each year, making it the leading cause of death among children aged 5–14 [2]. Many children are dependent on their parents or other adults to accompany them to the school [3–5], but some parents often feel inconvenienced when transporting their children to and from school because parents' working hours often overlap with their children's school hours, leading to time conflicts [6,7]. But to ensure their children's safety, many parents drive them to and from school [8,9]. On the other hand, commuters and young couples are more interested in autonomous vehicles (AVs) than couples with children. This is likely because couples with children are more concerned about the safety of their children and the lack of clear policies and regulations governing AVs [10,11]. Therefore, a reliable and safe school travel mode that provides affordable, accessible, independent, and sustainable mobility could be an appealing solution for parents.

AVs provide parents with greater convenience and freedom when it comes to transporting their children to and from school due to the development of autonomous driving technology. Many studies have shown that AVs could actually enhance passenger safety by eliminating human error, which is the cause of 90% of crashes [12–16]. However, the price of AVs will be higher than conventional vehicles, and they would increase travel distance and trips made by single-occupancy vehicles, which is in conflict with sustainable transportation development [17,18]. Therefore, policymakers may not generally support this type of motorized alternative, and shared use of this technology, called shared autonomous vehicles (SAVs), are likely to become more popular in the future. Dynamic ride-sharing (DRS) services, a service form of SAVs, enable children to share their trips to and from school with other students, which leads to a reduction in travel costs [19].

Even though SAVs offer many benefits [20,21], many concerns still remain regarding their safe operation, security, privacy, and design [22–25]. For example, as a result of an international adult sample, only 11% were inclined to use AVs for transporting their children, far less than the 22% who would not use AVs at all [26]. Therefore, the identification of factors motivating individuals to embrace new technologies and lessen their resistance to change is a crucial step in guaranteeing the successful implementation of any new policy [27–30]. In light of this, the authors of this paper sought to gain a better understanding of the factors influencing the intentions of parents to transport their children to and from school using SAVs. Moreover, based on the results, several implications for policy and practice are suggested to develop a suitable SAV service as a school travel mode. In this study, the authors aimed to answer questions such as "What factors significantly influence parents' usage intention to permit their children to travel unaccompanied to school?", "Among these significant factors, which ones have the greatest impact on parents' intentions?", and "Which factors show the heterogeneity among the intention levels of respondents?". One of the main novelties of our research is that, unlike previous studies, we have examined the effects of socioeconomic, travel-related, and attitudinal factors, and their interactions as a form of systematic heterogeneity. Furthermore, to the best of our knowledge, the integration of SAVs and DRS as a school travel mode has received less attention in previous studies, as most of them have focused on AVs. In addition, we used an intricate method (i.e., generalized ordered logit model) to consider the ordinal nature of the dependent variable (i.e., intention to use) and the heterogeneity among the levels of usage intention. Finally, most of the previous studies have been conducted in developed countries, and less attention has been paid to this topic in developing countries.

The remainder of paper is organized as follows: prior studies in the context of using SAVs/AVs as a school travel mode are presented in Section 2. Section 3 introduces the methodology, including the research method, survey process, and preliminary data analysis. Section 4 presents the estimation results, a critical discussion of the findings, and recommendations for policy and practice. In Section 5, the authors provide a conclusion, as well as some suggestions for future research.

2. Literature Review

There have been many studies investigating the determinants behind school travel modes. In this section, we focus on the factors affecting parents' usage intentions of AVs to send their children to school, which are very limited in the literature (Table 1). Prior studies can be classified based on the considered methodology and factors affecting parents' willingness/intentions to use automation technology for their children's school travel mode. For instance, in the following paragraphs, we review the studies that considered any combination of socioeconomic, travel-related, or psychological factors. Then, the gaps in each study are mentioned, followed by our formulated research hypotheses. Finally, our research contribution to the existing body of literature regarding the role of SAVs in school travel modes is presented.

In terms of the effects of socioeconomic characteristics, contradictory findings have been observed. For instance, in several studies, it has been found that the age of parents is positively associated with their intention to allow their children to travel with SAVs [7,31]. While Lee et al. [32] concluded that the parents' age has a negative effect on their intention, no significant effect has been observed in the studies conducted by Ayala et al. [33] or Lee and Mirman [34]. In addition to parents' age, the impact of the children's age has also been investigated in previous studies, such as that by Lee and Mirman [34], and a positive association between the children's age and the parents' intentions to allow their children to use SAVs has been observed. Parents' gender is also another factor that has been considered in previous studies. In most prior studies, it has been concluded that fathers are more likely to allow their children to use autonomous mobility as a school travel mode [7,31,32,35]; however, no significant relation was found by Ayala et al. [33] or Anania [36]. Parents' education level is another demographic factor that positively influenced their intentions to use AVs in most studies [9,31,32,34]; however, Mao et al. [7] found no significant relationship between these two factors. Different and contradictory findings have also been reported for the effect of income on parents' intention. Although Lee et al. [32] found a positive impact of income on parents' intention, Lee and Mirman [34] and Mao et al. [7] found a negative and not significant correlation, respectively. The effect of familiarity with automation technology is another demographic factor that has been examined by prior studies. In most of the studies, a significant and positive relation has been found [7,9,31]. The number of children and owning a driving license were other explanatory variables that were considered by Mao et al. [7], but no significant associations were found between these variables and parents' intentions to use automation.

Travel behavior characteristics is another category of explanatory variables that have been considered in previous studies. Experiencing any type of accident has been considered by Lee and Mirman [34] and Lee et al. [32]. However, two contradictory findings, including a negative and positive relationship, have been reported, respectively. In other words, they found that experiencing an accident could increase or decrease the likelihood of parent's intentions to use AVs for their children's school travel mode. In addition, Mao et al. [7] examined the influence of travel distance and driving frequency, and a positive and not significant correlation were found, respectively.

Attitudinal factors play a key role in exploring the intentions of individuals [17]. Consumer innovativeness, as an attitudinal construct, is the tendency of consumers to be early adopters of new products and services [37]. This factor has been considered as one of the main factors affecting individuals' intention to use AVs. A well-established positive relationship has been reported in prior research [9,31–34], indicating that individuals with higher innovativeness are more likely to adopt autonomous mobility for their children's school travel mode. Perceived usefulness refers to how much a person believes that using a particular technology will be beneficial to them [38]. Many studies have observed that perceived usefulness is one of the most critical factors in the acceptance of AVs [39–41]. The positive and significant relationship between perceived usefulness and parents' intentions to allow their children to use SAVs as a school travel mode has also been acknowledged in previous studies [1,7,34,42,43]. Perceived risk is a latent construct that explains how consumers make decisions when they are faced with uncertainty [44]. Consumers weigh the potential benefits and risks of an intention before making a decision. In the previous studies outlined in Table 1, perceived risk leads to a lower intention of parents to use AVs as the school travel mode for their children [1,7,42,43]. Safety measures refer to the tasks that are performed to reduce the risk of harm to people, property, and the environment [45]. Despite the numerous benefits of AVs, there are still many uncertainties regarding their acceptance. Therefore, parents were asked about their preferences for route control, assurance features, and other safety measures in the vehicles to help them trust AVs. Previous studies have also confirmed the positive impact of these measures on parents' usage intentions [9,31,32,43]. Attitude is another factor that refers to a person's overall liking or disliking for a particular behavior [46]. Mao et al. [7] and Jing et al. [1] have concluded that attitude is a factor positively affecting parents' intentions to use AVs.

Table 1. An overview of previous studies in parents' usage intentions of AVs as their children's school travel mode.

Author	Variable High			Mode [4]	Model [5]
	Demographic [1]	Travel-Related [2]	Attitudinal [3]		
Ayala et al. [33]	P.S.; P.A.	-	C.I.	AV	Chi-square
Lee and Mirman [34]	P.A.; In; C.A.; Edu; P.S.	E.A.	C.I.; P.U.;	AV	D.A.
Anania [36]	P.S.; Na.	-	-	AB	ANOVA
Hand et al. [35]	P.S.	-	-	AV	ANOVA
Mao et al. [7]	P.A.; P.S.; In; Edu; N.C.; D.L.; Kn.	D.F.; T.D.	Att.; P.R.; P.U.	AV	HCM
Ma et al. [42]	-	-	P.R.; P.U.	AV	SEM
Jing et al. [1]	-	-	P.U.; Att.; P.R.	AV	SEM
Lee et al. [32]	P.A.; Edu; In.	E.A.	S.M.; C.I.	AV	R.F.
Koppel et al. [9]	Edu.; Kn.	-	C.I.; S.M.	AV/R.S.	L.R.
Koppel et al. [31]	P.A.; P.S.; Edu.; Kn.	-	C.I.; S.M.	AV/R.S.	L.R.
Tremoulet et al. [43]	-	-	P.R.; P.U.; S.M.	AV	Interview and F.G.

Abbreviation: (1) ***P.S.***: parent sex; ***P.A.***: parent age; ***In***: income; ***C.A.***: child age; ***Edu***: education; ***Na.***: nationality; ***N.C.***: number of children; ***D.L.***: driving license; ***Kn.***: knowledge. (2) ***E.A.***: experience of accident; ***D.F.***: driving frequency; ***T.D.***: travel distance. (3) ***C.I.***: consumer innovativeness; ***P.U.***: perceived usefulness; ***Att.***: attitude; ***P.R.***: perceived risk; ***S.M.***: safety measure. (4) ***AV***: autonomous vehicle; ***AB***: autonomous bus; ***R.S.***: ridesharing. (5) ***HCM***: hybrid choice model; ***SEM***: structural equation modeling; ***R.F.***: random forest; ***L.R.***: logistic regression; ***F.G.***: focus group; ***D.A.***: descriptive analysis.

Based on the literature review, it was hypothesized that the following have negative effects on parents' high usage intention: respondents who have experienced any type of accident [34] (H1); those who are less educated [32] (H2); those who have a low income [34] (H3); those who are unemployed (H4); and those who are car-dependent [7] (H5). In contrast, it was hypothesized that the following have positive effects on parents' high usage intention: being a female student (H6); those who are innovative [33] (H7); those who are pro-environmental (H8); those who are well-educated [32] (H9); and those who find that SAVs have the possibility of monitoring students online [31] (H10) as well as a travel mode with safe operation and rapid response during emergency situations [43] (H11).

The limitation of each prior study (listed in Table 1) indicates that only a few studies [7,32,34] have considered all the categories of explanatory variables, while each still has its own limitations. For example, in the study by Lee and Mirman [34], only a descriptive analysis (correlation) was conducted, which does not provide much insight regarding the causality relationship and is also limited to only two variables. Although Mao et al. [7] provided insightful findings, their model (HCM) ignored the ordinal nature of the dependent variable (intention to use) and the existence of heterogeneity among the usage intention behavior levels. The latter limitation is also true for the study by Lee et al. [32], who used random forest. Other studies (e.g., Koppel et al. [9,31]) have neither considered the effects of all categories of explanatory variables nor the integration of AVs and ridesharing services.

Based on the aforementioned research gaps, it can be concluded that the majority of prior studies have focused on identifying the factors influencing parents' decisions about their intentions to use AVs [1,7,32,34–36,42,43], or to choose between AVs and rideshare services [9,31] to transport their unaccompanied child(ren) to and from school. However, little is known about the integration of automation and ridesharing services, which are called SAVs. This study contributes the following to the existing body of literature about children's school travel modes: (1) To our understanding, no studies have explored the integration of ridesharing services and automation, which could provide affordable, accessible, independent, and sustainable mobility for vulnerable groups, particularly students under the supervision of their parents. (2) Few studies [7,9,31] have taken into account socioeconomic characteristics, travel behavior, and psychological factors simultaneously. (3) Although most of the prior studies have estimated regression, logistic regression, and a multinomial logit model, the current authors propose a generalized ordered logit (GOL) model to consider the ordinal nature of respondents' usage intentions to understand the heterogeneity among the key factors associated with parents' intentions to use SAVs to transport their unaccompanied children (for further advantages of GOL, please refer to Section 3.1).

3. Methodology

In this section, firstly, an introduction to Kerman as the study area is presented (Section 3.1). Then, the research method, a generalized ordered logit model, is presented (Section 3.3), followed by the sample characteristic (Section 3.2). Finally, the questionnaire design and survey administration (Section 3.4) is presented.

3.1. Study Area

The city of Kerman, as one of the metropolises of Iran, has an area of over 130 km^2 and a population of over 738,000 people, located at a latitude 30.29 and a longitude 57.06 (Figure 1). Some major transportation-related problems of this city are car dependency, lack of suitable active travel mode infrastructure, and lack of public transportation networks, such as a subway, bus rapid transit, or light rail transit. The only public transportation system in Kerman is an insufficient and inefficient bus network, which leads to a high share of private cars (near 50%) in daily trips and an increasing trend in car ownership and dependency. In addition, the share of different school travel modes in Kerman (Figure 2) shows that almost 80% of school trips have been conducted using private cars and school vans, whereas transit accounts for only 1% of trips.

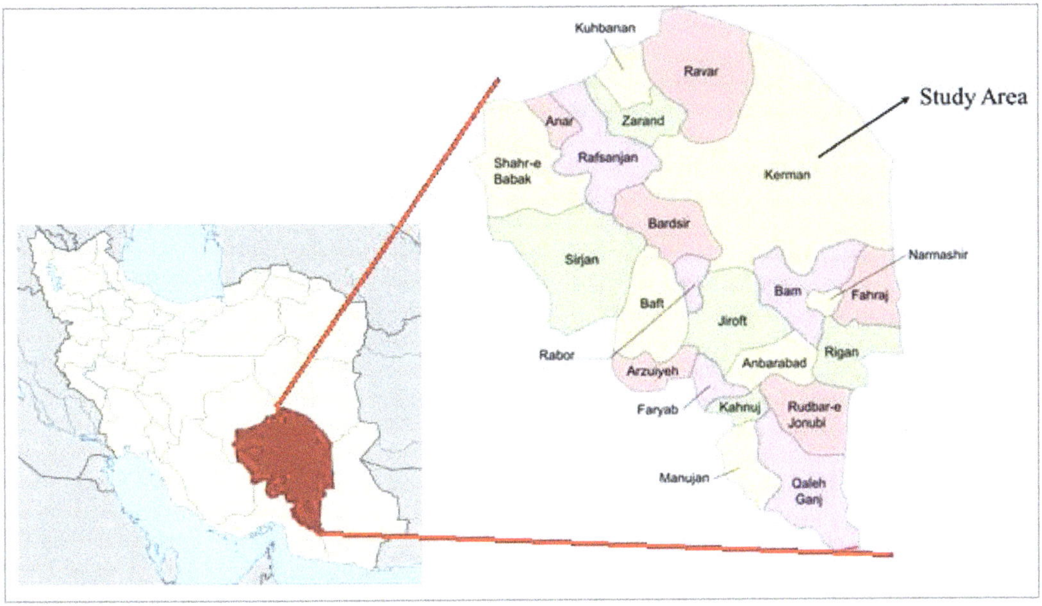

Figure 1. The location of Kerman city [47].

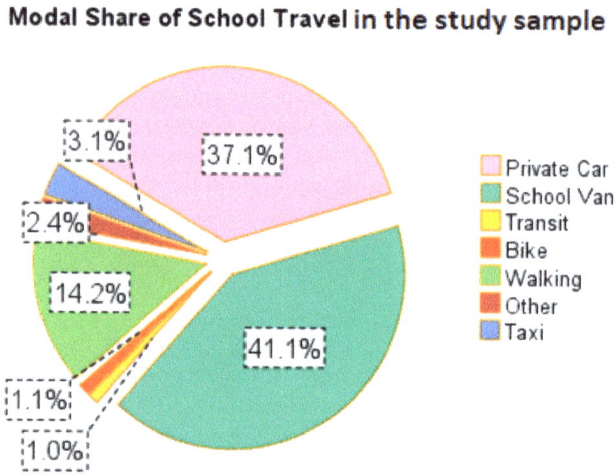

Figure 2. Shares of different school travel modes in Kerman.

3.2. Features of the Collected Sample

The total number of collected responses was 1468, and after removing invalid or incomplete responses, 1435 valid responses were used for further analysis. The incomplete questionnaires, the ones containing outlier answers, and questionnaires with an unusual completion time were the candidates for further screening or removal. The participants, on average, took about 10 min to complete the questionnaire, with a standard deviation of 2 min. This suggests that the participants answered the questions carefully.

The frequency analysis of the socio-demographic characteristics of the respondents (Table 2) shows that more than half (67.1%) of the parents were within the age range of 55–64 years. Regarding the number of children, about 794 (55.3%) respondents had two children. The majority of participants (69.9) had two driving licenses in their household. In terms of household size, 55% of the respondents belonged to four-member households. The majority of respondents (73.1%) owned at least one car. In terms of household income, the majority of respondents (about 90%) were, at most, a middle income. According to the characteristics of the children, girls comprised a slightly higher share of our sample (53%) than boys. In terms of school grades, the authors tried to equally collect data from different grades to prevent the biases of our findings for any specific group. Due to the existence of an unreliable, unsafe, old, and poor accessible public transportation system in Kerman, only 7.6% of parents send their children to school by transit/paratransit mode. On the other hand, about 80% of the parents choose a private car or school van (a van that is shared among students for transporting them to and from school) as the most frequent modes for their children's school trips.

Table 2. Descriptive analysis of explanatory variables: socio-demographic and travel-related characteristics of the respondents from Kerman.

Variable Description	Value *	Frequency	
		Absolute	Relative (%)
	Parents' Socio-Demographic Characteristics		
Age group (years)	35–44	12	0.8
	45–54	160	11.1
	55–64	963	67.1
	≥65	300	20.9

Table 2. *Cont.*

Variable Description	Value *	Frequency	
		Absolute	Relative (%)
	Parents' Socio-Demographic Characteristics		
Number of children	1	180	12.5
	2	794	55.3
	3	369	25.7
	≥4	92	6.4
Household size	2	14	1.0
	3	187	13.0
	4	790	55.1
	5	374	26.1
	≥6	70	4.9
Household car ownership	0	87	6.1
	1	1049	73.1
	2	272	19.0
	≥3	27	1.9
Father's education level	At most, high school diploma	707	49.2
	Apprenticeship diploma	190	13.2
	Bachelor's	340	23.6
	Master's	155	10.8
	PhD	43	3.0
Number of household members possessing a driving license	0	16	1.1
	1	261	18.2
	2	1003	69.9
	3+	155	10.8
Household income level	Very low	210	14.6
	Low	434	30.2
	Medium	646	45.0
	High	132	9.2
	Very high	13	0.9
Household bike ownership	0	1118	77.9
	1	302	21.0
	≥2	15	1.0
	Child characteristics		
Gender	Male (Boy)	674	47.0
	Female (Girl)	761	53.0
School grade (age)	4th (10)	289	20.1
	5th (11)	259	18.0
	6th (12)	270	18.8
	7th (13)	213	14.8
	8th (14)	202	14.1
	9th (15)	202	14.1
Most frequent transportation mode to school	School van	590	41.1
	Private car	532	37.1
	Walking	204	14.2
	Transit and paratransit	109	7.6

* The prevalent categories are highlighted in grey.

Moreover, the importance of various factors, such as environmental concerns, travel cost, accident risk, the possibility of online monitoring of children at any moment, and the safety of children on the way to school (Figure 3) were explored based on the most frequent children school modes (i.e., walking, private car, transit, and school van). It can be seen that the lowest accident risk, the possibility of monitoring children online, and their safety are the most important factors for parents who use their private car to transport their children to school. In contrast, environmental concerns and costs are of greater concern to parents whose children travel to school by walking or using transit.

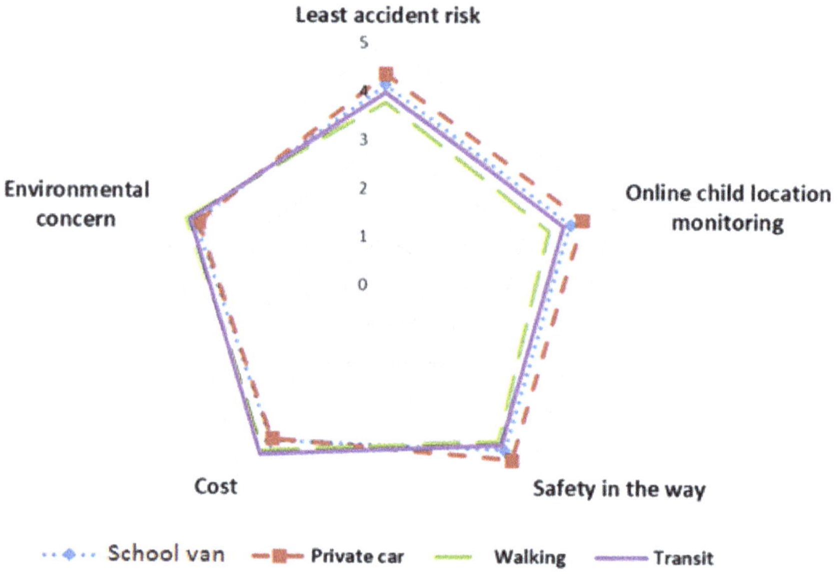

Figure 3. Importance of various factors in choosing school travel mode.

3.3. Generalized Ordered Logit (GOL)

This study models the intention to use SAVs as a school travel mode in children's educational trips by applying the prominent approach of discrete choice models.

In this study, parents' intention to use SAVs was measured as an ordinal variable. The authors have encoded this variable into three levels, namely low = 1, moderate = 2, and high = 3, to better understand the influential factors of parents' usage intentions of SAVs.

Given the discrete and ordinal nature of the dependent variable, the authors first estimated an ordered logit model (OLM) [48,49]. The probability of parents' usage intention can be calculated as Equation (1):

$$P(y_i > j) = \frac{\exp(X_i\beta' - \phi_j)}{1 + \exp(X_i\beta' - \phi_j)} \quad j = 1, 2, \ldots, M - 1 \tag{1}$$

where j is the parents' intention category, X_i is a vector containing the observed explanatory variables, β represents the vector of parameters that need to be estimated, ϕ_j are the cut-off points that determine the thresholds in the OLM, and M signifies the number of categories within the ordered-response variables.

One of the assumptions associated with the OLM is proportional odds, implying equal relationships between each pair of intention categories [49]. However, this assumption often has been violated and highlights the necessity of GOL model estimation. Using the OLM was required prior to utilizing the GOL model so that the authors could determine if the parallel lines assumption (or the proportional odds assumption) was violated. To confirm whether a GOL model should be employed instead of an OLM, the Brant test was utilized to examine the parallel assumption [50]. Using the Brant test, it is possible to check whether all the estimated coefficients are the same across different levels of the ordered dependent variable. When the GOL model works appropriately, the variables that satisfy the proportional odds assumption will be interpreted in the same way as in the OLM. As an advantage of GOL model, it can also handle independent variables that vary in size and direction of their effects. Furthermore, it can help researchers avoid major mistakes related to statistical significance, which could lead them to incorrectly assume that an explanatory variable has minimal or no impact on the outcome variable. Moreover, it can sometimes tackle issues of reporting heterogeneity, where different groups respond to questions in different ways. This can prevent differences in measurement across groups from being misinterpreted as differences in effects [51].

If the data satisfy the proportional odds assumption, then the OLM is considered an appropriate method for data description; otherwise, the GOL model should be used. Therefore, the coefficients of independent variables vary across different cut points of the dependent variable. The probability of parents' intention for allowing their children to use SAVs is expressed as Equation (2) [52]:

$$P(y_i > j) = \frac{\exp(X_i\beta_j' - \phi_j)}{1 + \exp(X_i\beta_j' - \phi_j)} \quad j = 1, 2, \ldots, M-1 \tag{2}$$

where β_j' represents a set of parameters that are adjusted based on the specific threshold point of the model.

3.4. Questionnaire

A web-based questionnaire was designed in Google Forms to assess parents' intentions to use SAVs as a school travel mode for their children in the city of Kerman, Iran. Before the main survey administration, 45 parents were randomly selected to participate in a pilot study, which resulted in a more comprehensible and less ambiguous questionnaire, based on their comments and feedback.

The final survey link was distributed among parents through the administration of schools in their social media groups from 12 April to 13 May 2021. Since the one-month data collection process was during the pandemic period, all the classes were held online, and all the important notifications from school were reported to the parents in Telegram groups. As a further explanation, using clustered sampling, the survey was distributed in different districts of Kerman. To avoid duplicate responses, we asked parents to fill out the school name and student ID.

The questionnaire included three main sections: (1) the travel behavior of parents and their children; (2) psychological factors; and (3) the socioeconomic characteristics of parents and their children. All the questions in the survey were derived from the literature or were customized for our study. It is worth mentioning that the objective of the survey was clearly announced to the participants, and they were assured that their responses would be anonymous and confidential. In the first section, we asked about the travel-related characteristics of the respondents using questions that have been used in previous studies (each question is cited to the corresponding paper(s)) or that were customized for this research. The respondents were asked about their travel behavior (five questions) including their most frequent school travel mode [36], monthly frequency of using a private car [7], accident experience and the corresponding severity [32,34], and frequency of using a seat belt while traveling [53]. In the second section, since SAV technology is new to the city of Kerman, a video clip preceded before asking parents' intention to use SAVs as a school travel mode for their unaccompanied children trip to go to school using a five-point Likert scale ranging from (1) strongly disagree to (5) strongly agree [17,19]. The following is an elaboration of the video: "*A self-driving vehicle, also known as an autonomous vehicle (AV), has the capability to operate and perform essential tasks independently without the need for human intervention, thanks to its ability to perceive its environment. The vehicle is equipped with a fully automated driving system that enables it to react to external circumstances in the same way a human driver would. Users can no longer drive themselves in such a vehicle, as there are no steering wheels or pedals. In the future, because of the high purchase costs of AVs, their use is expected to be more widely shared. In other words, there is no ownership of the vehicle and it is rented to the user on a temporary basis or on a predetermined schedule at specific times. Students will share their travel plans with others, and other students with similar origins and destinations join them to share their trips".* Then, psychological factors, such as attitude toward technology (six questions), attitude toward driving (five question), perceived usefulness (six question), and safety benefits (five questions) of SAVs were asked. The corresponding references of each question are presented in Table 3. Finally, the third section asked about the socio-economic characteristics (twelve questions) of the parents, such as their age, number of children, education, occupation, household car and bike ownership level, household size, status of driving license, and income, as well as their children's characteristics (seven questions), such as gender and school grade. It should be noted that these factors were derived from the previous studies outlined in Table 1.

4. Estimation Results and Discussion

In Section 4.1, using a maximum likelihood estimator (MLE) and AMOS, the authors provide an analysis of the results obtained from a confirmatory factor analysis (CFA) to identify the significant items in the corresponding latent constructs as a prerequisite of using psychological factors in the GOL model. Following this, the estimation result of GOL as a discrete choice modeling approach (Section 4.2) is presented to determine the effect of each exogenous variable on parents' intentions to use SAVs as their children's school travel mode. In Section 4.3, a critical discussion of our findings is presented. Finally, a number of policy and practice recommendations are provided in Section 4.4 to increase the likelihood of using SAVs as a school travel mode.

4.1. Confirmatory Factor Analysis

To determine the relationships between latent constructs and indicators, the authors conducted a CFA using AMOS software version 24 [54]. Table 3 provides a summary of the CFA results, which include the factor loadings of items, averages, and other measures of model fit, as well as indices of validity and reliability. In order to assess the validity and reliability of the latent constructs, the authors considered the content, convergent, and discriminant validity measures. Since the items were derived from the literature, their content validity is supported. Convergent validity was examined through item reliability, average variance extracted (AVE), and composite construct reliability (CR). According to Hair et al. [55], values greater than 0.5 and 0.7 are required to demonstrate convergent validity for AVE and CR, respectively [56]. Ideally, a Cronbach's alpha (CA) value should be 0.8 or above; however, a value between 0.5 and 0.7 is considered acceptable. In addition to CA, McDonald's Omega (ω) was also calculated since, unlike CA, ω operates under fewer and more realistic assumptions [57]. The acceptable threshold of ω is similar to that of CA. Based on Table 3, it can be concluded that the all the aforementioned measures are greater than the acceptable thresholds. Hence, the convergent validity is also supported. According to Table 4, discriminant validity describes how constructs differ from each other. The square root of the AVEs, which is in bold font, is greater that the correlation coefficient in the corresponding rows and columns. This suggests that the test for discriminant validity is also confirmed [58].

To determine whether the hypothetical model is supported by the data, the authors used model identification indices, including chi-squares (X^2), normed chi-squares (*CMIN/df*), normed fit indexes (*NFI*), comparative fit indexes (*CFI*), and the root mean square error of approximation (*RMSEA*) as part of the model goodness-of-fit evaluation [17]. Based on the estimated values, it can be concluded that all the necessary criteria have been satisfied.

Table 3. The results of reliability and validity evaluation of the measurement model.

Construct	Item	Mean	Factor Loadings	CA (ω)	CR	AVE
Attitude toward technology	The disadvantages of technologies are more than their advantages [59].	3.04	0.958	0.880 (0.889)	0.865	0.572
	I am not open-minded toward new products [60].	3.21	0.749			
	I am excited about the possibilities offered by new technologies [61]	3.36	−0.848			
	If I heard about new technologies, I would look for ways to try it [59].	3.49	−0.626			
	Among my peers, I am usually the first to try out new technologies [59].	3.39	−0.522			
Attitude toward driving	I prefer not to be responsible for driving a car [62].	2.76	0.951	0.805 (0.823)	0.821	0.545
	In a car, I prefer being the driver rather than the passenger [62].	3.40	−0.742			
	I like driving a car [62].	3.17	−0.683			
	I prefer not to drive in a regular path [63].	2.89	0.509			
Perceived usefulness	SAVs will reduce traffic congestion [64].	3.74	0.798	0.905 (0.905)	0.906	0.616
	SAVs would enable me to save time [64].	3.66	0.807			
	SAVs will reduce emissions [64].	3.81	0.774			
	Using SAVs would increase my productivity [65].	3.99	0.785			
	SAVs will reduce parking spaces [63].	3.82	0.766			
	SAVs will enhance my well-being [63].	3.97	0.779			
$\chi^2 = 1194.866\ (p < 0.001);\ \frac{CMIN}{df} = 3.112;\ GFI = 0.946;\ NFI = 0.958;\ CFI = 0.971;\ RMSEA = 0.038\ (CI90*: [0.036, 0.41])$						

* CI90 indicate the confidence interval at a 10% significance level.

Table 4. Correlation matrix and discriminant validity results.

	Construct	1	2	3
1	Attitude toward technology	0.756 *		
2	Attitude toward driving	−0.081	0.738	
3	Perceived usefulness	0.108	−0.063	0.784

* Bold numbers are the square root of AVE.

4.2. Estimation Results of GOL Model

Initially, an OLM was estimated to investigate how various explanatory variables affect parents' intentions to use SAVs as a school travel mode for their children. For the purpose of checking the parallel slope assumption, the Brant test was performed. The authors performed the Brant test on all coefficients, utilizing the estimated coefficients and their variances for the calculation. In addition, a test was conducted separately for each individual coefficient. According to the Brant test, the proportional odds hypothesis is rejected (χ^2 = 20.77, p-value = 0.049), suggesting that at least one variable rejects the hypothesis. The OLM was thus respecified using a GOL model, which allowed each outcome of the endogenous variable to have a unique coefficient (refer to Table 5). In order to assess the model's goodness of fit, pseudo-R^2 was used [66].

Based on the GOL estimation results, a series of attitudinal, travel-related, and socioeconomic characteristics significantly affect parents' intentions to use SAVs as a school travel mode for their unaccompanied children trips. The marginal effect values were also estimated, which allows for a more detailed interpretation of the results. Based on the marginal effect values, the critical factors are the interaction effect of the father's education level (MSc degree) and environmental concern, extremely strong belief in SAV safety for passengers, the capability of SAV in quick assistance during emergency situations (e.g., accidents), experiencing property damage crash as a pedestrian, the possibility of monitoring children online in the school path (particularly girls), and the interaction effect of parents' innovativeness and strong belief in the safe operation of SAVs, respectively. Meanwhile, in descending order, the interaction effect of education level (i.e., illiterate) and perceived usefulness, driver property damage crash experience, father's occupation (i.e., unemployed), the interaction effect of low income and importance of travel cost, fatal crash experience, and interaction effect of attitude toward driving and using a private car as a school travel mode significantly decrease the high intentions of using SAVs as a school travel mode among parents, respectively.

Table 5. Estimation results of GOL model of parents' intentions to use SAVs as a school travel mode.

Variable	Moderate		High		Marginal Effect		
	Coef.	P > |z|	Coef.	P > |z|	Low	Moderate	High
Experiencing property damage crash several times as a driver	−0.592	0.067	−0.521	0.020	0.043	0.066	−0.109
Experiencing fatal crashes more than two times	−0.670	0.014	−0.282	0.151	0.048	0.010	−0.059
Father's occupation: Unemployed	−0.654	0.086	−0.523	0.062	0.047	0.062	−0.109
Attitude toward technology × Strong belief in the safe operation of SAVs	0.034	0.142	0.024	0.081	−0.002	−0.003	0.005
Attitude toward driving × Use a private car as school travel mode	−0.155	0.053	−0.137	0.010	0.011	0.017	−0.029
$\frac{Illiterate\ Father}{\sqrt{Perceived\ Usefulness}}$	−0.576	0.280	−0.950	0.023	0.042	0.157	−0.198
Strongly perceive SAVs as enhanced emergent-assisted technology during accidents	−0.082	0.582	0.195	0.028	0.006	−0.047	0.041
Strong and very strong belief in SAV as a safe travel mode for passengers	0.373	0.009	0.406	0.000	−0.027	−0.058	0.085
Father with M.Sc. degree × Having high environmental concern	0.662	0.367	0.655	0.086	−0.048	−0.089	0.137
Having low income × High importance of cost in choosing school travel mode	−0.711	0.138	−0.422	0.102	0.051	0.037	−0.088
Girl student × High importance of online monitoring of student location	0.071	0.100	0.042	0.144	−0.005	−0.004	0.009
Constant	1.332	0.000	−1.581	0.000			
Model statistics							
Number of observations			1435				
Log-likelihood at convergence			−1078.663				
Restricted log-likelihood			−1162.22				
Pseudo R^2			0.0711				

4.3. Discussion

Marginal effects indicate that parents who have previously experienced property damage crash as drivers are less likely to allow their children to use SAVs. In the same way, experiencing fatal crashes two times among parents significantly decreases their high likelihood of allowing their children to be transported to and from school using SAVs. In other words, being such a parent reduces the likelihood of having a high intention of using SAVs by 5.9%. In accordance with previous studies [26], this may be due to the lower trust level of respondents to automation. The increase in innovativeness among parents who have a strong belief in the safe operation of SAVs is another factor that is positively correlated with the high intention of parents to use SAVs as their children's school transportation mode. This is due to the fact that pro-technology parents have higher knowledge and familiarity with technologies, which has been demonstrated in previous studies [32], too. Based on the marginal effect value, it can be seen that a one-unit increase in this systematic heterogeneity is associated with a 0.5% increase in the high intention of parents. This finding is also in accordance with previous studies.

As an indication of car dependency among parents, it can be seen that increasing the attitude toward driving among parents who use a private car as their children's school travel mode increases the likelihood of lower and moderate intentions rather than high ones. This is also aligned with the findings of Aboutorabi Kashani et al. [19]. Based on the marginal effects, an increase of one unit in this systematic heterogeneity reduces the probability of high intention by 2.9%, while it increases the likelihood of low and moderate usage intentions by 1.1% and 1.7%, respectively.

A decrease in the perception of the potential benefits of SAVs among fathers who are illiterate leads to lower and moderate intentions of letting their children travel with SAVs to school. This is due to the lower knowledge and income of these individuals, which is in accordance with the findings of Jing et al. [1] and Ma et al. [42]. However, the marginal effect value indicates that a one-unit increase in this interactive variable decreases the high intentions of parents to use SAVs as a service for their children by 19.8%.

A strong belief in the quick response of SAVs during accidents is positively related to high intentions of letting their children utilize SAVs as a school travel mode. This might be due to the facilitation conditions that are provided by SAVs, leading to a higher level of trust. According to the marginal effect value, a one-unit increase in this interactive variable increases high intentions by 4.1%. People who strongly believe that SAVs are a safe travel mode for passengers are less likely to have lower and moderate intentions, which is line with previous studies [1,42]. When this variable rises by a single unit, there are decreases in the likelihood of low and moderate intentions to use SAVs by 2.7% and 5.8%, respectively. However, the likelihood of high usage intention rises by 8.5%.

Fathers with M.Sc. degrees who are environmentally friendly are less likely to have lower and moderate intentions. This is due to the fact that SAVs are well aligned with sustainable development, and parents who are worried about the environment are more likely to use SAVs, as has been documented in prior research [19]. If this variable goes up by one unit, the likelihood of high usage intention sees a rise of 13.7% based on marginal effects. On the other hand, the probabilities of low and moderate usage intentions experience drops of 4.8% and 8.9%, respectively.

A lower likelihood of high usage intention was observed among the low-income parents who care about their children's travel cost to school, which is in accordance with the findings of Bansal et al. [67]. In other words, being such respondents increases low and moderate intentions by 5.1% and 3.7%, respectively. Parents with daughters who prefer to continuously monitoring their student's location are less likely to have lower and moderate intentions. Due to their fears of abuse, imaginary harassment, and social halo associated with their gender, women may be less likely to use shared mobility services [17]. Considering the marginal effect value, it should be noted that a one-unit increase in this systematic heterogeneity will increase high usage intentions by 0.9%.

4.4. Implications for Policy and Practice

Parent's intentions to use SAVs for their unaccompanied children trips are significantly affected by their accident experiences. The manufacturer should issue a statement explaining how AVs are designed to protect passengers in the following circumstances, in order to reduce parental concerns regarding the functional risk of SAVs. Additionally, manufacturers should support the research and development of automatic driving technologies in order to ensure the safety of SAVs in a variety of situations.

Parents' innovativeness is a psychological factor that positively influences their intentions to use SAVs as their children's school travel mode. As an implication, by inspiring a sense of innovation among individuals, policymakers may be able to encourage them to use SAVs as a school travel mode. A free trial can be offered to first-time users as a means to encourage them to try out the service. In this way, they will be more inclined to continue to use the service in the future.

Attitudes toward driving and car dependency among parents decrease the likelihood of letting their children use SAVs. It is recommended to establish travel demand management (TDM) policies to reduce the dependency on private cars of parents and encourage them to use more sustainable travel modes, reducing their dependence on private cars. In addition to increasing taxes and related costs associated with households' additional cars, congestion pricing, and parking management may also be implemented to encourage car users to use shared vehicles.

To encourage parents to use SAVs for the transportation of their children to and from school, it is critical to improve their overall evaluation of SAVs. A positive introduction to the usefulness, attributes, and performance of AVs could be made by SAV manufacturers and relevant government agencies. As a result, parents may be able to better appreciate the advantages of using SAVs for child transportation, such as an improvement in parents' time utilization value and an enhancement in the mobility of their children's travel. A public awareness campaign about the benefits of using SAVs for child transportation might have a positive effect on parents' attitudes and intentions toward using them.

It is critical to note that respondents' intention to use SAVs is significantly influenced by their belief that SAVs could provide emergency services during accidents. As an implication, it is recommended that SAVs are equipped with enhanced audio and video assistant systems that can quickly assist passengers in emergency situations. Also, it is recommended that parents are invited to try SAVs through simulators and to compare how SAVs and parents react in emergency situations. When parents perceive SAVs as quick and responsive technologies, they are more likely to allow their children to use them as a mode of transportation to and from school.

Having a high environmental concern among fathers with MSc degrees is positively correlated with their usage intentions. To encourage parents to consider SAVs as a means for their children's school commute, it is suggested that policymakers highlight the environmental advantages, such as the reduction of noise and air pollution.

The authors propose that by giving passengers the choice to select the desired number of children to share the journey with at different travel costs, service providers can potentially broaden the use of SAVs among individuals of various income levels. This approach could also contribute to more sustainable development. An additional critical factor in shaping parents' high usage intention is the possibility of tracking their children online. It is suggested that manufacturers consider equipping SAVs with cameras to monitor the interior. This would allow parents to use their mobile phones to ensure their children's safety while in the car. If an unusual situation arises, parents could use the camera feed to assess the situation and activate the car's alarm via their mobile phones, potentially attracting the attention of passersby or police officers.

5. Conclusions

5.1. Findings

The estimation results of the generalized ordered logit (GOL) model reveal the significant influence of several factors on parents' intentions to use shared autonomous vehicles (SAVs) for their unaccompanied children's trips to school, including accident experience, parent occupation, education level, income level, the children's gender, the possibility of monitoring the children online, the importance of cost and safety, perceiving SAVs as a quick responsive technology, attitude toward driving, perceived usefulness, and parents' innovativeness. Among the aforementioned factors, we found that individuals who have encountered any form of accident, are less educated, have a low income, are jobless, or are car-dependent are less likely to consider SAVs for their children's school commute. However, females, innovative individuals, environmental enthusiasts, and well-educated individuals, with the addition of the possibility of online student monitoring and ensuring the safe operation of SAVs, along with quick emergency responses, had higher intentions of using SAVs. Here is a list of some important findings:

- Experiencing property damage/fatal accidents decreases the parents' intentions to use SAVs;

- Fathers with no occupation/education are less likely to allow their children to use SAVs;
- Innovative parents who strongly believe in the safe operation of SAVs are more likely to use this technology for their children's school travel mode;
- Car-dependent parents have a lower intention of using SAVs for their children;
- Parents who find that SAVs are a beneficial technology are more likely to allow their children to use SAVs;
- Providing safety measures (e.g., online monitoring of student's location, and rapid assistance during emergency situations) by SAVs' service providers increases the usage intention of parents; and
- Being a pro-environmental parent increases the likelihood of using SAVs as a school travel mode.

In terms of the most influential factors in each intention level (low, moderate, and high), the variable of low-income parents concerned about their children's school travel costs has the greatest marginal effect value at the low intention level, indicating their lowest intentions of using SAVs. Furthermore, among the various factors that influence the moderate usage intentions of SAVs, the decrease in the perception of the potential benefits of SAVs among illiterate fathers has the greatest impact. However, fathers with an M.Sc. degree who are environmentally friendly have the highest intentions of letting their children use SAVs.

Taking a practical perspective, the research results provide us with additional insights into parents' intentions and provide manufacturers and government officials with valuable ideas for promoting the use of SAVs for school travel.

5.2. Limitations and Suggestions for Future Studies

This study contributes to the existing body of literature about the usage intentions of SAVs for school transportation by filling previous gaps. In addition to providing insight into some of the underlying issues, this study also raises some new questions that require further investigation.

Although the modeling framework of our research can be used as the basis of future studies in other contexts, the generalization of our findings to other contexts is limited. Therefore, it is recommended to conduct a comparative study with other cities to understand the effects of context on the impact of explanatory variables.

This study used psychological, sociodemographic, and travel-related variables to examine factors associated with the intentions to use SAVs for school trips. However, the model could not take into account certain travel-related attributes found in stated preference surveys, such as travel time, waiting time, and travel cost. It is, therefore, recommended that future research examines the impact of these variables on parent's usage intention of SAVs.

Due to the fact that the authors provided a description of SAVs, there is some uncertainty regarding how they affected the respondents' judgments in this study. The respondents' subjective judgments about SAVs may have been influenced to some extent by the description of SAV concepts, even though it may have provided an overview of the research topic and made them more familiar with it. Other methods can be used to help respondents better understand SAVs, such as using simulators to experience autonomous vehicles.

Finally, since SAVs are not still available in Iran's market, this research employed a stated preference (SP) survey. Since the SP approach is based on a hypothetical situation, it is recommended to explore the revealed preferences of respondents in the future when SAVs are available on the market to better understand the determinants behind the acceptance of SAVs.

Author Contributions: Conceptualization, M.A.K., S.K., A.R.M. and G.S.; methodology M.A.K., S.K. and A.R.M.; software, M.A.K. and S.K.; formal analysis, M.A.K., S.K., A.R.M. and G.S.; investigation, M.A.K., S.K. and A.R.M.; resources M.A.K., S.K. and A.R.M.; data curation, M.A.K. and S.K.; writing—original draft preparation, M.A.K., S.K., A.R.M. and G.S.; writing—review and editing, M.A.K., S.K., A.R.M. and G.S.; visualization, M.A.K. and S.K.; supervision, A.R.M. and G.S.; project administration, M.A.K. and A.R.M. All authors have read and agreed to the published version of the manuscript.

Funding: This research received no external funding.

Institutional Review Board Statement: All methods were carried out in accordance with relevant guidelines and regulations. Furthermore, written informed consent was obtained from every participant, which detailed, at their level of understanding, the aims, methods, any possible conflicts of interest, institutional affiliations of the researcher, the expected benefits and potential risks of the study, and their responsibilities and rights to refuse responding to the questionnaire. It is also necessary to mention that the Ethical Committee in the Deputy Research Directorate (DRD) of Tarbiat Modares University, Tehran, Iran, approved ethical clearance prior to the beginning of the data collection.

Informed Consent Statement: Not applicable.

Data Availability Statement: The data presented in this study are available upon request from the corresponding author. The data are not publicly available due to restrictions, e.g., privacy or ethical.

Conflicts of Interest: The authors declare no conflict of interest.

References

1. Jing, P.; Du, L.; Chen, Y.; Shi, Y.; Zhan, F.; Xie, J. Factors that influence parents' intentions of using autonomous vehicles to transport children to and from school. *Accid. Anal. Prev.* **2021**, *152*, 105991. [CrossRef] [PubMed]
2. Wang, S.; Wang, J.; Li, J.; Wang, J.; Liang, L. Policy implications for promoting the adoption of electric vehicles: Do consumer's knowledge, perceived risk and financial incentive policy matter? *Transp. Res. Part A Policy Pract.* **2018**, *117*, 58–69. [CrossRef]
3. Mehdizadeh, M.; Fallah Zavareh, M.; Nordfjaern, T. School travel mode use: Direct and indirect effects through parental attitudes and transport priorities. *Transp. A Transp. Sci.* **2019**, *15*, 749–775. [CrossRef]
4. Mehdizadeh, M.; Nordfjaern, T.; Mamdoohi, A. Environmental norms and sustainable transport mode choice on children's school travels: The norm-activation theory. *Int. J. Sustain. Transp.* **2020**, *14*, 137–149. [CrossRef]
5. Ho, S.S.; Tan, W. Mapping mental models of parents' risk perceptions of autonomous public transport use by young children: A social representations theory approach. *J. Risk Res.* **2023**, *26*, 989–1005. [CrossRef]
6. He, S.Y. Will you escort your child to school? The effect of spatial and temporal constraints of parental employment. *Appl. Geogr.* **2013**, *42*, 116–123. [CrossRef]
7. Mao, Y.; Mei, Q.; Jing, P.; Zha, Y.; Xue, Y.; Huang, J.; Shao, D.; Luo, P. Factors Affecting the Parental Intention of Using AVs to Escort Children: An Integrated SEM–Hybrid Choice Model Approach. *Sustainability* **2022**, *14*, 11640.
8. McMillan, T.E. The relative influence of urban form on a child's travel mode to school. *Transp. Res. Part A Policy Pract.* **2007**, *41*, 69–79. [CrossRef]
9. Koppel, S.; McDonald, H.; Peiris, S.; Zou, X.; Logan, D.B. Parents’ Willingness to Allow Their Unaccompanied Children to Use Emerging and Future Travel Modes. *Sustainability* **2022**, *14*, 1585.
10. Webb, J.; Wilson, C.; Kularatne, T. Will people accept shared autonomous electric vehicles? A survey before and after receipt of the costs and benefits. *Econ. Anal. Policy* **2019**, *61*, 118–135. [CrossRef]
11. Jing, P.; Xu, G.; Chen, Y.; Shi, Y.; Zhan, F. The determinants behind the acceptance of autonomous vehicles: A systematic review. *Sustainability* **2020**, *12*, 1719. [CrossRef]
12. Anderson, J.M.; Nidhi, K.; Stanley, K.D.; Sorensen, P.; Samaras, C.; Oluwatola, O.A. *Autonomous Vehicle Technology: A Guide for Policymakers*; Rand Corporation: Santa Monica, CA, USA, 2014.
13. Pudāne, B.; Rataj, M.; Molin, E.J.; Mouter, N.; van Cranenburgh, S.; Chorus, C.G. How will automated vehicles shape users' daily activities? Insights from focus groups with commuters in the Netherlands. *Transp. Res. Part D Transp. Environ.* **2019**, *71*, 222–235. [CrossRef]
14. Fagnant, D.J.; Kockelman, K. Preparing a nation for autonomous vehicles: Opportunities, barriers and policy recommendations. *Transp. Res. Part A Policy Pract.* **2015**, *77*, 167–181. [CrossRef]
15. Paddeu, D.; Tsouros, I.; Parkhurst, G.; Polydoropoulou, A.; Shergold, I. A study of users' preferences after a brief exposure in a Shared Autonomous Vehicle (SAV). *Transp. Res. Procedia* **2021**, *52*, 533–540. [CrossRef]
16. Farzin, I.; Mamdoohi, A.R.; Abbasi, M.; Baghestani, A.; Ciari, F. Determinants behind the acceptance of autonomous vehicles in mandatory and optional trips. In Proceedings of the Institution of Civil Engineers—Engineering Sustainability; Emerald Publishing Limited: Bingley, UK, 2023; pp. 1–10. [CrossRef]

17. Abbasi, M.; Mamdoohi, A.R.; Sierpiński, G.; Ciari, F. Usage Intention of Shared Autonomous Vehicles with Dynamic Ride Sharing on Long-Distance Trips. *Sustainability* **2023**, *15*, 1649. [CrossRef]
18. Golbabaei, F.; Yigitcanlar, T.; Paz, A.; Bunker, J. Understanding Autonomous Shuttle Adoption Intention: Predictive Power of Pre-Trial Perceptions and Attitudes. *Sensors* **2022**, *22*, 9193. [CrossRef]
19. Aboutorabi Kashani, M.; Abbasi, M.; Mamdoohi, A.R.; Sierpiński, G. The Role of Attitude, Travel-Related, and Socioeconomic Characteristics in Modal Shift to Shared Autonomous Vehicles with Ride Sharing. *World Electr. Veh. J.* **2023**, *14*, 23. [CrossRef]
20. Meyer, J.; Becker, H.; Bösch, P.M.; Axhausen, K.W. Autonomous vehicles: The next jump in accessibilities? *Res. Transp. Econ.* **2017**, *62*, 80–91. [CrossRef]
21. Adnan, N.; Nordin, S.M.; bin Bahruddin, M.A.; Ali, M. How trust can drive forward the user acceptance to the technology? In-vehicle technology for autonomous vehicle. *Transp. Res. Part A Policy Pract.* **2018**, *118*, 819–836. [CrossRef]
22. Li, M.; Holthausen, B.E.; Stuck, R.E.; Walker, B.N. No risk no trust: Investigating perceived risk in highly automated driving. In Proceedings of the 11th international conference on automotive user interfaces and interactive vehicular applications, Utrecht, The Netherlands, 21–25 September 2019; pp. 177–185.
23. Menon, N.; Pinjari, A.; Zhang, Y.; Zou, L. *Consumer Perception and Intended Adoption of Autonomous-Vehicle Technology: Findings from A University Population Survey*; TRID: Ludwigsfelde, Germany, 2016.
24. Nees, M.A. Acceptance of self-driving cars: An examination of idealized versus realistic portrayals with a self-driving car acceptance scale. In Proceedings of the Human Factors and Ergonomics Society Annual Meeting, Washington, DC, USA, 19–23 September 2016; pp. 1449–1453.
25. Ayoub, J.; Mason, B.; Morse, K.; Kirchner, A.; Tumanyan, N.; Zhou, F. Otto: An autonomous school bus system for parents and children. In Proceedings of the Extended Abstracts of the 2020 Chi Conference on Human Factors in Computing Systems, Honolulu, HI, USA, 25–30 April 2020; pp. 1–7.
26. Kyriakidis, M.; Happee, R.; de Winter, J.C. Public opinion on automated driving: Results of an international questionnaire among 5000 respondents. *Transp. Res. Part F Traffic Psychol. Behav.* **2015**, *32*, 127–140. [CrossRef]
27. Farzin, I.; Abbasi, M.; Macioszek, E.; Mamdoohi, A.R.; Ciari, F. Moving toward a More Sustainable Autonomous Mobility, Case of Heterogeneity in Preferences. *Sustainability* **2023**, *15*, 460. [CrossRef]
28. Acheampong, R.A.; Cugurullo, F.; Gueriau, M.; Dusparic, I. Can autonomous vehicles enable sustainable mobility in future cities? Insights and policy challenges from user preferences over different urban transport options. *Cities* **2021**, *112*, 103134. [CrossRef]
29. Nordhoff, S.; De Winter, J.; Kyriakidis, M.; Van Arem, B.; Happee, R. Acceptance of driverless vehicles: Results from a large cross-national questionnaire study. *J. Adv. Transp.* **2018**, *2018*, 5382192. [CrossRef]
30. Nordhoff, S.; Kyriakidis, M.; Van Arem, B.; Happee, R. A multi-level model on automated vehicle acceptance (MAVA): A review-based study. *Theor. Issues Ergon. Sci.* **2019**, *20*, 682–710. [CrossRef]
31. Koppel, S.; Lee, Y.-C.; Mirman, J.H.; Peiris, S.; Tremoulet, P. Key factors associated with Australian parents' willingness to use an automated vehicle to transport their unaccompanied children. *Transp. Res. Part F Traffic Psychol. Behav.* **2021**, *78*, 137–152. [CrossRef]
32. Lee, Y.-C.; Hand, S.H.; Lilly, H. Are parents ready to use autonomous vehicles to transport children? Concerns and safety features. *J. Saf. Res.* **2020**, *72*, 287–297. [CrossRef]
33. Ayala, A.; Hickerson, K.; Lettie, H.; Lee, Y.-C. The impact of peer influence on parent willingness to transport children in autonomous vehicles. In Proceedings of the Human Factors and Ergonomics Society Annual Meeting, Atlanta, GA, USA, 10–14 October 2022; pp. 692–696.
34. Lee, Y.-C.; Mirman, J.H. Parents' perspectives on using autonomous vehicles to enhance children's mobility. *Transp. Res. Part C Emerg. Technol.* **2018**, *96*, 415–431. [CrossRef]
35. Hand, S.; Lee, Y.-C. Who would put their child alone in an autonomous vehicle? Preliminary look at gender differences. In Proceedings of the Human Factors and Ergonomics Society Annual Meeting, Philadelphia, PA, USA, 1–5 October 2018; pp. 256–259.
36. Anania, E.C.; Rice, S.; Winter, S.R.; Milner, M.N.; Walters, N.W.; Pierce, M. Why People Are Not Willing to Let Their Children Ride in Driverless School Buses: A Gender and Nationality Comparison. *Soc. Sci.* **2018**, *7*, 34. [CrossRef]
37. Roehrich, G. Consumer innovativeness: Concepts and measurements. *J. Bus. Res.* **2004**, *57*, 671–677. [CrossRef]
38. Venkatesh, V.; Morris, M.G.; Davis, G.B.; Davis, F.D. User acceptance of information technology: Toward a unified view. *MIS Q.* **2003**, *27*, 425–478. [CrossRef]
39. Yuen, K.F.; Choo, L.Q.; Li, X.; Wong, Y.D.; Ma, F.; Wang, X. A theoretical investigation of user acceptance of autonomous public transport. *Transportation* **2022**, *50*, 545–569. [CrossRef]
40. Smyth, J.; Chen, H.; Donzella, V.; Woodman, R. Public acceptance of driver state monitoring for automated vehicles: Applying the UTAUT framework. *Transp. Res. Part F Traffic Psychol. Behav.* **2021**, *83*, 179–191. [CrossRef]
41. Zhang, T.; Tao, D.; Qu, X.; Zhang, X.; Zeng, J.; Zhu, H.; Zhu, H. Automated vehicle acceptance in China: Social influence and initial trust are key determinants. *Transp. Res. Part C Emerg. Technol.* **2020**, *112*, 220–233. [CrossRef]
42. Ma, Y.; Li, S.; Qin, S.; Qi, Y. Factors affecting trust in the autonomous vehicle: A survey of primary school students and parent perceptions. In Proceedings of the 2020 IEEE 19th International Conference on Trust, Security and Privacy in Computing and Communications (TrustCom), Guangzhou, China, 29 December 2020; pp. 2020–2027.

43. Tremoulet, P.D.; Seacrist, T.; Ward McIntosh, C.; Loeb, H.; DiPietro, A.; Tushak, S. Transporting children in autonomous vehicles: An exploratory study. *Hum. Factors* **2020**, *62*, 278–287. [CrossRef]
44. Dowling, G.R.; Staelin, R. A model of perceived risk and intended risk-handling activity. *J. Consum. Res.* **1994**, *21*, 119–134. [CrossRef]
45. Zhang, P.; Guan, Z.; Sun, J. Performance Evaluation of New Energy Vehicles with Human-Machine Interaction in Intelligent Transportation Systems. *Strateg. Plan. Energy Environ.* **2021**, *40*, 331–362. [CrossRef]
46. Davis, F.D. User acceptance of information systems: The technology acceptance model (TAM). *Int. J. Man-Mach. Stud.* **1993**, *38*, 475–487. [CrossRef]
47. Mirshekar, A.; Madjdzadeh, S.M.; Khayrandish, M. Spider wasps (Hymenoptera, Pompilidae) from the Southeastern Iran, Kerman. *J. Insect Biodivers. Syst.* **2020**, *6*, 9–19. [CrossRef]
48. Greene, W.H.; Hensher, D.A. *Modeling Ordered Choices: A Primer*; Cambridge University Press: Cambridge, UK, 2010.
49. Washington, S.; Karlaftis, M.G.; Mannering, F.; Anastasopoulos, P. *Statistical and Econometric Methods for Transportation Data Analysis*; CRC press: Boca Raton, FA, USA, 2020.
50. Brant, R. Assessing proportionality in the proportional odds model for ordinal logistic regression. *Biometrics* **1990**, *46*, 1171–1178. [CrossRef]
51. Williams, R. Understanding and interpreting generalized ordered logit models. *J. Math. Sociol.* **2016**, *40*, 7–20. [CrossRef]
52. Long, J.S. *Regression Models for Categorical and Limited Dependent Variables*; Advanced Quantitative Techniques in the Social Sciences; SAGE: Thousand Oaks, CA, USA, 1997; Volume 7, p. 219.
53. Beck, L.F.; Nguyen, D.D. School transportation mode, by distance between home and school, United States, ConsumerStyles 2012. *J. Saf. Res.* **2017**, *62*, 245–251. [CrossRef] [PubMed]
54. AArbuckle, J.L. *Amos*, version 23.0; [computer program]; IBM SpSS: Chicago, IL, USA, 2014.
55. Hair, J.F.; Page, M.; Brunsveld, N. *Essentials of Business Research Methods*; Routledge: London, UK, 2019.
56. Hair, J.F.; Ringle, C.M.; Sarstedt, M. PLS-SEM: Indeed a silver bullet. *J. Mark. Theory Pract.* **2011**, *19*, 139–152. [CrossRef]
57. McKay, M.T.; Perry, J.L.; Cole, J.C.; Magee, J. Adolescents consider the future differently depending on the domain in question: Results of an exploratory study in the United Kingdom. *Personal. Individ. Differ.* **2017**, *104*, 448–452. [CrossRef]
58. Fornell, C.; Larcker, D.F. Evaluating structural equation models with unobservable variables and measurement error. *J. Mark. Res.* **1981**, *18*, 39–50. [CrossRef]
59. Keszey, T. Behavioural intention to use autonomous vehicles: Systematic review and empirical extension. *Transp. Res. Part C Emerg. Technol.* **2020**, *119*, 102732. [CrossRef]
60. Leicht, T.; Chtourou, A.; Youssef, K.B. Consumer innovativeness and intentioned autonomous car adoption. *J. High Technol. Manag. Res.* **2018**, *29*, 1–11. [CrossRef]
61. Chen, J.; Li, R.; Gan, M.; Fu, Z.; Yuan, F. Public acceptance of driverless buses in China: An empirical analysis based on an extended UTAUT model. *Discret. Dyn. Nat. Soc.* **2020**, *2020*, 4318182. [CrossRef]
62. Garidis, K.; Ulbricht, L.; Rossmann, A.; Schmäh, M. Toward a user acceptance model of autonomous driving. In Proceedings of the 53rd Hawaii International Conference on System Sciences, Wailea-Makena, HI, USA, 7–10 January 2020; pp. 1381–1390.
63. Mason, J.; Classen, S.; Wersal, J.; Sisiopiku, V.P. Establishing face and content validity of a survey to assess users' perceptions of automated vehicles. *Transp. Res. Rec.* **2020**, *2674*, 538–547. [CrossRef]
64. Liu, H.; Yang, R.; Wang, L.; Liu, P. Evaluating initial public acceptance of highly and fully autonomous vehicles. *Int. J. Hum.–Comput. Interact.* **2019**, *35*, 919–931. [CrossRef]
65. Acheampong, R.A.; Cugurullo, F. Capturing the behavioural determinants behind the adoption of autonomous vehicles: Conceptual frameworks and measurement models to predict public transport, sharing and ownership trends of self-driving cars. *Transp. Res. Part F Traffic Psychol. Behav.* **2019**, *62*, 349–375. [CrossRef]
66. McKelvey, R.D.; Zavoina, W. A statistical model for the analysis of ordinal level dependent variables. *J. Math. Sociol.* **1975**, *4*, 103–120. [CrossRef]
67. Bansal, P.; Kockelman, K.M.; Singh, A. Assessing public opinions of and interest in new vehicle technologies: An Austin perspective. *Transp. Res. Part C Emerg. Technol.* **2016**, *67*, 1–14. [CrossRef]

sustainability

MDPI

Article

Households' Willingness to Pay for Interactive Charging Stations for Vehicle to Grid System in South Korea

Ju-Hee Kim [1], Min-Ki Hyun [2] and Seung-Hoon Yoo [1,*]

1 Department of Future Energy Convergence, College of Creativity and Convergence Studies, Seoul National University of Science & Technology, Seoul 01811, Republic of Korea; jhkim0508@seoultech.ac.kr

2 Department of Energy Policy, Graduate School of Convergence Science, Seoul National University of Science & Technology, Seoul 01811, Republic of Korea; hmkii098@seoultech.ac.kr

* Correspondence: shyoo@seoultech.ac.kr; Tel.: +82-2-970-6802

Abstract: The South Korean government intends to construct interactive charging stations (ICSs) for a vehicle to grid (V2G) system that uses electric vehicles as a type of energy storage system. This article employs contingent valuation (CV) to examine households' willingness to pay (WTP) to construct the ICSs. To this end, a CV survey of 1000 people was performed using the one-and-one-half-bound dichotomous choice questioning as the elicitation method for the WTP. To check if the response effect incurred by the questioning exists, a single-bound model, which partly uses the responses from the questioning, was also applied. Furthermore, a spike model which can model WTP observations with lots of zeros was adopted. The single-bound spike model, finally chosen for further analysis, produced some results securing statistical significance. The average household WTP is estimated as KRW 4017 (USD 3.51) per annum. A national version of the yearly WTP is derived as KRW 85.48 billion (USD 74.65 million). Considering that the occurrence period of the annual WTP is 10 years, the total present value as of the end of 2022 is computed as KRW 676.4 billion (USD 590.7 million). This study has significance in two aspects. First, quantitative information on household WTP is explicitly provided. Second, to the best of the authors' knowledge, the household WTP is empirically dealt with for the first time in the literature. In addition, the implications of this value from a policy perspective, as well as four challenges to be solved for constructing ICSs for a V2G system, are discussed.

Keywords: interactive charging station; vehicle to grid; willingness to pay; energy storage system; contingent valuation

1. Introduction

The number of electric vehicles (EVs) is scheduled to rise from 0.23 million in 2021 to 3.62 million by 2030, to reduce air pollutants and greenhouse gas (GHG) emissions in South Korea [1]. As of the end of June 2021, 12,789 fast EV chargers and 59,316 slow EV chargers were in operation, and the number of EVs per charger was 2.4. The number of EVs per charger in China, Japan, France, and the United States was 6, 10, 10, and 16, respectively [2]. Therefore, compared to other countries, South Korea seems to have an excellent level of EV charging infrastructure. Moreover, the government plans to build 12,000 fast EV charging stations and 500,000 slow EV charging stations by 2025 [3].

South Korea has been trying to increase renewable energy (RE). Recently, the country has established a plan to significantly augment the share of RE in total power generation from 5% in 2019 to 30.6% by 2036 [4]. Solar and wind power will account for most of the RE. However, they are typically variable RE (VRE) sources in the country [5]. Currently, the ratio of solar to wind power is approximately nine to one, and this situation is expected to continue for the time being. Thus, energy storage systems (ESSs) to help overcome the problem of variability and intermittence involved in VRE are widely needed [6–10]. The frequent curtailment of RE generation will emerge as a social issue [11–13].

Two representative types of ESSs under discussion in the country are pumping-up power generation and batteries. The government decided to build three new pumping-up power plants with a 1.8 GW capacity, and is considering further construction of such plants. However, they not only incur high costs, but also require a suitable large-scale site. Local residents and environmental groups are bitterly opposed to the additional construction of pumping-up power plants. This is because the construction of a pumping-up power plant damages forests. Moreover, a battery ESS (BESS) is quite vulnerable to fire. In fact, many installed BESSs have not been operable in the country due to frequent BESS fires.

Consequently, the government is considering another kind of ESS, an interactive vehicle to grid (V2G) system that use EVs. An interactive V2G means supplying power stored in the EV battery to consumers through the grid [14,15]. EVs are charged during times when power demand is low or curtailment of VRE generation is needed. Power is transmitted from the EVs to the grid at times when power demand is high. According to an empirical study, 100,000 EVs can play a role as great as a power plant with a capacity of 1 GW [16].

As of 2021, the average daily mileage of passenger cars in the country is 37.2 km per car [17]. According to the Ministry of Land Infrastructure and Transport [18], a total of 1.59 million eco-friendly cars, including EVs, were registered in South Korea as of 2022, accounting for 6.2% of the total number of registered cars. It would be more useful if there were specific performance data related to the driving of EVs, but the data have not yet been found in South Korea. Therefore, in this study, it is assumed that the driving performance of EVs will be similar to the statistics of general passenger cars in South Korea.

Assuming that the average driving speed of a passenger car is 50 km per hour, the average driving time per day is 0.74 h. In other words, most cars are parked for more than 90% of the day. Thus, if an infrastructure that can transmit the power stored in EVs to the grid is established, parked EVs can be used as a power supply source. For the interactive V2G system to succeed, two things are required. First, active participation of EV owners in the system is needed. The government intends to introduce a system that differentiates EV charging rates according to changes in power demand during the day, and pay a certain amount when returning power from EVs to the grid. The charging rates should be high when power demand is high, and the rates should be low when power demand is low. Second, interactive chargers that connect EVs and the grid should be constructed.

The establishment of a government-led interactive charging station (ICS) infrastructure demands a large investment of money. The investment will eventually be financed by the general public. For this reason, information on the public acceptance of bearing the cost incurred by constructing an ICS infrastructure for V2G is needed. Thus, the main objective of this article is to examine households' willingness to pay (WTP) to construct an ICS infrastructure for a V2G system using contingent valuation (CV). This research has significance in two aspects. First, quantitative information on household WTP is explicitly provided to the government for policy making and decisions. Second, to the best of the authors' knowledge, the household WTP is empirically dealt with for the first time in the literature.

In other words, the purpose of this paper is not to develop innovative methodologies or creative ideas, but to secure and present evidence to carry out government-led ICS construction. The subsequent composition of this article consists of four sections. The method applied in this study is explained in Section 2. Section 3 describes the results, and Section 4 presents a discussion. Conclusions and policy implications are presented in Section 5.

2. Methodology

2.1. Method: CV

One representative method that can be applied to obtaining households' WTP for a commodity is CV. Of course, there are other methods that can measure WTP in addition to CV, but this study applies CV for two reasons. Firstly, CV is the most frequently employed technique in evaluating WTP to consume a good not traded in the market. This kind of good is called a nonmarket good; the ICS for a V2G system is such a case. There are many studies applying CV [5,19–22], among which, the main results of research cases related to renewable energy or eco-friendly vehicles similar to this study are summarized in Table 1. Secondly, the reliability and validity of the CV technique using observations gathered from a survey of people are well-verified in previous research [23,24]. In short, CV is not only well-established in theory, but is also useful and frequently applied in empirical studies.

Table 1. Summary of literature review of studies evaluating eco-friendly vehicle infrastructure or renewable energy using contingent valuation.

Sources	Countries	Main Results
Bigerna and Polinori [19]	Italy	The additional willingness to pay (WTP) for the introduction of the hydrogen bus was EUR 2.01 to 2.44 per single-trip bus fare.
Yang et al. [20]	South Korea	The public's WTP for expanding hydrogen stations from 20 in 2016 to 100 by 2020 was KRW 2258 (USD 2.04) per household per year.
Lee et al. [21]	South Korea	The mean willingness-to-accept for vehicle-to-grid service was KRW 9821 (USD 8.83) per month, per vehicle.
Kim et al. [5]	South Korea	The mean WTP for introducing power-to-heat was KRW 4348 (USD 3.59) per household per year.
Kowalska-Pyzalska [22]	Poland	The mean WTP for green electricity was PLN 2.83 (USD 3.5 USD) per month.

The application of CV is largely standardized in the literature. The process of applying CV consists of three main steps, which are shown in Figure 1. A questionnaire is prepared in the first step with great caution. The employment of CV requires a survey of individuals with a well-prepared questionnaire. Thus, a CV questionnaire should be adequately prepared and modified so that respondents can read it without difficulty, and can easily understand it [25–28]. In the second step, data are collected from conducting the CV survey. For the purpose of obtaining reliable observations in applying CV, the guidelines given in the Korea Development Institute [29], Johnston et al. [30], and Sajise et al. [31] must be carefully followed. The collected data are statistically analyzed to derive some necessary information in the third step. Among the various models to deal with CV data, a model widely accepted and applied in the literature should be adopted [32].

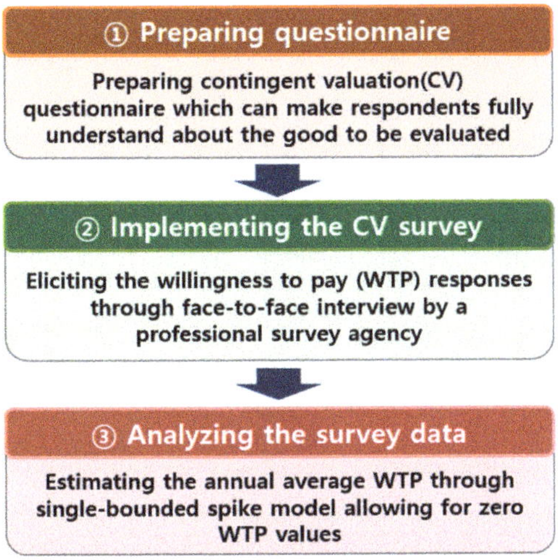

Figure 1. Process of applying contingent valuation in this study.

2.2. Step 1: Questionnaire Preparation

The important points in the first stage, questionnaire design, are summarized in three ways. First, the goods to be evaluated must be clearly presented. In this study, the reference state (Q_0) is that there are no ICSs, and only 512,000 EV charging stations (12,000 fast EV charging stations and 500,000 slow ones) by 2025. The target state (Q_1) is that all 512,000 EV charging stations will be built with ICSs by 2025. The goods to be evaluated should be explicitly a move from Q_0 to Q_1. In particular, sufficient explanations of constructing ICSs for a V2G system should be provided to the respondents so that they can understand the goods without any difficulties. Consequently, the goods to be evaluated were described to the respondents in the questionnaire as follows:

"For compensating the intermittence and variability involved in the VRE, the government has the intention of introducing a V2G system that utilizes EVs as a type of ESS. With the introduction, there is no need to construct additional ESS, power plants, and transmission lines in preparation for suspension of RE supply, which can save costs and prevent social conflicts such as resident opposition. In addition, since there is a lot of traffic in the downtown area, the V2G system has an advantage as a distributed power source. However, to introduce the V2G system, it is necessary to build an ICS infrastructure that connects EVs and the grid, which has a high cost".

Moreover, four effects expected from constructing ICSs for a V2G system were explained to the respondents. First, VRE resources can be efficiently utilized because it is not necessary to curtail the power from VRE, or at least the frequency of such curtailments is drastically reduced. Second, this reduces the demand for additional construction of fossil fuel-based power plants, which in turn can reduce emissions of GHGs and air pollutants. Third, energy security can be enhanced by curbing the importation of fuel needed for fossil fuel-based power plants. Fourth, the stability of the supply of power may be improved by shaving the peak demand for power.

Second, the WTP induction method and the payment vehicle are determined. The open-ended question format asks respondents directly about the magnitude of the WTP. Respondents may be embarrassed when an open-ended question is asked, which may increase their objection to the survey itself or the probability of reporting protest responses. On the other hand, the closed-ended question format, which asks respondents about whether they are willing to pay a presented bid, can avoid this problem. A dichotomous choice (DC) questioning method was applied to induce the WTP. That method has the advantage of preventing protest bid responses and inducing an incentive-compatible response, as it asks whether or not the respondent has the intention of paying a predetermined bid amount instead of plainly asking the size of the WTP [33].

Specifically, of several kinds of DC questioning methods, the one-and-one-half-bound DC (OBDC) one was chosen to elicit the respondents' WTP. As mentioned earlier, the government tries to cover the costs associated with the construction of an ICS infrastructure with taxes, which are public resources. Therefore, the income tax was selected as a payment vehicle, considering that it is concerned with the goods under investigation and is familiar to the respondents. In addition, the unit of payment was clearly presented as a household, the frequency of payment was annual, and the payment period was set as 10 years.

Third, the entire contents of the questionnaire should be easily understood and well-accepted by the respondents. To this end, a preliminary draft questionnaire prepared by the authors was reviewed by the supervisors from the polling company and a focus group comprising 10 people. As a result, all parts that were considered incomprehensible or inaccurate could be adequately corrected. In addition, color photographic data were also provided to the respondents to help them figure out the goods to be valued. In other words, color visual cards were used in the survey along with the final revised questionnaire.

2.3. Step 2: Survey Implementation

The decisions to be made for the second step, the survey implementation, can be summarized in four points. The first point concerns the method of conducting the survey.

In this study, the face-to-face interview method was selected. Other interview methods using the Internet, telephone, or mail have difficulty avoiding sample selection bias, and it is not easy to explain to respondents the purpose of the survey and the main survey contents sufficiently. The second point is the unit of survey. This was determined as the household rather than the individual, considering that the payment vehicle selected was income tax per household. Therefore, the number of households, not the number of individuals, should be used when calculating a national version of the WTP.

The third point is who will extract the respondents and interview the extracted respondents in the CV survey. The first option, of course, is the authors. However, they lack experience in sampling and interviews. Regarding sampling and interviews, it is reasonable to obtain help from a company specializing in public opinion polls. Accordingly, the authors entrusted the entire process of the survey to a specialized polling company. The final point is the sample size. This was determined to be 1000. Ordinary public opinion polls in South Korea are urged to use this figure by the Korea Development Institute [29]. Moreover, this size was supported by Arrow et al. [34].

2.4. Step 3: Data Analysis

The utility difference model given by Hanemann [35] was basically adopted to analyze the DC CV data. The model notes that if the response to the WTP question is "yes," the utility ensuing from paying the bid amount is greater than that ensuing from not paying it. That is, the difference between these two utilities is greater than zero. This point enables us to model the probability of answering "yes" to a bid amount, P. Two bid amounts, P^U and P^L, have to be determined beforehand in applying the OBDC model. P^L is lower than P^U. Half of all respondents were presented with a lower bid amount, P^L, first, and the other half were presented with an upper bid amount, P^U, first. If the answer to P^L is "no" or the answer to P^U "yes," no other questions are required. However, if the answer to P^L is "yes" or the answer to P^U is "no," P^U or P^L are additionally presented, respectively. Thus, some people face two WTP questions, while others face only one WTP question. As a result, there are a total of six possible cases, as follows:

(i) If the response is "no", WTP $\leq P^L$,
(ii) If the response is "yes-no", $P^L <$ WTP $\leq P^U$,
(iii) If the response is "yes-yes", $P^U <$ WTP.
(iv) If the response is "no-no", WTP $\leq P^L$,
(v) If the response is "no-yes", $P^L <$ WTP $\leq P^U$,
(vi) If the response is "yes", $P^U <$ WTP.

From (i) to (iii) are cases where the lower bid (P^L) is presented first, and from (iv) to (vi) are answers derived when the higher bid (P^U) is presented first.

Furthermore, respondents who answer "no" to the lower amount, P^L, should be identified if they have positive or zero WTP. To this end, a question of whether or not they intended to pay a penny was additionally offered to them. Finally, the WTP of each respondent, W, corresponds to one of the following four cases.

$$\begin{cases} W = 0 \\ 0 < W < P^L \\ P^L \leq W < P^U \\ P^U \leq W \end{cases} \quad (1)$$

where $G_W(\cdot)$ is the cumulative distribution function of W.

The probability of accepting the payment of P is $Pr(P \leq W) = 1 - G_W(P;\theta)$ where θ is a parameter vector of $G_W(\cdot)$. Looking at Equation (1), the first case means point data, and the other three cases imply interval data. Therefore, it is required to apply a specially designed model to deal with point data and interval data in one configuration. Taking this into account, the spike model developed by Kriström [36] is applied in this study. This is because it has been found in empirical studies that the model is useful, as well as most widely applied to dealing with data on WTP with many zero observations [37]. Let θ be (θ_0, θ_1). In the spike model, $G_w(P;\theta)$ is often formulated as:

$$G_w(P;\theta) = \begin{cases} 0 & \text{if } W < 0 \\ [1+\exp(\theta_0)]^{-1} & \text{if } W = 0 \\ [1+\exp(\theta_0 - \theta_1 P)]^{-1} & \text{if } W > 0 \end{cases} \quad (2)$$

If covariates concerning the characteristics of a respondent are additionally penetrated into Equation (2), θ_0 can be replaced by $\beta\prime c$ where β is a parameter vector that matches the covariate vector, c. The spike, meaning the probability of $W = 0$, is $[1+\exp(\theta_0)]^{-1}$. The average of W is derived as $\left(\frac{1}{\theta_1}\right)\ln[1+\exp(\theta_0)]$.

3. Results

During March 2021, the CV survey was conducted nationwide. Respondents had no particular difficulty in responding to the WTP questions. If some items were not answered, or it was judged that they were not sincere in filling out the questionnaire, the questionnaire was boldly discarded and further investigation was conducted on other respondents. Finally, the survey of 1000 households was completed. Table 2 summarizes the answers to the bid amounts offered to the respondents. The upper and lower panels of the table show the answers when P^L and P^U are presented first, respectively. Overall, it can be seen that as the bid presented first increases, the number of "yes" answers decreases. In the second column from the right, "no-no" and "no-no-no" answers indicate a WTP of zero. That is, the number of respondents with $W = 0$ is 590 (= 281 + 309).

Table 2. Summary of answers to each set of bids.

Bids [a]		Number of Answers				
First	Second	"Yes-Yes"	"Yes-No"	"No-Yes"	"No-No"	Totals
1000	3000	20	20	3	29	72
2000	4000	8	23	8	33	72
3000	6000	9	14	9	39	71
4000	8000	5	17	5	44	71
6000	10,000	3	13	11	44	71
8000	12,000	4	8	13	46	71
10,000	15,000	4	11	11	46	72
	Totals	53	106	60	281	500
First	Second	"Yes"	"No-Yes"	"No-No-Yes"	"No-No-No"	Totals
3000	1000	19	8	3	41	71
4000	2000	24	13	5	30	72
6000	3000	12	6	4	50	72
8000	4000	12	4	10	45	71
10,000	6000	9	3	15	44	71
12,000	8000	9	2	14	46	71
15,000	10,000	7	2	10	53	72
	Totals	92	38	61	309	500

[a] The unit is Korean won (USD 1.0 = KRW 1145 at the time of the survey).

Although the survey was conducted in March 2021, about two years ago, the scenario set in this study is still significant in two respects. First, it is the stability of the government's EV supply plan. In South Korea, a plan to supply eco-friendly vehicles is being established every five years under the 'Eco-friendly Vehicle Act'. Electric cars are representative eco-friendly vehicles. In the future, the supply of eco-friendly vehicles will increase, and EVs will be responsible for most of the supply. As a result, the deployment of V2G systems will increase, and the government is still very interested in public WTP for ICSs for V2G. Second, ICS construction for the introduction of V2G systems is still in the empirical research stage due to technical limitations. In other words, since there is no ICS established so far, and the policy for introducing the V2G system is still in the preimplementation stage, the current state and target state scenarios set in the CV survey of this study are still valid. Therefore, the authors think that the analysis results of the data collected two years ago still provide useful implications for policy officials.

One more thing that needs to be emphasized is that the beneficiaries of the benefits arising from the ICSs for V2G covered in this study are not limited to EV owners, but to the entire population. This is because ICSs for V2G will primarily benefit EV owners, but ultimately create positive effects such as efficient use of surplus electricity produced from VRE, reducing greenhouse gas and air pollutant emissions, improving energy security, and improving power supply stability. Therefore, EV owners were not only considered in selecting survey subjects. Respondents from all over the country were randomly selected. Of course, the construction of the ICSs for V2G allows EV owners to gain financial benefits from power transactions, but this is not the focus of this study. The subject of evaluation of this study is the various benefits to the entire people due to the construction of ICSs for V2G.

It is necessary to check if a response effect involved in the OBDC questioning method takes place or not. The response effect is related to the second question. Thus, a model that does not use the answers to the second question can be considered. To this end, the single-bound DC (SBDC) model using the response to only the first question was also applied. If there is no significant distinction between the results from OBDC and SBDC models, it can be accepted that the response effect does not occur. Otherwise, since there exists the response effect, it is desirable to adopt the estimation results from the SBDC model. Table 3 shows the estimation results of both models.

Table 3. Results from estimating the models.

Variables	**One-and-One-Half Bounded Model** [d]	**Single-Bound Model** [d]
Constant	−0.3651 (−5.71) [#]	−0.3719 (−5.80) [#]
Bid amount [a]	−0.1810 (−17.68) [#]	−0.1305 (−13.47) [#]
Spike	0.5903 (38.15) [#]	0.5919 (38.25) [#]
Yearly average household willingness to pay	KRW 2912 (USD 2.54)	KRW 4017 (USD 3.51)
t-values	15.32 [#]	12.52 [#]
95% CI [b]	KRW 2574 to 3316 (USD 2.25 to 2.90)	KRW 3462 to 4715 (USD 3.02 to 4.12)
Log-likelihood	−1133.67	−928.76
Wald statistics (*p*-values) [c]	422.49 (0.000)	261.66 (0.000)
Sample size	1000	1000

[a] The unit is KRW 1000 (USD 0.87). [b] It means confidence interval computed using the method given by Krinsky and Robb [38]. [c] The null hypothesis is that the model is mis-specified. [d] The values shown in the parentheses next to the coefficient estimates are *t*-values. [#] Implies that the estimate holds statistical significance at the 5% level.

Statistical significance at the 1% level is secured for the estimated coefficients for the constant term and the bid amount of the two models. Particularly, the coefficient for the bid amount has a negative sign. This implies that as the size of the bid increases, the probability of accepting the payment of it diminishes. Thus, the respondents appropriately provided WTP responses in the CV survey. The null hypothesis for the Wald statistic is that the coefficients are all zero. The hypothesis can be rejected in both models. The estimates found in the two models for the spikes, 0.5903 and 0.5919, respectively, were statistically significant. As mentioned above, the sample proportion of zero WTP observations was 59.0%, which is almost equal to the spike values. Consequently, the spike models were reasonably estimated.

By putting the estimates of the two models into the average WTP formula presented by Kriström [36], KRW 2912 (USD 2.54) and KRW 4017 (USD 3.51), respectively, were statistically significantly discovered for the yearly average WTP per household in the two models. Interestingly, the latter is much greater than the former. The gap is KRW 1105 (USD 0.96), accounting for 37.9% of the average WTP from the OBDC model. We are faced with the challenge of choosing one of these two and using it in the further analysis.

In this regard, an overlap test is adopted to examine whether the gap is significant. The test takes two steps. First, the 95% confidence intervals (CIs) are constructed for the average WTP estimates. In this research, the technique given by Krinsky and Robb [38] is used. Second, it is checked if the 95% CIs overlap each other or not. As shown in Table 2, the CIs do not overlap. It appears that a response effect as reported by Bateman et al. [39] exists in the OBDC model. Furthermore, the response effect significantly lowers the average WTP estimate. Consequently, the SBDC model is preferred to the OBDC model in this research. The SBDC model forms the basis of the following discussion.

4. Discussion of the Results

Three aspects will be discussed in relation to the above results. First, the effect of several variables related to the respondent on the WTP response can be explored. Five individual characteristics were selected for the exploration. Table 4 presents information about these. Specifically, there are two variables, Education and Knowledge, related to the individual characteristics of the respondent, and three variables, Income, Metro, and Solar, related to the respondent's household.

Table 4. Description of variables used in the model.

Variables	Definitions	Mean	Standard Deviation
Education	The respondent's education level in years.	14.24	2.13
Income	The respondent household's monthly income (unit: million Korean won).	4.88	1.98
Metro	The respondent's residence (0 = non-Seoul Metropolitan area; 1 = Seoul Metropolitan area).	0.53	0.50
Knowledge	Whether the respondent recognizes the vehicle-to-grid system before seeing the questionnaire (0 = no; 1 = yes).	0.05	0.22
Solar	Whether there is a solar power facility in the respondent's house (0 = no; 1 = yes).	0.03	0.18

Table 5 reports the estimation results of the SBDC model including the covariates. All of the coefficient estimates have statistical significance. Incidentally, their signs are all positive. For instance, the respondent's level of education or income is positively correlated to the probability of responding "yes" to a bid. Coefficients for other variables can be interpreted similarly. The estimated yearly average WTP for the construction of an ICS infrastructure is KRW 3522 (USD 3.08) per household, of which, the t-value is 12.50. Therefore, it is distinguishable from zero at a significance level of 1%. The Wald statistic computed under the null hypothesis that all of the parameter estimates are jointly zero is 156.21, and its p-value is 0.000. Thus, the hypothesis can be rejected at a significance level of 1%. The estimate for the spike is 0.5977, which is close to the sample proportion of 0.5900 and has statistical meaningfulness.

Second, the sample version of the average WTP can be extended to the population version of WTP. In the extension, two points are important. The first point is to determine the appropriate size of the population. When the survey was performed, South Korea had 21,278,321 households [40]. This number is used for the size of the population. The second point is the representativeness of the sample. In this respect, a specialized polling company implemented the sampling while maintaining consistency with the characteristics of the population. Thus, the extension can be attempted without causing much controversy. In this study, the population becomes the whole of South Korea. The average household WTP for the construction of an ICS infrastructure obtained above was KRW 4017 (USD 3.51) per annum. The yearly national WTP becomes KRW 85.48 billion (USD 74.65 million).

Table 5. Results from estimating the single-bound model with covariates.

Variables [a]	Coefficient Estimates [e]
Constant	−2.9602 (−6.12) #
Bid amount [b]	−0.1461 (−13.49) #
Education	0.1105 (3.26) #
Income	0.0670 (2.05) #
Metro	1.0598 (7.33) #
Knowledge	1.2989 (4.32) #
Solar	0.9878 (2.74) #
Spike	0.5977 (36.40) #
Yearly average household willingness to pay	KRW 3522 (USD 3.08)
t-value	12.50 #
95% CI [c]	KRW 3028 to 4142 (USD 2.64 to 3.62)
Log-likelihood	−862.37
Wald statistic (*p*-value) [d]	156.21 (0.000)
Sample size	1000

[a] They are described in Table 4. [b] The unit is KRW 1000 (USD 0.87). [c] It means confidence interval computed using the method given by Krinsky and Robb [38]. [d] The null hypothesis is that the model is mis-specified. [e] The values shown in the parentheses next to the coefficient estimates are *t*-values. # Implies that the estimate holds statistical significance at the 5% level.

Third, the present value of the population version of the annual WTP needs to be computed. The duration of the occurrence of the annual WTP was suggested to be ten years from 2022 to 2031 in the CV survey. Currently, an appropriate social discount rate is recommended to be 4.5% by the government. With these two pieces of information, the total present value as of the end of 2022 can be computed as KRW 676.4 billion (USD 590.7 million). To check the economic feasibility of the construction of an ICS infrastructure, this value can be compared with the cost corresponding to it. However, it is difficult to obtain specific information on the total cost of this construction project. This is because the V2G system is still in the empirical research stage and there is no case where an ICS has been built.

Thus, the government budget invested in related projects in the past can be considered as a proxy for the cost. The relationship between the government budget and the WTP derived in this study can be explained in two main aspects. First, since the construction of the ICSs for V2G is promoted by the government, not by the private sector, all of the costs related to its construction are filled by the government budget. In other words, ICSs for V2G are built by the government with financial resources raised by taxes. Second, it is desirable that the appropriate level of budget allocated to a specific project set by the government does not exceed the economic benefits generated by the project. The public WTP for ICS construction derived from this study can be interpreted as an economic benefit arising from the construction. Therefore, it would be desirable for the government budget for ICS construction projects not to exceed the total WTP amount derived in this study. Although the construction cost of ICSs for V2G has not yet been accurately calculated, it can be compared to the results of this study if the cost is calculated in the future.

The total budget invested by the government for the ten years from 2011 to 2020 to build an EV charging station infrastructure in the early stage of introducing EVs was KRW 668.9 billion (USD 584.2 million). Since constructing an ICS infrastructure means attaching an interactive charger to an existing EV charger, the construction is expected to cost less than the cost invested in building the infrastructure for the initial EV charging stations. The total present value is larger than the total budget invested by the government to build the infrastructure for EV charging stations over the past ten years. It seems that constructing the ICS infrastructure is socially beneficial. Of course, after obtaining accurate cost information, the work of more strictly comparing the value obtained in this study with it should be followed.

However, there are still issues to be resolved for the successful introduction of a V2G system, and especially, the opinions collected in this survey can be largely summarized into four categories. First, it is necessary to introduce a charging and discharging rate system exclusively for V2G. There is still only an EV charging rate system in South Korea, and no discharge rate. In addition, it is necessary to prepare a differentiated charging and discharging rate system by time and season. Second, specific methods of participating in the V2G system should be disclosed. Since the public's overall understanding of the V2G concept is still at a low level, various information such as ICS location, real-time price, and mandatory access time must be provided to users, including potential users, easily and clearly for the successful introduction of the V2G system.

Third, appropriate incentives should be provided to V2G system participants. The benefits of providing power reserves due to the V2G system will be enjoyed by the entire nation. Therefore, in consideration of the social benefits, V2G system participants should be provided with incentives for participating in the system, in addition to the profits from energy arbitrage. This can be an effective inducement to sufficiently secure EVs required to activate the V2G system. Fourth, it is necessary to resolve public concerns and distrust about the possibility of fire and shortening the life of EVs due to frequent battery charging and discharging. It is necessary to promote and improve public awareness that the fire and life shortening problems can be prevented through several means such as a battery management system.

As mentioned earlier, the purpose of this paper is not to develop innovative methodologies or creative ideas, but to evaluate the benefits of certain policies that the South Korean government intends to pursue. In other words, in a situation where an ESS is needed due to the expansion of VRE, this paper attempts to quantitatively analyze the economic benefits of ICS construction for the introduction of V2G, one of the ESS alternatives, using CV. CV has been widely applied for the valuation of nonmarket goods, and Table 6 summarizes the results of this study and the results of similar previous studies. Although the same methodology was applied, it is difficult to compare the results of each study with each other because the previous research cases and this study differ in five main aspects: evaluation target, sample size, survey method, survey time, and model.

Table 6. Summary of the results of previous studies and this study using contingent valuation.

Sources	Countries	Object to Be Valued	Mean Willingness to Pay
Bigerna and Polinori [19]	Italy	Introduction of hydrogen bus	EUR 2.01 to 2.44 per single-trip bus fare
Yang et al. [20]	South Korea	Expanding hydrogen stations	KRW 2258 (USD 2.04) per household per year
Lee et al. [21]	South Korea	Vehicle-to-grid service	KRW 9821 (USD 8.83) per month, per vehicle
Kim et al. [5]	South Korea	Introducing power-to-heat	KRW 4348 (USD 3.59) per household per year
Kowalska-Pyzalska [22]	Poland	Green electricity	PLN 2.83 (USD 3.5 USD) per month
This study	South Korea	Construction of interactive charging stations	KRW 3522 (USD 3.08) per household per year

Since the methodology used in this study has already been developed a long time ago and its framework has been established to some extent, it is difficult to say that it is innovative. Nevertheless, the implications of this paper can be useful in three aspects. First, to the extent that the authors know, this study is the first attempt in the literature to evaluate the public WTP for ICS construction for the introduction of V2G. There have been quite a few research cases in which CVs have been applied to energy issues, but it is difficult to find cases in which CVs have been applied in evaluating the economic value of infrastructure construction for V2G. Second, policy officials were provided with basic data to determine the budget for the ICS construction project or to evaluate the economic feasibility of the project. The government was demanding information on economic benefits for the construction of ICSs, which is preemptively necessary for the introduction of V2G systems, but this information has not been available so far. Third, other countries that are planning policies similar to South Korea may refer to their findings.

5. Conclusions

The government is considering introducing a V2G system to overcome the problem of variability and intermittence involved in VRE. For its introduction, an ICS connecting EVs and the power grid is required, and the government intends to construct the infrastructure by investing public funds. In this regard, this research attempted to quantitatively examine household WTP for the construction by using CV. For the sake of eliciting the WTP response, the OBDC model was adopted. For comparison, the SBDC model, which uses only the response to the first question in the OBDC model, was also applied. A spike model with a considerable plausibility that could explicitly deal with zero WTP responses was employed. Consequently, the estimation results of the SBDC model were used for further policy analysis. The average household WTP was statistically significantly derived as KRW 4017 (USD 3.51) per year. Its population value was KRW 85.48 billion (USD 74.65 million) per annum. The present value corresponding to ten years' payment was KRW 676.4 billion (USD 590.7 million) as of the end of 2022. This value is clearly significant, and appears to be greater than the cost involved in constructing the ICSs for V2G.

Of course, the construction of an ICS infrastructure does not guarantee the successful introduction of a V2G system. As can be seen from the four issues discussed above, a systematic system necessary for operating the V2G system and a support policy for participants should be established. If the 30.6% RE supply target is achieved by 2036, it will be a key task to stably include intermittent RE in the power system. Accordingly, various types of flexible resources are required. The EV resources can be used in downtown areas with a high demand for electricity, and can contribute to stabilizing RE output by being used as reserve power. Therefore, for the introduction of a V2G system, efforts such as system preparation, infrastructure construction, and public awareness improvement are necessary.

The results obtained from this research have three policy implications. The first implication is that the upper limit of the government budget invested in constructing an ICS infrastructure for a V2G system was quantitatively calculated. The estimated total benefit can be interpreted as the upper limit, and the government can refer to this information when allocating the budget. The second implication is that among the individual characteristics of the respondents, factors affecting the WTP response were identified. When the government selects target areas for constructing ICSs, it can give priority to places where a group with a high preference resides. The third implication is that 59.0% of all respondents had no intention of paying for the construction. This indicates that more than half of all households are negative or indifferent to the construction. Therefore, in order for the government to successfully promote the construction project and introduce a V2G system, it is necessary to persuade the public through publicity about its effectiveness or institutional improvement and support.

To complement the implications of this study more abundantly, three further studies can be conducted in the future. First, it is necessary to evaluate the public's acceptability of the charging and discharging rate system exclusively for a V2G system. Second, as the WTP for constructing ICSs can vary from region to region, it can be estimated by region by obtaining a greater budget for the survey and increasing the sample size. Third, it would also be meaningful to try experiments related to CV data analysis. For example, a comparison of results using various DC questioning methods. In this study, the OBDC model and the SBDC model were compared, but the double-bound DC model may be considered additionally.

Author Contributions: Conceptualization, S.-H.Y.; methodology, S.-H.Y. and M.-K.H.; software, J.-H.K. and M.-K.H.; validation, J.-H.K., M.-K.H. and S.-H.Y.; formal analysis, J.-H.K.; investigation, M.-K.H.; resources, S.-H.Y.; data curation, J.-H.K.; writing—original draft preparation, J.-H.K.; writing—review and editing, M.-K.H. and S.-H.Y.; visualization, M.-K.H.; supervision, S.-H.Y.; project administration, S.-H.Y. All authors have read and agreed to the published version of the manuscript.

Funding: This research did not receive any specific grant from funding agencies in the public, commercial, or not-for-profit sectors.

Institutional Review Board Statement: All subjects gave their informed consent for inclusion before they participated in this study. This study was conducted in accordance with the Declaration of Helsinki, and the survey was approved by Institutional Review Board at Seoul National University of Science and Technology for exemption from deliberation.

Informed Consent Statement: Informed consent was obtained from all subjects involved in the study.

Data Availability Statement: Not applicable.

Conflicts of Interest: The authors declare no conflict of interest.

References

1. Korean Government. *2030 Nationally Determined Contribution*; Korean Government: Sejong, Republic of Korea, 2021.
2. International Energy Agency. *Global EV Outlook*; International Energy Agency: Paris, France, 2021.
3. Korean Government. *BIG 3 Key Tasks for Each Industry*; Korean Government: Sejong, Republic of Korea, 2021.
4. Korea Ministry of Trade, Industry and Energy. *The 10th Basic Plan for Power Demand and Supply (2022–2036)*; Korea Ministry of Trade, Industry and Energy: Sejong, Republic of Korea, 2023.
5. Kim, J.H.; Lim, S.Y.; Yoo, S.H. Public preferences for introducing a power-to-heat system in South Korea. *Renew. Sustain. Energy Rev.* **2021**, *151*, 111630. [CrossRef]
6. Zhao, H.; Wu, Q.; Hu, S.; Xu, H.; Rasmussen, C.N. Review of energy storage system for wind power integration support. *Appl. Energy* **2015**, *137*, 545–553. [CrossRef]
7. Guney, M.S.; Tepe, Y. Classification and assessment of energy storage systems. *Renew. Sustain. Energy Rev.* **2017**, *75*, 1187–1197. [CrossRef]
8. Yang, Y.; Bremner, S.; Menictas, C.; Kay, M. Battery energy storage system size determination in renewable energy systems: A review. *Renew. Sustain. Energy Rev.* **2018**, *91*, 109–125. [CrossRef]
9. Hanif, S.; Alam, M.J.E.; Fotedar, V.; Crawford, A.; Vartanian, C.; Viswanathan, V. Managing the techno-economic impacts of partial string failure in multistring energy storage systems. *Appl. Energy* **2022**, *307*, 118196. [CrossRef]
10. Le, T.S.; Nguyen, T.N.; Bui, D.K.; Ngo, T.D. Optimal sizing of renewable energy storage: A techno-economic analysis of hydrogen, battery and hybrid systems considering degradation and seasonal storage. *Appl. Energy* **2023**, *336*, 120817. [CrossRef]
11. Golden, R.; Paulos, B. Curtailment of renewable energy in California and beyond. *Electr. J.* **2015**, *28*, 36–50. [CrossRef]
12. Henriot, A. Economic curtailment of intermittent renewable energy sources. *Energy Econ.* **2015**, *49*, 370–379. [CrossRef]
13. Agbonaye, O.; Keatley, P.; Huang, Y.; Odiase, F.O.; Hewitt, N. Value of demand flexibility for managing wind energy constraint and curtailment. *Renew. Energy* **2022**, *190*, 487–500. [CrossRef]
14. Hildermeier, J.; Kolokathis, C.; Rosenow, J.; Hogan, M.; Wiese, C.; Jahn, A. Smart EV charging: A global review of promising practices. *World Electr. Veh. J.* **2019**, *10*, 80. [CrossRef]
15. United Kingdom Government. *Electric Vehicle Smart Charging*; United Kingdom Government: London, UK, 2021.
16. Hankyoreh. Are You Using ESS Beyond Electric Cars and Campsite Appliances? 2022. Available online: https://www.hani.co.kr/arti/economy/car/1033781.html (accessed on 10 February 2023).
17. Statistics Korea. Vehicle Mileage Statistics. 2022. Available online: https://kosis.kr/statHtml/statHtml.do?orgId=426&tblId=DT_426001_N004&conn_path=I2 (accessed on 10 March 2023).
18. Ministry of Land Infrastructure and Transport. Automobile Registration Status. 2023. Available online: http://stat.molit.go.kr (accessed on 14 July 2023).
19. Bigerna, S.; Polinori, P. Willingness to pay and public acceptance for hydrogen buses: A case study of Perugia. *Sustainability* **2015**, *7*, 13270–13289. [CrossRef]
20. Yang, H.J.; Cho, Y.; Yoo, S.H. Public willingness to pay for hydrogen stations expansion policy in Korea: Results of a contingent valuation survey. *Int. J. Hydrogen Energy* **2017**, *42*, 10739–10746. [CrossRef]
21. Lee, C.Y.; Jang, J.W.; Lee, M.K. Willingness to accept values for vehicle-to-grid service in South Korea. *Transp. Res. D Transp. Environ.* **2020**, *87*, 102487. [CrossRef]
22. Kowalska-Pyzalska, A. Do consumers want to pay for green electricity? A case study from Poland. *Sustainability* **2019**, *11*, 1310. [CrossRef]

23. Loomis, J.; González-Cabán, A.; Champ, J. Estimating the robustness of contingent valuation estimates of WTP to survey mode and treatment of protest responses. In *The International Handbook on Non-market Environmental Evaluation*; Bennett, J., Ed.; Edward Elgar: Cheltenham, UK, 2011.
24. Haab, T.; Lewis, L.; Whitehead, J. *State of the Art of Contingent Valuation. Oxford Research Encyclopedia of Environmental Science*; Oxford University Press: Oxford, UK, 2020.
25. Groves, R.M.; Fowler, F.J.; Couper, M.P.; Lepkowski, J.M.; Singer, E.; Tourangeau, R. *Survey Methodology*; Wiley: Hoboken, NJ, USA, 2004.
26. Boyle, K.J. Contingent valuation in practice. In *A Primer on Nonmarket Valuation*, 2nd ed.; Springer: Dordrecht, The Netherlands, 2017.
27. Champ, P.A.; Boyle, K.J.; Brown, T.C.; Peterson, L.G. *A Primer on Nonmarket Valuation*; Springer Science & Business Media: Amsterdam, The Netherlands, 2017.
28. Mariel, P.; Hoyos, D.; Meyerhoff, J.; Czajkowski, M.; Dekker, T.; Glenk, K.; Jacobsen, J.B.; Liebe, U.; Olsen, S.B.; Sagebiel, J.; et al. *Environmental Valuation with Discrete Choice Experiments: Guidance on Design, Implementation and Data Analysis*; Springer Nature: Berlin, Germany, 2021.
29. Korea Development Institute. *Guidelines for Applying Contingent Valuation Method to Pre-Evaluation of Feasibility*; Korea Development Institute: Sejong, Republic of Korea, 2012.
30. Johnston, R.J.; Boyle, K.J.; Adamowicz, W.; Bennett, J.; Brouwer, R.; Cameron, T.A.; Hanemann, W.M.; Hanley, N.; Ryan, M.; Scarpa, R.; et al. Contemporary guidance for stated preference studies. *J. Assoc. Environ. Resour. Econ.* **2017**, *4*, 319–405. [CrossRef]
31. Sajise, A.J.; Samson, J.N.; Quiao, L.; Sibal, J.; Raitzer, D.A.; Harder, D. *Contingent Valuation of Nonmarket Benefits in Project Economic Analysis: A Guide to Good Practice*; Asian Development Bank: Manila, Philippines, 2021.
32. Haab, T.C.; McConnell, K.E. *Valuing Environmental and Natural Resources*; Edward Elgar: Cheltenham, UK, 2002.
33. Mitchell, R.C.; Carson, R.T. *Using Surveys to Value Public Goods: The Contingent Valuation Method*; Resources for the Future: Washington, DC, USA, 1989.
34. Arrow, K.; Solow, R.; Portney, P.R.; Leamer, E.E.; Radner, R.; Schuman, H. Report of the NOAA panel on contingent valuation. *Fed. Regist.* **1993**, *58*, 4601–4614.
35. Hanemann, W.M. Welfare evaluations in contingent valuation experiments with discrete responses. *Am. J. Agric. Econ.* **1984**, *66*, 332–341. [CrossRef]
36. Kriström, B. Spike models in contingent valuation. *Am. J. Agric. Econ.* **1997**, *79*, 1013–1023. [CrossRef]
37. Yoo, S.H.; Kwak, S.J. Using a spike model to deal with zero response data from double bounded dichotomous choice contingent valuation surveys. *Appl. Econ. Lett.* **2002**, *9*, 929–932. [CrossRef]
38. Krinsky, I.; Robb, A.L. On approximating the statistical properties of elasticities. *Rev. Econ. Stat.* **1986**, *68*, 715–719. [CrossRef]
39. Bateman, I.J.; Day, B.H.; Dupont, D.P.; Georgiou, S. Procedural invariance testing of the one-and-one-half-bound dichotomous choice elicitation method. *Rev. Econ. Stat.* **2009**, *91*, 806–820. [CrossRef]
40. Statistics Korea. Korea Statistical Information Service. 2023. Available online: https://kosis.kr/statHtml/statHtml.do?orgId=101&tblId=DT_1BZ0503&conn_path=I2 (accessed on 10 March 2023).

sustainability

MDPI

Article

A Data-Driven Approach to Analyze Mobility Patterns and the Built Environment: Evidence from Brescia, Catania, and Salerno (Italy)

Rosita De Vincentis [1], Federico Karagulian [2], Carlo Liberto [2,*], Marialisa Nigro [1], Vincenza Rosati [1] and Gaetano Valenti [2]

1 Department of Engineering, Roma Tre University, Via Vito Volterra 62, 00146 Rome, Italy
2 ENEA Research Center Casaccia, Via Anguillarese 301, 00123 Rome, Italy
* Correspondence: carlo.liberto@enea.it

Abstract: Investigating the correlation between urban mobility patterns and the built environment is crucial to support an integrated approach to transportation and land-use planning in modern cities. In this study, we aim to conduct a data-driven analysis of these two interrelated parts of the urban environment through the estimation of a set of metrics to assist city planners in making well-informed strategic decisions. Metrics are computed by aggregating and correlating different types of data sources. Floating Car Data (FCD) are used to compute metrics on mobility demand and traffic patterns. The built environment metrics are mainly derived from population and housing census data, as well as by investigating the topology and the functional classification adopted in the OpenStreetMap Repository to describe the importance and the role of each street in the overall network. Thanks to this set of metrics, accessibility indexes are then estimated to capture and explain the interaction between traffic patterns and the built environment in three Italian cities: Brescia, Catania, and Salerno. The results confirm that the proposed data-driven approach can extract valuable information to support decisions leading to more sustainable urban mobility volumes and patterns. More specifically, the application results show how the physical shape of each city and the related street network characteristics affect the accessibility profiles of different city zones and, consequently, the associated traffic patterns and travel delays. In particular, the combined analysis of city layouts, street network distributions, and floating car profiles suggests that cities such as Brescia, which is characterized by a homogeneously distributed radial street system, exhibit a more balanced spread of activities and efficient mobility behaviors.

Keywords: mobility patterns; built environment; Floating Car Data; OpenStreetMap repository; road network; accessibility

1. Introduction

The relationship between land use and mobility patterns has been a research topic for years, due to its importance in traffic management and urban planning [1]. Indeed, the knowledge of this relationship supports the adoption of policies regarding sustainable planning for the increase in the efficiency of transportation systems [2], for supporting the street network design [3], and for the improvement of social equity and inclusion in terms of access to services and activities [4,5].

The several studies on the topic can be differentiated in terms of investigated mobility options, data adopted, and related metrics computed.

For example, there are studies concentrating on walking mode, as in the case of the "15 min city" [6], where the positive effect of mixed-land-use policy to walking accessibility has been investigated in the city of Milan.

Other studies have focused on car travel mode: in [7], the impact of the built environment and activities on private mobility has been analyzed by taking into consideration the "3 Ds" (Density of population and workers; Diversity of types of activities and land-use; Design for transportation supply). Others have concentrated on the case of public transportation: [8], for example, used both land-use variables (population density, service facility density, etc.) and transport variables (bus stops, road density, etc.) to evaluate the public transport ridership in each traffic zone. Finally, studies exist considering together cars, walking, and public transportation, such as [9], which underlined that population and workers density allow the decrease in travel times and car use for the benefit of public transport, or [10], where the distance to transit and destination accessibility have been added to the "3 Ds" analyzed in [7].

About data, in [11], open data such as Points Of Interest (POIs) and the road network from Open Street Map have been used as independent variables for a model for the fruition of car-hailing, also demonstrating the changes in the impact of land-use on mobility patterns during the day. Indeed, infrastructural indicators, such as the length of the different road types in each zone, their density and distribution, as well as the capacity and the performances of the transportation supply, could give information about accessibility [12,13] or about the classification of the land-use (i.e., residential areas are characterized by local streets with low capacity and low speeds; instead, industrial hubs are often served by road with high capacity).

Floating Car Data (FCD) are adopted even more for exploiting human mobility behaviors and to derive useful insights into urban and transportation planning. For example, Ref. [14] captured the spatiotemporal evolution of urban hot zones from people's arrive-stay-leave (ASL) behavior, which is fundamental where changes or improvements in urban networks must be planned. Moreover, it is underlined how much the stay time by FCD and its correlation with additional sources of data such as Points Of Interest (POIs) could stimulate the research on transport and urban planning. Ref. [15] explored the relationship between urban mobility by FCD and specific built environment factors (e.g., average house price, scenic spots, or shopping areas) in Shanghai through an ordinal logistic regression (OLR) analysis to help better understand their relationship. Ref. [16] built a mobility-related typology of territorial zones by investigating vehicle movements by FCD in the Great Paris region; they found that the derived mobility types of zones have a correspondence with the common recognition of their social functions. In addition, in [17], significant places have been localized in the Metropolitan Region of Paris by mining trajectories and activity durations from FCD.

Our work is a first step to computing data-driven metrics for the investigation of the mutual interaction between the transportation system and the land use, with a specific focus on private transport. However, despite the analyses matching big data from traffic with the built environment usually focusing on large urban networks, we decided to focus on three medium-size cities. Taking inspiration from previous studies in the literature, metrics are computed by adopting different layers of data (Floating Car Data, Census data, and Open Street Map data); FCD allow one to derive traffic performances as seen in [18] and to use these performances for the computation of the zone's accessibility (similarly to the case in [19]); the built environment derives from population and employees data, together with the physical structure of the road network [20]. The 3 medium-size cities are Brescia, Catania, and Salerno, located in Italy and characterized by different urban forms (in terms of road network structure) and land-use activities.

The proposed data-driven metrics as well as their relationship could be the base for the development of a Decision Support System module supporting planners and engineers to localize the critical elements (e.g., zones) and to investigate the impact of planning policies.

The paper is structured as follows: the method of analysis is described in the following section, and then the results of the metrics computation for the three medium-size Italian cities are presented and discussed. Finally, the conclusion and the future research ideas are given.

2. Methods

The analysis framework is applied with the same criteria to three medium-sized cities (Brescia, Catania, and Salerno), allowing for a direct comparison based on the proposed zonal metrics and accessibility scores.

The three cities are selected for their differences in terms of shape, extension, and density. Catania and Salerno are in the south of Italy, respectively, in Sicily and Campania regions. Salerno is a city economically based on the commercial and tertiary sectors, with a port in the north of the city, which is hub of several commercial and touristic flows (due to the proximity with the Amalfi coast). Catania is a medium-sized city and the core city of a metropolitan region. Catania has a strong agricultural sector, an important commercial and touristic port (both commercial and for ferries), and a large industrial district (Etna Valley). The street networks of Catania and Salerno, in fact, extend along the coastline. Brescia, located in the north of Italy, is an important industrial and commercial city. Brescia has a more regular morphology characterized by a street network having both a grid and a radial structure.

Catania has an extension of 183 km^2, about 296,000 inhabitants, and 130,000 employees. Salerno is the smallest one, covering 59 km^2, with about 132,000 inhabitants and 25,000 employees; finally, Brescia extends over 90 km^2 with 196,000 inhabitants and 63,000 employees.

According to Figure 1, each city is divided with a regular hexagonal grid (side of length 500 m) in order to obtain normalized maps without irregular shapes and to perform standardized comparisons between different urban contexts. Thanks to its flexibility and generalization feature, this approximation has been used in several research contexts, such as the spatial accessibility analysis [21–23], where the availability of transport data collections is still growing [15,24]. Despite its higher computational complexity, hexagonal cell tessellation is able to cover an entire geographical region without any gaps and overlapping, ensuring a greater mapping accuracy with respect to other methods (e.g., circular or square shapes). Moreover, the method allows a better distribution of centroids, an optimal management of nearest neighborhoods, an easy matching of curve patterns, and irregular territorial boundaries.

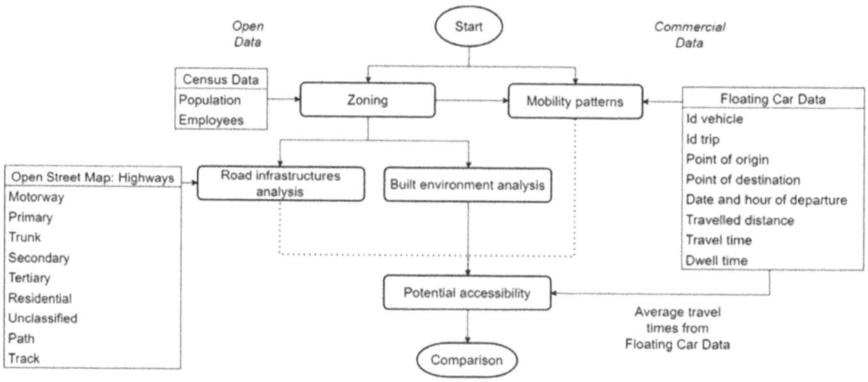

Figure 1. Workflow of the analysis framework.

The study adopts different sources and types of data for the analysis: open data (census data; Open Street Map database—OSM) and GPS traces data from a fleet of privately owned probe vehicles traveling across the three cities (Floating Car Data—FCD) (Figure 1).

Census data rely on the ISTAT database that provides information about inhabitants and employees in each city census zone [25]. The assignment of census data to the hexagonal grid is applied through a procedure. It is based on the hypothesis that inhabitants and employees are homogeneously distributed within the census zone; thus, the percentage of overlapping of each census zone and each hexagon is computed; the shares of population and employees are proportionally assigned to the hexagon through these percentages.

OSM data rely on the road network in terms of road link classification (Figure 1) and the related extension. Only infrastructures where cars are allowed to travel are selected. Particularly, the selected street types have different capacity and speed limits, starting from motorways to residential streets.

Census data and OSM allow us to compute metrics to fully characterize the built environment.

FCD, as a sample representing the whole mobility in the city, are used to deduce mobility and traffic patterns. FCD derive from probe vehicles, in particular, private cars, tracked by an On-Board Unit every 30–60 s. FCD are filtered to remove measurement errors and then aggregated by trip chains. Information extracted from the trip datasets includes the id of the vehicle, the starting GPS coordinates, the departure time, the destination GPS coordinates, the traveled distance, the travel time, and the dwell time. The resulting datasets are:

1. Brescia: 65,720 trips of 3445 vehicles tracked during November 2019;
2. Catania: 182,701 trips of 7393 vehicles tracked during November 2019;
3. Salerno: 178,545 trips of 7479 vehicles tracked during October 2019.

The whole set of metrics computed through the data previously reported are shown in Table 1.

Table 1. Mobility patterns and built environment metrics.

Mobility Patterns		
Metrics	**Unit**	**Note**
Origin-Destination matrices	(passenger car equivalent/time period)	FCD Demand Matrices for different time periods (whole month of data; average of weekdays; average of weekend days).
Generated trips	(passenger car equivalent/time period)	Average number of vehicles generated by each zone for different time periods (average of the whole period; average of weekdays during 7:00 a.m.–2:00 p.m.).
Attracted trips	(passenger car equivalent/time period)	Average number of vehicles attracted by each zone for different time periods (average of the whole period; average of weekdays during 7:00 a.m.–2:00 p.m.).
Time matrices	(Minutes)	Average travel times between origin and destination, during the whole period, during weekdays and during weekend days.
Travelled distance matrices	(km)	Average distances between origin and destination zone, for different time periods (during the whole period, during weekdays and during weekend days).
Distributions of trips during the day	(Vehicles/h)	Number of vehicles departing in the hour (average of weekdays; average of weekend days).
Travel times	(Minutes)	Average value, median value, and standard deviation for different time periods (the whole period; weekdays and weekend days).

Table 1. *Cont.*

Mobility Patterns		
Metrics	**Unit**	**Note**
Distribution of travel times	(%)	Statistical distribution of travel times during the whole period.
Distribution of average travel times during the day	(Minutes)	Average travel times of vehicles departing in each hour of the day (weekdays and weekend days).
Traveled distances	(km)	Average value, median value, and standard deviation for different time periods (the whole period; weekdays and weekend days).
Distribution of traveled distances	(%)	Statistical distribution of distances traveled during the whole period.
Distribution of average distances traveled during the day	(km)	The average traveled distances in each hour during weekdays and during weekend days.
Distributions of classes of dwell times	(%)	Statistical distributions of dwell times (short dwell times and long dwell times) during the whole month.
Mean speeds	(km/h)	Average value, median value, and standard deviation for different time periods (the whole period; weekdays and weekend days).
Distribution of mean speeds	(%)	Statistical distribution of mean speeds during the whole period.
Distribution of average mean speeds during the day	(km/h)	Average mean speeds in each hour during weekdays and during weekend days.
Built Environment		
Metrics	**Unit**	**Note**
Zone Population	(Number)	Number of inhabitants living in zone *z*.
Zone Employees	(Number)	Number of employees working in zone *z*.
Population distribution	-	Computed adopting the Gini index reported in (1) to Zone Population.
Employees distribution	-	Computed adopting the Gini index reported in (1) to Zone Employees.
Total road network of type *i*	(km)	Total length of road type *i* in the city.
Total road network extension	(km)	Total length of all the roads in the city.
Total road network of type *i* in zone *z*	(km)	Total length of road type *i* in zone *z*.
Total road network in zone *z*	(km)	Total length of all the roads in zone *z*.
Percentage of road type *i*	(%)	The percentage of length of road type *i* in the city.
Percentage of road type *i* in zone *z*	(%)	The percentage of length of road type *i* in the zone *z*.
Road network distribution of type *i*	-	Computed adopting the Gini index reported in (1) to Total length of road type *i* in zone *z*.
Road network distribution	-	Computed adopting the Gini index reported in (1) to Total length of all the roads in zone *z*.

Table 1. *Cont.*

Interaction between Mobility Patterns and the Built Environment		
Metrics	**Unit**	**Note**
Active potential accessibility of zone z	(Employees/second)	Computed according to (2).
Passive potential accessibility of zone z	(Inhabitants/second)	Computed according to (3).

The spatial distribution of data is computed adopting the Gini index [26]. It is usually adopted to quantify the concentration of an economic variable and it assumes a value between 0 (homogeneous distribution of the variable) and 1 (high concentration of the variable), according to:

$$G = \frac{N-1}{N} * \frac{\sum_{i=1}^{N-1} (P_i - L_i)}{\sum_{i=1}^{N-1} P_i} \tag{1}$$

where:

1. N is the total number of the elements of the sample (i.e., zones of the cities);
2. $P_i = i/N$ is the rank of the *i*-th element;
3. $L_i = \frac{\sum_{k=1}^{i} x_k}{\sum_{k=1}^{N} x_k}$ is the concentration ratio to the *i-th* element of the variable x_k (e.g., in the case of computation of the Gini index for the distribution of population, x_k is the population assigned to the zone k).

As reported in Table 1, accessibility measures are also computed as a metric of the interaction between the mobility patterns and the built environment.

The concept of accessibility is well-known [27]. Accessibility can be defined as the easiness to reach activities from a starting zone or to be reached from several zones through transportation infrastructures and services. Several formulations have been defined for accessibility as a function of the selected variables [28,29], for example, based exclusively on the performance of the transportation supply or based on the spatial distribution of activities and on the cost to reach them. The proposed formulation for accessibility is inspired by the gravitational law [30], considering socioeconomic characteristics such as the attractive mass and the travel time between zones [31]. Particularly, two potential accessibility metrics are computed: (i) the active one (Figure 2a) that measures the possibility to reach several destinations and using their activities for a given zone; (ii) the passive one (Figure 2b) that measures the possibility for a given zone to be easily reached from the inhabitants of several origins.

(a) (b)

Figure 2. Example of potential accessibility: (**a**) active; (**b**) passive.

The proposed formulation for the computation of active potential accessibility is:

$$AA_i = \sum_{j=1} \frac{empl_j}{\bar{t}_{ij}^{FCD}} \tag{2}$$

where:

1. i is the origin zone;
2. j is the destination zone;
3. $empl_j$ is the number of employees in zone j;
4. $\bar{t}_{ij}^{FCD}$ is the average travel time between zones i and j as derived by FCD.

In a comparable way, the passive potential accessibility assumes the following formulation:

$$PA_j = \sum_{i=1} \frac{inhab_i}{\bar{t}_{ij}^{FCD}} \quad (3)$$

where $inhab_i$ is the number of inhabitants of zone i.

3. Results

3.1. Mobility Patterns

Considering the 338 zones of Brescia, 522 of Catania, and 209 of Salerno, the trips, as derived by FCD datasets, cover, respectively, 12%, 6%, and 20% of the O-D pairs, thus obtaining very sparse demand matrices. However, the zones covered by at least one generated trip are 77%, 59%, and 78%, respectively, for the three cities; similar coverage has been obtained in terms of attracted trips. Comparing attracted trips and generated trips with the respective employees and inhabitants, a high correlation index is obtained (Pearson coefficient between 0.777 for Brescia and 0.908 for Salerno), demonstrating an adequate representativeness of FCD with respect to the built environment despite the low coverage in terms of O-D trips.

In terms of trip distribution during the daytime, Catania and Salerno show similar trends in both the weekends and the weekdays; Brescia shows an evening peak just one hour earlier with respect to Catania and Salerno during weekdays (two hours earlier on weekend), thus reporting some differences in terms of mobility behaviors (activity duration and/or departure time intervals).

Moving to traveled distances and travel times, the average of traveled distances per trip in the whole month (thus, despite the time interval and the specific day, if weekday or weekend) is equal to 4.64 km for Brescia (median 3.7 km, standard deviation 3.53 km), 5.17 km for Catania (median 3.9 km, standard deviation 4.27 km), and 5.05 km for Salerno (median 3.8 km, standard deviation 4.11 km). According to these metrics and as shown in Figure 3a, Brescia has 34% of trips with a traveled distance between 2 and 4 km, while datasets of Catania and Salerno have, respectively, 31% and 30% of trips belonging to this class, showing a stronger trend of Brescia for low distances than in the other two cities. This confirms a more compact city structure for Brescia or at least a more mixed one.

The average travel times in the whole period (thus, despite the time interval and the specific day, if weekday or weekend) are 14 min for Brescia (median 12 min, standard deviation 8 min), 18 min for Catania (median 15 min, standard deviation 10 min), and 16 min for Salerno (median 13 min, standard deviation 10 min). In the dataset of Brescia, the classes between 5 and 15 min represent 68% of the total trips, those in the dataset related to Salerno represent 60%, and those for Catania represent half of the total trips sample (Figure 3b). For travel time values over 15 min, the percentage of trips for the database of Brescia considerably drops, while the percentages in the datasets of Catania and Salerno decrease slowly. Again, as underlined in terms of distances traveled, it seems that Catania and Salerno have a sprawled structure with respect to Brescia.

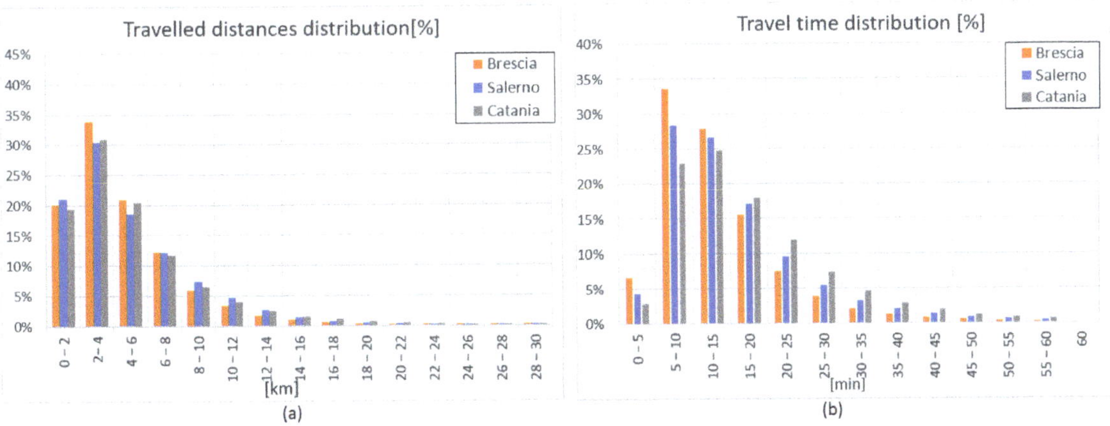

Figure 3. Distribution of traveled distances (**a**) and distribution of travel times (**b**).

The average mean speed is 20 km/h for the database of Brescia (median 19 km/h, standard deviation 8 km/h), 18 km/h for the database of Catania (median 16 km/h, standard deviation 9 km/h), and 20 km/h for the database of Salerno (median 18 km/h, standard deviation 9 km/h).

According to the distribution of Figure 4, mean speeds in the database of Brescia are concentrated in a medium speed range, 30% of trips experience an average speed between 15 and 20 km/h, while trips in the database of Catania are concentrated in lower-speed classes (between 10 and 15 km/h). If the observed speeds are analyzed according to the peak hour in the morning, previous values are amplified as they reach 16 km/h for Catania, while higher ones (about 19 km/h) are reached for both Brescia and Salerno.

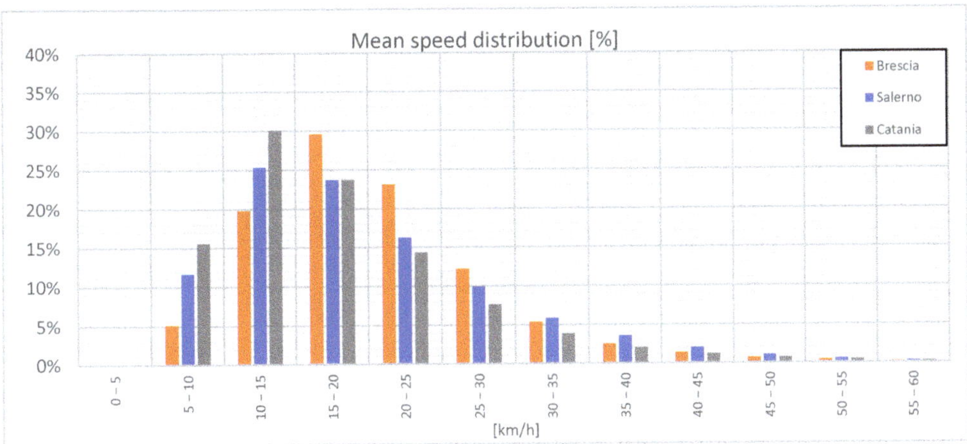

Figure 4. Mean speed distribution.

3.2. Built Environment Analysis

The analysis of the built environment is conducted through the computation of density of inhabitants (Figure 5) and employees in each hexagonal zone and of the Gini index for quantifying the spatial distribution.

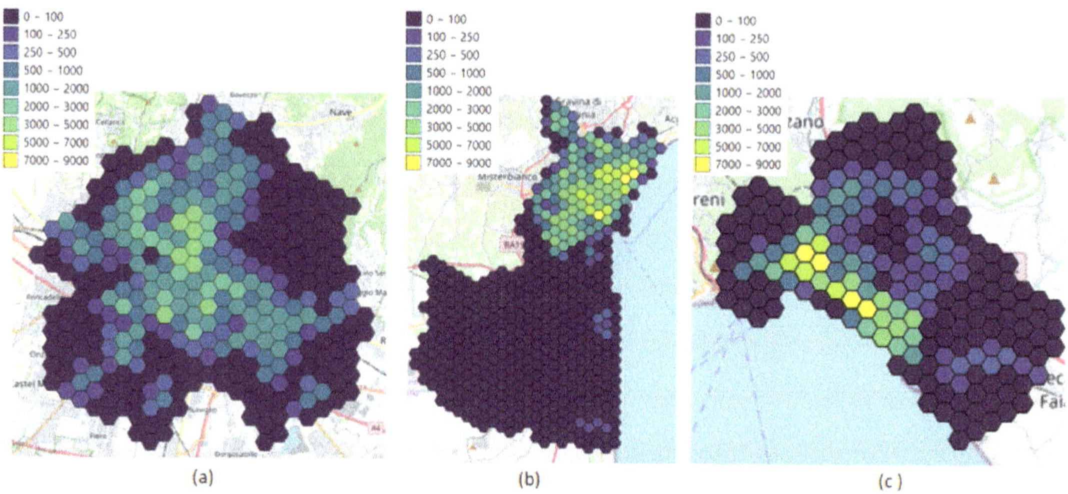

Figure 5. Distribution of inhabitants in each zone: Brescia (**a**), Catania (**b**), and Salerno (**c**).

The city of Brescia has a density of 2094 inhabitants/km^2 and 700 employees/km^2. The city shows a Gini index equal to 0.68 and 0.69 for the population distribution and for the employees, respectively.

The city of Catania is the largest one, with a density of 1602 inhabitants/km^2 (lower than Brescia). The Gini index for population is 0.86, with the concentration in few zones (in this case, in the northern part of the city, where its "core" center is located). Despite the density of activities being similar to Brescia (711 employees/km^2), it is also concentrated in the "core" center, with a Gini index of 0.88.

The city of Salerno is the smallest one, with a density of 2251 inhabitants/km^2 and a Gini index for population of 0.83, with a high concentration along the coast and very low values in the hinterland. Employees density is 423 employees/km^2 (lower than Brescia and Catania) with a Gini index of 0.87, showing a higher concentration near the harbor.

It must be underlined that, while both Catania and Salerno are city ports, the port of Salerno is more remotely located with respect to the downtown than Catania. The location of the port of Catania near the center of the city involves a higher concentration of employees, whereas the position of the port of Salerno allows a slight minor concentration of employees (in fact, the Gini Index is slightly lower than the one of Catania).

It should also be noticed that in Brescia, there are no zones with the highest-class value of inhabitants (i.e., no yellow zones are reported in Figure 5).

Moving to the road network structure (Figure 6), in the three cities, the main road type is the residential street covering more than 35% of the total network kilometers. Catania and Salerno have higher percentages of high-speed roads (motorway and primary) than Brescia. The Gini index of the road network is 0.26 for Brescia, demonstrating a more uniform distribution of the street network, while the Gini indexes for Catania and Salerno are, respectively, 0.57 and 0.44. Brescia has, in general, more roads for penetration in the city center than the other two cities. Salerno has in its municipality an important motorway—the A2 Salerno-Reggio Calabria.

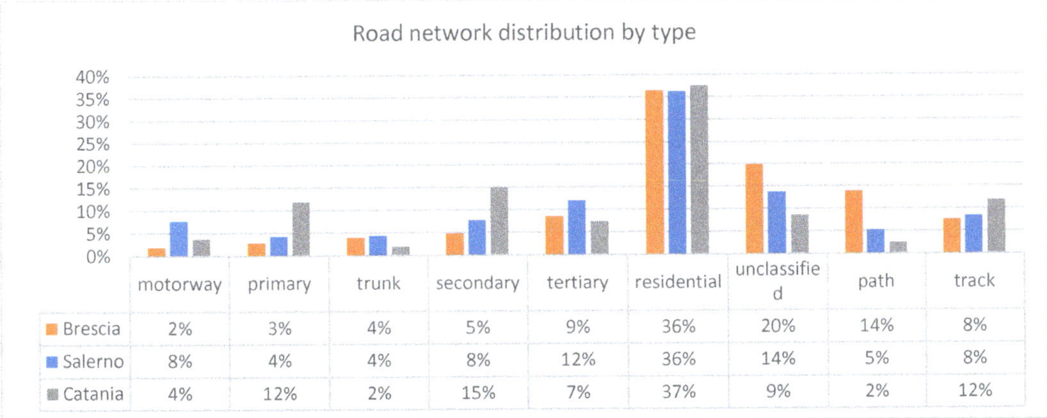

	motorway	primary	trunk	secondary	tertiary	residential	unclassified	path	track
Brescia	2%	3%	4%	5%	9%	36%	20%	14%	8%
Salerno	8%	4%	4%	8%	12%	36%	14%	5%	8%
Catania	4%	12%	2%	15%	7%	37%	9%	2%	12%

Figure 6. Distribution of road network by type.

3.3. *Integrating Mobility Patterns and the Built Environment: The Potential Accessibility*

The active potential accessibility, computed as reported in (2), is 22 employees/s for Brescia (standard deviation 46 employees/s) with a Gini index of 0.79, 26 employees/s for Catania (standard deviation 91 employees/s) with a Gini index of 0.91, and 13 employees/s for Salerno (standard deviation 41 employees/s) with a Gini index of 0.90. Gini indexes for active potential accessibility are slightly higher than the Gini indexes of employees.

Instead, for the passive potential accessibility (Figure 7), computed as reported in (3), the average value is 68 inhabitants/s for Brescia (standard deviation 121 inhabitants/s) with a Gini index of 0.77, 62 inhabitants/s for Catania (standard deviation 160 inhabitants/s) with a Gini index of 0.88, and 69 inhabitants/s for Salerno (standard deviation 186 inhabitants/s) with a Gini index of 0.87.

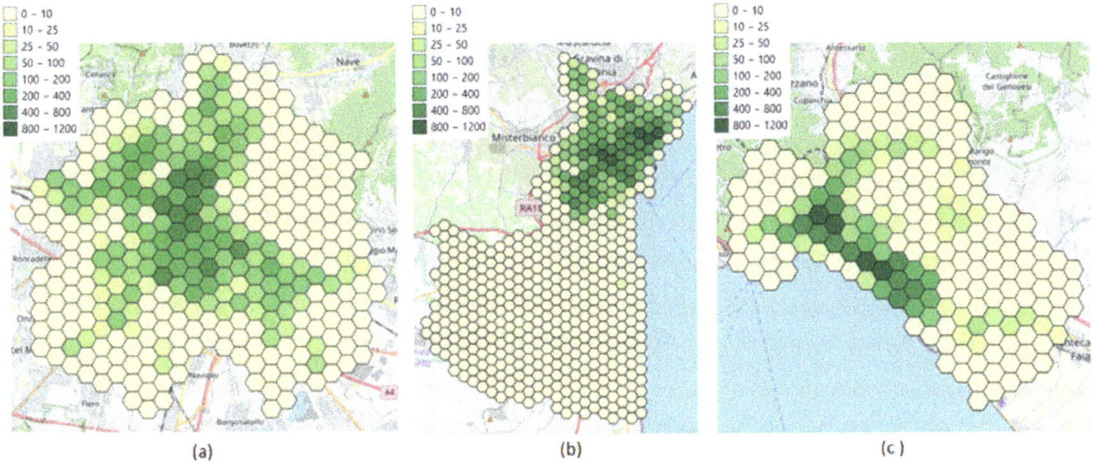

Figure 7. Distribution of passive potential accessibility: Brescia (**a**), Catania (**b**), and Salerno (c).

4. Discussion

In this work, we promoted a data-driven framework regarding the interrelation between mobility patterns, the built environment, and related accessibility in order to analyze the correlation between the private transportation system and the land-use in three cities.

As seen in the results section and summarized in Figure 8, potential accessibility indicators have slightly higher values of Gini indexes with respect to those of their respective variable (e.g., passive accessibility with population and active accessibility with employees). If the potential accessibility is higher than 0.8 (see Catania and Salerno, Figure 8), approximately the same values are obtained for population and employees' distribution. Brescia shows a reduction of approximately 0.1 between the Gini index of the potential accessibility and the related socioeconomic factor. This result can be summarized, postulating that more homogeneous conditions in terms of population and employees imply a higher possibility to reach a high level of accessibility on the whole city.

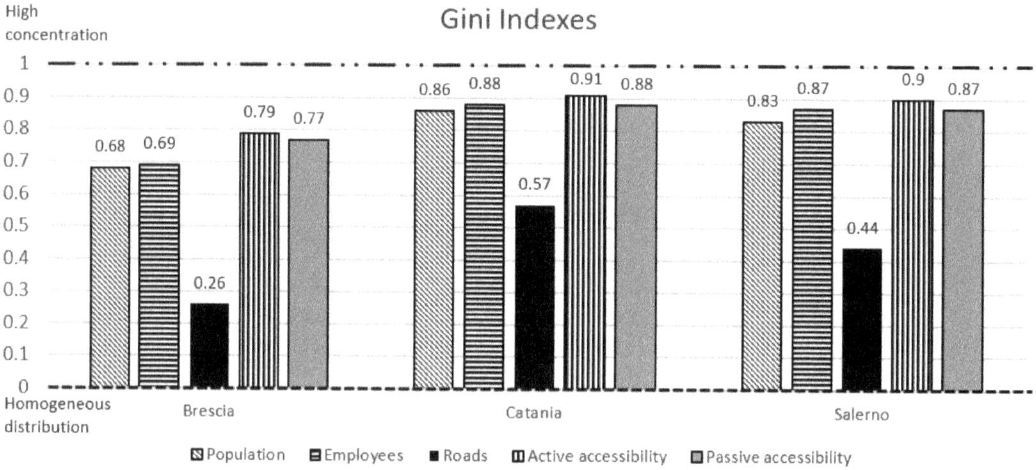

Figure 8. Summary of Gini Indexes for each city.

Instead, observing the differences between the Gini index for the potential accessibility and for the road network, the differences are higher where homogenous road structure conditions can be assured. Thus, a very uniform and distributed road network, as in the case of Brescia, can help in improving accessibility values, but it seems to be not sufficient alone.

The spatial distribution computed through the Gini index is demonstrated to be essential for this type of analysis: Salerno, in fact, shows the same passive potential accessibility of Brescia on average, but in Salerno, the higher values of accessibility (and population) are concentrated in few zones, while Brescia shows lower accessibility values (and population values) at the zone level but is more evenly distributed.

The layout of the city (compact for Brescia, sprawled for the other two cases) impacts the road infrastructures: Brescia has a radial structure that allows for the diffusion of the built environment (Figure 5a) and the transportation supply in all the directions; in fact, this city has the lowest Gini indexes for the distribution of population and employees and for the distribution of the road network. Instead, Catania and Salerno have been developed along the coast concentrating the built environment (Figure 5b,c) and the local road network in their "cores". The concentration of roads and activities impacts on the revealed mobility patterns: the database of Catania has lowest average mean speeds and highest trip times (Figures 3b and 4), while the database of Brescia reports the lowest average travel time and highest average mean speed. These findings could involve the hypothesis that Catania results as more congested than Brescia due to the concentration of traffic flows to specific destinations and on few corridors. Thus, the layout of the city, also dependent on its geographical location (in the hinterland for Brescia and along the coastline for Salerno and Catania), must be considered in the integrated planning of transportation systems and land-use. The structure of the city indeed allows the localization of potential areas to maximize the benefits of sustainable policies related to the growth of the city itself, along with the identification of the critical elements that must be solved. Moreover, the knowledge of the layout of the city and the metrics described above support the several best practices usually applied in urban planning (e.g., a deep analysis of the road infrastructures could avoid the construction of new buildings, directing the efforts toward sustainable management policies of the existing transportation supply).

5. Conclusions

The present work is focused on the evaluation of the mutual impact between the transportation systems and the land-use based on several metrics estimated from different data sources and applied to three Italian medium-size cities: Brescia, Catania, and Salerno.

The results highlight this well-known interrelation showing how the layout of the city and the geographical constraints impact the road infrastructures. In Brescia, the diffusion of the built environment and the transportation supply follow a radial structure, due to the lack of natural constraints in its potential territorial development, as confirmed by the low Gini indexes for the distribution of population and employees and for the distribution of the road network. Instead, Catania and Salerno have been developed along the coast and, thus, present a concentrated built environment and road network. In addition, the low mean speed and high travel times estimated for the municipality of Catania suggest that this city is more congested than Brescia due to the concentration of traffic flows to specific destinations and on few corridors. Catania also shows a high average value for potential active accessibility with respect to the other cities, with a maximum concentration along the coast, while Brescia has a lower value that is more distributed within the municipality, demonstrating an extended use of its whole territory.

This work represents a first step toward the development of an extended analytical framework aimed at investigating the complex relationship between private transportation and the built environment.

In this context, future research will focus mainly on two topics.

First, the framework could be improved by estimating mean travel speeds and delays in each city zone to obtain the spatial distribution of congestion [32,33]. Spatial distribution and duration of the dwell-time could also be exploited by FCD to classify the land-use of each zone (e.g., residential and commercial areas, industrial districts, and playgrounds) or to evaluate the effects of parking policies and traffic restrictions.

Moreover, the framework can be extended to include other modes of transport, such as public transport, in order to further capture the interaction between mobility and land-use [34,35].

Author Contributions: Conceptualization, C.L., M.N. and G.V.; data curation, R.D.V., F.K. and V.R.; formal analysis, R.D.V., M.N. and V.R.; investigation, R.D.V. and V.R.; methodology, C.L., M.N. and G.V.; supervision, F.K., C.L., M.N. and G.V.; writing—original draft, R.D.V., M.N. and G.V.; writing—review and editing, R.D.V., C.L., M.N. and G.V. All authors have read and agreed to the published version of the manuscript.

Funding: This research received no external funding.

Institutional Review Board Statement: Not applicable.

Informed Consent Statement: Not applicable.

Data Availability Statement: Not applicable.

Acknowledgments: We are extremely grateful to Fabio Carapellucci for his continuous support in the management of IT systems and resources.

Conflicts of Interest: The authors declare no conflict of interest.

References

1. Gkania, V.; Dimitriou, L. A back-engineering approach to explore human mobility patterns across megacities using online traffic maps. In *Mobility Patterns, Big Data and Transport Analytics: Tools and Applications for Modeling*; Elsevier: Amsterdam, The Netherlands, 2018; pp. 345–363.
2. Vulevic, A. Accessibility concepts and indicators in transportation strategic planning issues: Theoretical framework and literature review. *Logist. Sustain. Transp.* **2016**, *7*, 58–67. [CrossRef]
3. Tong, L.; Zhou, X.; Miller, H.J. Transportation network design for maximizing space–time accessibility. *Transp. Res. Part B Methodol.* **2015**, *81*, 555–576. [CrossRef]
4. Hellervik, A.; Nilsson, L.; Andersson, C. Preferential centrality–A new measure unifying urban activity, attraction and accessibility. *Environ. Plan. B Urban Anal. City Sci.* **2019**, *46*, 1331–1346. [CrossRef]
5. Habib, M.A.; McCarthy, S. Development of an Integrated Transportation and Land-Use Microsimulation Model on a flexible modeling platform. *Transp. Res. Rec. J. Transp. Res. Board* **2021**, *2675*, 355–369. [CrossRef]
6. Abdelfattah, L.; Deponte, D.; Fossa, G. The 15-minute city: Interpreting the model to bring out urban resiliencies. *Transp. Res. Procedia* **2022**, *60*, 330–337. [CrossRef]
7. Cervero, R.; Kockelman, K. Travel Demand and the 3Ds: Density, Diversity, and Design. *Transp. Res. Part D* **1997**, *2*, 199–219. [CrossRef]
8. Ma, X.; Zhang, J.; Ding, C.; Wang, Y. A geographically and temporally weighted regression model to explore the spatiotemporal influence of built environment on transit ridership. *Comput. Environ. Urban Syst.* **2018**, *70*, 113–124. [CrossRef]
9. Frank, L.D.; Pivo, G. *Relationships between Land-Use and Travel Behavior in the Puget Sound Region*; Elsevier: Amsterdam, The Netherlands, 1994. [CrossRef]
10. Choi, K. The influence of the built environment on household vehicle travel by the urban typology in Calgary, Canada. *Cities* **2018**, *75*, 101–110. [CrossRef]
11. Li, T.; Jing, P.; Li, L.; Sun, D.; Yan, W. Revealing the Varying Impact of Urban Built Environment on Online Car-Hailing Travel in Spatio-Temporal Dimension: An Exploratory Analysis in Chengdu, China. *Sustainability* **2019**, *11*, 1336. [CrossRef]
12. Brandi, A.; Gori, S.; Nigro, M.; Petrelli, M. Activities Relocation for a Sustainable Mobility System. *Transp. Res. Procedia* **2015**, *5*, 4–12. [CrossRef]
13. Ibáñez, J.N.; Rotoli, F. Measuring the Impact of the Trans-European Road Transport Network on the Accessibility of European Urban Agglomerations. *Transp. Res. Rec. J. Transp. Res. Board* **2017**, *2605*, 72–82. [CrossRef]
14. Xiao, Z.; Xiao, H.; Jiang, H.; Chen, W.; Chen, H.; Regan, A.C. Exploring Human Mobility Patterns and Travel Behavior: A Focus on Private Cars. *IEEE Intell. Transp. Syst. Mag.* **2022**, *14*, 129–146. [CrossRef]
15. Chen, C.; Feng, T.; Shao, M.; Yao, B. Understanding the determinants of spatial-temporal mobility patterns based on multi-source heterogenous data. *Transp. Res. Procedia* **2021**, *52*, 477–484. [CrossRef]
16. Sun, D.; Leurent, F.; Xie, X. Uncovering mobility typologies of territorial zones based on Floating Car Data mining. *Transp. Res. Procedia* **2021**, *52*, 708–715. [CrossRef]
17. Sun, D.; Leurent, F.; Xie, X. Mining Vehicle Trajectories to Discover Individual Significant Places: Case Study using Floating Car Data in the Paris Region. *Transp. Res. Rec.* **2021**, *2675*, 1–9. [CrossRef]
18. Ferrara, M.; Liberto, C.; Nigro, M.; Trojani, M.; Valenti, G. Multimodal choice model for e-mobility scenarios. *Transp. Res. Procedia* **2019**, *37*, 409–416. [CrossRef]
19. Pirra, M.; Carboni, A.; Deflorio, F. Freight delivery services in urban areas: Monitoring accessibility from vehicle traces and road network modelling. *Res. Transp. Bus. Manag.* **2019**, 100680. [CrossRef]
20. Laviolette, J.; Morency, C.; Waygood, O.D. Car Ownership and the Built Environment: A Spatial Modeling Approach. *Transp. Res. Rec. J. Transp. Res. Board* **2022**, *2676*, 125–141. [CrossRef]
21. Burdziej, J. Using hexagonal grids and network analysis for spatial accessibility assessment in urban environments – a case study of public amenities in Toruń. *Misc. Geogr.* **2019**, *23*, 99–110. [CrossRef]
22. Cabrera-Barona, P.; Wei, C.; Hagenlocher, M. Multiscale evaluation of an urban deprivation index: Implications for quality of life and healthcare accessibility planning. *Appl. Geogr.* **2016**, *70*, 1–10. [CrossRef]
23. Zhu, X.; Liu, S.; Yeow, M.C. Accessibility analysis for housing development in Singapore with GIS and multicriteria analysis methods. *Appl. GIS* **2006**, *2*, 13.1–13.12. [CrossRef]
24. Chmielewski, J.; Kempa, J. Hexagonal Zones in Transport Demand Models. *KnE Eng.* **2020**, 103–116. [CrossRef]
25. ISTAT Census Data. Available online: https://www.istat.it/it/archivio/104317 (accessed on 18 July 2022).
26. Pellegrino, S. *The Gini Coefficient: Its Origins*; Working paper No. 70. 2020. Available online: https://ideas.repec.org/p/tur/wpapnw/070.html (accessed on 2 October 2022).
27. Dalvi, M.Q.; Martin, K.M. The Measurement of Accessibility: Some Preliminary Results. *Transportation* **1976**, *5*, 17–42. [CrossRef]
28. Geurs, K.T.; Van Wee, B. Accessibility evaluation of land-use and transport strategies: Review and research direction. *J. Transp. Geogr.* **2004**, *12*, 127–140. [CrossRef]
29. Liu, M.; Jiang, Y. Measuring accessibility of urban scales: A trip-based interaction potential model. *Adv. Eng. Inform.* **2021**, *48*, 101293. [CrossRef]
30. Simini, F.; González, M.; Maritan, A.; Barabási, A.L. A universal model for mobility and migration patterns. *Nature* **2012**, *484*, 96–101. [CrossRef] [PubMed]

31. Yoshida, N.; Deichmann, U. Measurement of Accessibility and Its Applications. *J. Infrastruct. Dev.* **2009**, *1*, 1–16. [CrossRef]
32. Altintasi, O.; Tuydes-Yaman, H.; Tuncay, K. Detection of urban traffic patterns from Floating Car Data (FCD). *Transp. Res. Procedia* **2017**, *22*, 382–391. [CrossRef]
33. Wolniak, M.J.; Mahapatra, S. Data- and Performance-Based Congestion Management Approach for Maryland Highways. *Transp. Res. Rec. J. Transp. Res. Board* **2014**, *2420*, 23–32. [CrossRef]
34. Nakshi, P.; Debnath, A.K. Impact of Built Environment on mode choice to major destinations in Dhaka. *Transp. Res. Rec. J. Transp. Res. Board* **2021**, *2675*, 281–296. [CrossRef]
35. Hardy, D.; Bell, A.; McCahill, C. Accessibility Measurement for Project Prioritization in Virginia. *Transp. Res. Rec. J. Transp. Res. Board* **2019**, *2673*, 266–276. [CrossRef]

sustainability

MDPI

Review

Mapping the Path to Sustainable Urban Mobility: A Bibliometric Analysis of Global Trends and Innovations in Transportation Research

Enver Cenan İnce

Department of City and Regional Planning, Faculty of Architecture, Fine Arts and Design, Sivas Cumhuriyet University, Sivas 58140, Türkiye; eince@cumhuriyet.edu.tr

Abstract: This article presents a bibliometric analysis of urban transportation research, focusing on trends in urban mobility from 2018 to 2023. By analyzing 2000 articles from 617 journals in the Scopus database, this study identifies key themes, such as sustainable transportation, smart mobility, and technology integration into transit systems. Emerging terms, such as public health, walkability, Simulation of Urban Mobility (SUMO), and Network Simulator 3 (ns-3) highlight the growing intersection of transportation with environmental and societal factors. China has emerged as a leader in urban transportation research, excelling in publication volume and international collaboration. Notable figures and journals, such as Liu Y. and "Sustainability (Switzerland)", emphasize the field's interdisciplinary nature. The use of tools like the Analytic Hierarchy Process (AHP), simulation, and a Geographic Information System (GIS) underscores a comprehensive approach to sustainable urban mobility planning, helping planners make data-driven decisions. These findings add vital insights into worldwide urban transportation research and provide recommendations for future studies aimed at improving sustainable transportation policies and practices.

Keywords: urban transportation; urban mobility; sustainability; bibliometric analysis; Biblioshiny; VOS (visualization of similarities) viewer

1. Introduction

Transportation in urban areas affects key aspects of modern societies, such as the quality of life, economic development, and environmental sustainability [1]. Rapid urbanization and population growth have exerted significant pressure on the planning and management of urban transportation systems. Urban transportation research is critical for comprehending these complex environmental and social interactions and developing effective solutions [1].

In this context, bibliometric analysis is a powerful tool for understanding the quantitative and qualitative characteristics of the scientific literature on urban transportation. Bibliometrics helps visualize the body of knowledge by examining the relationships between scientific publications, publication trends, most frequently cited works, and interdisciplinary collaboration [2]. This type of analysis can help researchers and policymakers identify important issues and trends in the field of urban transportation, shaping future research and policy agenda.

Historically, urban transportation research has concentrated on the physical aspects of transportation infrastructure, as well as the performance of public transportation networks.

However, in recent years, technological improvements and shifting societal requirements have prompted researchers and politicians to adopt broader approaches to this topic. This larger approach encompasses issues such as sustainable transportation regulations, Intelligent Transportation Systems, travel behavior analysis, environmental implications, and social injustice [1]. Another key aspect of sustainable urban mobility is tackling gender disparities, social exclusion, and equity challenges, all of which are critical to meeting Agenda 2030 goals. According to research, women and underprivileged communities frequently experience severe impediments in accessing transportation systems, resulting in restricted mobility, economic disadvantages, and social marginalization [3]. Urban transport research has traditionally focused on the physical characteristics of transport infrastructure and the performance of public transport systems. However, in recent years, technological advances and changing societal needs have led researchers and policymakers to adopt a broader perspective in this field. This broader perspective includes topics such as sustainable transport policies, intelligent transport systems, travel behavior analysis, environmental impacts, and social inequality [1]. Another critical dimension of sustainable urban mobility involves addressing gender gaps, social exclusion, and equity issues, which are pivotal in achieving the goals of Agenda 2030. Research highlights that women and marginalized communities often face significant barriers in accessing transportation systems, resulting in reduced mobility opportunities, economic disadvantages, and social exclusion [4]. By taking into account the unique travel habits, security concerns, and accessibility requirements of various groups, gender-responsive and -inclusive transportation design seeks to close these gaps. For example, initiatives such as improving last-mile connectivity, integrating affordable and safe public transportation, and fostering pedestrian-friendly environments can significantly reduce these inequalities. Incorporating these aspects into urban transportation systems not only promotes equity, but also ensures that sustainable mobility solutions are socially inclusive and aligned with broader Sustainable Development Goals (SDGs). Future urban mobility policies must prioritize inclusivity as a cornerstone for fostering a just and equitable transition to sustainable cities.

Bibliometric analysis provides an opportunity to examine the scientific literature on a topic in detail. It serves as a means to provide a comprehensive overview of past and current studies, visualize and map data, identify trends, shed light on future studies, and offer a distinct advantage to researchers in these respects [5]. In this study, Biblioshiny, a web-based bibliometric analysis platform, was used to collect and process numerous academic publications in the field of urban transportation. Biblioshiny facilitates data collection, evaluation, and visualization by providing information on the chronological development of research in urban transportation, leading authors, journals, and keywords. In addition, VOSviewer, an advanced network visualization tool, has been used to create network maps showing keywords and their evolution over the years to reveal emerging trends and research themes for urban transportation researchers [6,7].

This thorough bibliometric analysis aims to provide readers with a comprehensive understanding of the scholarly evolution of urban transportation research. Tracing significant individuals and charting the changing boundaries of urban transportation studies reveals scholarly milestones. Beyond satisfying the curiosity of academics, this effort is poised to provide practical insights to decision-makers, policy practitioners, and researchers to solve urban transportation problems and address complex real-world challenges in urban transportation. By embarking on this bibliometric survey, the intention was to introduce readers to the rich tapestry of research in the field of urban transportation and explore the threads that connect the past, present, and future of the ever-evolving field of transportation studies.

Novelty of This Study

This study makes a unique contribution to the field of urban transportation research by conducting a comprehensive bibliometric analysis that focuses exclusively on the recent surge in research output between 2018 and 2023. Unlike previous bibliometric studies that often cover broader or more generalized timeframes, this paper highlights emerging themes, such as the integration of public health, advanced simulation tools (e.g., SUMO and ns-3), and interdisciplinary approaches to urban mobility. By combining Biblioshiny and VOSviewer to map the evolution to visualize the evolution of keywords and collaborative networks, this study provides fresh insights into global trends, research hotspots, and technological innovations that shape urban transportation research. These findings not only fill a gap in the literature, but also offer direction for researchers, policymakers, and urban planners.

The existing literature on urban transportation and sustainable mobility has primarily focused on specific regional or temporal contexts, with few comprehensive reviews providing a global perspective on recent trends and innovations. This study fills this gap by conducting a bibliometric analysis of 2000 articles published between 2018 and 2023, offering both a broad temporal overview and insights into spatial variations in the research focus.

This research paper is structured as follows. Section 2 introduces the bibliometric analysis, research gaps, and research topics, as well as a brief overview of urban transportation. Section 3 introduces the research strategy in terms of materials and methods. Section 4 summarizes and evaluates the research findings, and identifies relevant research issues and trends in the field of urban transport. Section 5 contains the discussion, while Section 6 outlines recommendations for future research.

2. A Brief Overview of Urban Transportation

Urban transportation and urban mobility are critical areas of study that address the complex interactions between infrastructure, societal needs, and technological advancements within urban settings. The evolving dynamics of urban transportation systems and mobility patterns are influenced by various factors including population growth, environmental concerns, and technological innovation.

Urban transportation systems encompass a range of modes including public transit (buses, trams, and subways), private vehicles, bicycles, and pedestrian pathways. The efficiency and sustainability of these systems are paramount in addressing urban congestion, pollution, and accessibility issues.

Newman and Kenworthy [8] highlighted the importance of integrating land use and transportation planning to promote sustainable urban development. They contend that compact city layouts, when backed by effective public transportation, may drastically reduce the use of private automobiles, which will reduce greenhouse gas emissions and ease traffic.

Public transit systems are also essential for ensuring equitable access to urban amenities. Litman [9] emphasized the role of public transportation in enhancing social equity by providing affordable mobility options for all socioeconomic groups. Furthermore, comprehensive public transit networks can mitigate the adverse environmental impacts associated with urban transportation by reducing the overall number of vehicle miles traveled (VMT). The effectiveness of public transit is often measured based on its coverage, frequency, and reliability. Cervero [10] explored the concept of transit-oriented development (TOD), where residential and commercial areas are designed to maximize access to public transportation, thereby encouraging higher transit ridership and reducing dependence on private cars.

In addition, the advent of technological innovation has significantly transformed urban mobility. The rise of smart mobility solutions, such as electric vehicles (EVs), autonomous vehicles (AVs), and shared mobility services (ride hailing and bike sharing), presents both opportunities and challenges for urban transportation systems. Gonzalez et al. [11] highlight the potential of EVs to reduce urban air pollution and dependence on fossil fuels. However, they also highlight the need for an adequate charging infrastructure and the integration of renewable energy sources to maximize the environmental benefits of EVs. However, the transition to autonomous vehicles (AVs) is not without challenges, particularly during the mixed-use phase, when both AVs and conventional vehicles coexist on roads. This transitional state can limit the realization of AV-related benefits, such as reduced congestion and enhanced safety, owing to the potential conflicts between human-driven and autonomous vehicles. Effective policies, infrastructure adaptations, and public awareness initiatives are necessary to mitigate these challenges and facilitate the integration of AVs into existing transportation systems.

Xie, Gou, and Gui [12] underscore the positive impact of EV adoption on urban air quality improvement, particularly in regions like Wuhan, where electrification policies have been implemented. Similarly, Tao [13] highlighted the dynamic relationship between EV uptake and green development, arguing that local government competition plays a crucial role in accelerating EV adoption while ensuring its alignment with sustainable urban policies. Autonomous vehicles are expected to revolutionize urban mobility by enhancing safety, reducing traffic congestion, and improving the efficiency of transportation networks. However, their widespread adoption requires substantial advancements in technology, regulatory frameworks, and public acceptance [14]. In their systematic review, Olayode et al. [15] examined the applications, impacts, and public perceptions of AVs, noting that, while AVs hold significant promise, their deployment must address challenges related to infrastructure, ethical considerations, and societal readiness. Shared mobility services have gained popularity because of their convenience and cost-effectiveness. Shaheen et al. [16] examine the impacts of ride-hailing and car-sharing services on urban transportation, noting that, while these services can complement public transit and reduce the need for private car ownership, they can also contribute to increased VMT and traffic congestion if not properly managed. Furthermore, the environmental and social impacts of urban transportation systems are significant considerations in planning and policymaking. Banister [17] emphasizes the need for a paradigm shift towards sustainable mobility, which prioritizes walking, cycling, and public transit over private car use. This shift could lead to reduced emissions, improved public health, and enhanced urban livability. Equitable access to transportation is crucial for fostering inclusive urban environments. Grengs [18] explores the concept of "transport justice", advocating for transportation policies that address the needs of marginalized communities and ensure the fair distribution of mobility resources.

Moreover, the future of urban transportation and mobility lies in the integration of innovative technologies, sustainable practices, and inclusive policies. The overall effectiveness and user experience of urban mobility systems can be improved by the creation of multimodal transportation networks that seamlessly link different forms of transportation [19].

Recent studies have introduced graph neural networks (GNNs) as a promising approach to urban transportation modeling. Li et al. [20] proposed an end-to-end heterogeneous graph neural network for traffic assignment, which enables the better prediction and allocation of traffic flow in complex urban networks. Similarly, Zhang et al. [21] utilized GNNs for nationwide human mobility prediction and demonstrated the efficacy of these methods in capturing human movement patterns and predicting future mobility trends.

These advancements reflect the increasing use of deep learning techniques to address the challenges of urban mobility and enhance the accuracy and efficiency of transportation systems [20,21]. These research papers represent state-of-the-art (SOTA) methodologies in urban transportation studies, highlighting cutting-edge approaches for managing complex traffic and mobility data with high precision.

Recent advances in neural network applications have further transformed urban mobility research by enabling more accurate traffic prediction, route optimization, and real-time decision making. In particular, deep learning approaches such as graph neural networks have proven effective in capturing the complex spatial and temporal dependencies within urban transportation networks. For instance, Liu and Meidani [22] demonstrated that end-to-end heterogeneous graph neural networks can significantly enhance traffic assignment accuracy by modeling intricate urban network dynamics, thereby offering promising improvements in efficiency and sustainability. This integration of neural network methodologies into urban mobility planning represents a pivotal step toward smarter transportation systems that can dynamically respond to real-world challenges.

In summary, urban transportation and mobility are dynamic fields influenced by technological advancements, environmental considerations, and social equity issues. Effective planning and policymaking, supported by continuous research and innovation, are essential for the development of sustainable and inclusive urban transportation systems. Thus, conducting a thorough bibliometric analysis of urban transportation and mobility is pivotal for understanding the underlying mechanisms and trends that shape these concepts.

2.1. Bibliometric Analysis

Bibliometric analysis is a valuable method for monitoring research trends and tracking discoveries across various fields. For instance, Demir et al. [5] utilized the Scopus search engine to review studies on sensitivity analysis in multi-criteria decision making (MCDM). They analyzed 1374 articles using VOSviewer (version 1.6.18), Biblioshiny (version 4.1.3), and Citespace (version 6.3.R1 Basic). Twinomurinzi and Mageto [23] conducted a bibliometric review of smart mobility research by examining 3223 documents from the Web of Science Core Collection with the help of VOSviewer and CiteSpace. They emphasized the integration of advanced technologies, such as IoT and AI, in transportation systems. Additionally, Ceccato et al. [24] analyzed 245 articles on passenger safety in public transportation systems using VOSviewer, providing insights into safety concerns in railway transportation. Moreover, Mirabi and Noorid [25] recently carried out a bibliometric study to examine current research trends and patterns in the spheres of urban mobility and transportation planning. Utilizing the Scopus database, they accumulated a dataset comprising 1255 publications spanning from 1995 to 2021. VOSviewer and the Bibliometrix R-package were employed to conduct an analysis, visualizing and charting co-authorship networks, citation trends, and keyword co-occurrences. Their research underscores the substantial increase in publications and identifies the key authors, institutions, and collaborative networks operating in these areas. While Mirabi and Noorid [25] provided valuable insights into urban mobility and transportation planning, their analysis primarily spans from 1995 to 2021. By contrast, this study focuses on the more recent period of 2018–2023, capturing emerging trends and innovations in the field, particularly in the context of sustainable mobility and smart transportation solutions. Additionally, this study delves deeper into thematic clusters and research gaps, providing a more targeted perspective on advancements that shape contemporary urban mobility research. Finally, Zhiyou Long [26] performed a comprehensive bibliometric analysis on how urban-built environments impact travel behavior using 1745 publications from the Web of Science database, identifying key research clusters and trends, and highlighting the significant contributions from institutions in the USA and China. The author used network analysis and content analysis tools, such as VOSviewer.

These studies showcase the diverse applications of bibliometric analysis for understanding and advancing urban mobility and transportation planning. Each study utilized comprehensive datasets and sophisticated software to reveal trends, identify research gaps, and suggest future research directions.

2.2. Research Gaps and Research Questions

A comprehensive review of existing academic papers in the field of urban transportation revealed critical research needs.

- There is a need for a generally accepted model that facilitates the evaluation of urban transportation studies.

In this study, quantitative tools, such as Bibliometrix, Biblioshiny, R software with a web-based interface, and VOSviewer, were used to design a bibliometric analysis for articles related to urban transportation in the field of social sciences, engineering, and environmental sciences in the Scopus database. This study established a conceptual framework for urban transportation publications and identified the most influential articles and prolific contributors to the field. These results provide valuable insights for transportation researchers, helping them understand current and future research trends. The main objective of this research was to find answers to the following questions.

- What is the trajectory of the growth and citation trend status of articles on urban transportation?
- Who are the most cited major authors in the field?
- What are the most cited publications, journals, organizations, and countries?
- What are the priority research points and key topics in related applications in the field of urban transportation?

This study can help urban transportation researchers understand current and future research trends and, thus, design more innovative and creative research by revealing the conceptual richness of urban transportation studies.

3. Materials and Methods

In the first step, the Scopus database was chosen for the research. In the second step, the keywords used were urban transportation, document type: article, publication language: English, publication type: journal, and publication year: 2018–2023. Reviews, conference proceedings, book chapters, articles or reviews, and commentaries published on preprint websites, editorials, and letters were excluded. Step three involved conducting performance and mapping analyses. In the fourth step, Biblioshiny and VOSviewer were utilized.

Evaluation Metrics

We employed several standard evaluation metrics commonly used in bibliometric studies to ensure the robustness and validity of the bibliometric analysis. These include the following.

Annual Growth Rate (AGR): We calculated the AGR of publications from 2018 to 2023 to assess the overall growth in research output in the field of urban transportation. AGR was computed as the percentage change in the number of publications each year compared to the previous year, providing insights into the trends and dynamics of the field.

Citation Analysis: We assessed the impact of individual articles, journals, and authors by analyzing citation counts. This metric helped helps evaluate the influence and relevance of the research within the field. We also identified the most cited articles to highlight their key contributions to the literature.

Co-authorship and Collaboration Patterns: The level of international collaboration was evaluated by tracking co-authorship across different countries and institutions. This metric revealed the extent of interdisciplinary and international cooperation in urban transportation research.

Keyword Co-occurrence and Clustering: We used VOSviewer to analyze the co-occurrence of keywords, allowing us to identify research clusters and emerging themes. This approach aids in visualizing the evolving focus areas within urban transportation research, such as sustainable mobility, smart cities, and transportation efficiency.

On the other hand, considering the strengths of different approaches, bibliometric analysis tools, such as Biblioshiny, VOSviewer, and keyword analysis, were employed.

Biblioshiny: This platform was utilized to collect and analyze a large dataset of urban transportation publications. It provides a user-friendly interface for data visualization and in-depth bibliometric analysis, including citation analysis, author and journal impacts, and publication trends.

VOSviewer: This tool enabled the creation of network maps of co-occurring keywords, which allowed us to identify thematic clusters in urban transportation research. It also enabled us to track the evolution of these themes over time, highlighting the emerging focus on public health, walkability, and simulation-based approaches, such as SUMO and ns-3.

Keyword Analysis: Using VOSviewer, we conducted a keyword co-occurrence analysis to explore the relationships between terms. We categorized the keywords into two primary clusters: sustainable urban mobility (e.g., urban mobility, smart cities) and motor transportation efficiency (e.g., travel time, vehicles, and intelligent systems). This clustering approach revealed the dual focus of current urban transportation research: promoting sustainability and optimizing transportation efficiency.

By providing further details on these evaluation metrics and methodological approaches, we hope to strengthen the overall analysis in this study.

4. Results of Bibliometric Analysis of Publications on Urban Transportation

4.1. Performance Analysis

A comprehensive descriptive analysis of publications was conducted, as well as an assessment of the annual publication growth rate and an estimate of the average number of citations per publication. A graph, commonly referred to as a Sankey diagram, was used to compare three separate areas, such as determining which journal has the most publications and citations and which organization and document have the most publications and citations. In addition to data performance analysis, this study also identified and evaluated authors and countries with the highest output in terms of publications and citations.

4.1.1. General Overview of the Database

A descriptive analysis of studies in the field of urban transportation was conducted using Biblioshiny. Figure 1 shows the main information in the data file.

Between 2018 and 2023, 2000 articles on urban transportation from 617 sources (journals) in the Scopus database were evaluated. Publications increased by 31.32% annually; the average age of publications was 2.92 years, and each received an average of 12.19 citations.

Among the 5609 authors, 204 published single-author papers. The proportion of international co-authorships in the documents was 30.15%.

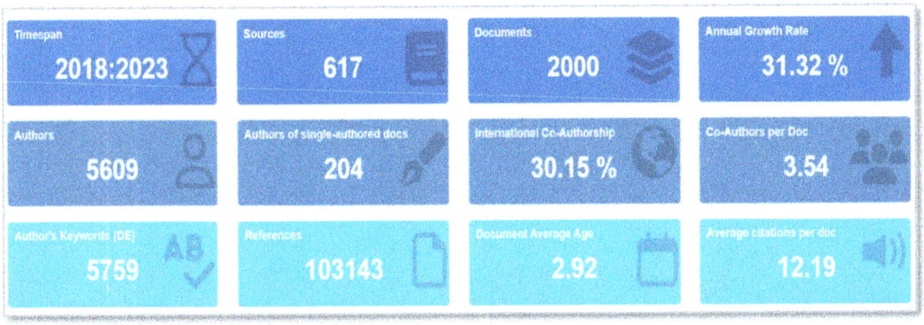

Figure 1. Main information.

4.1.2. Annual Increase in Publications

Figure 2 illustrates shows the evolution of documents over time in the bibliometric analysis of research in the field of urban transportation.

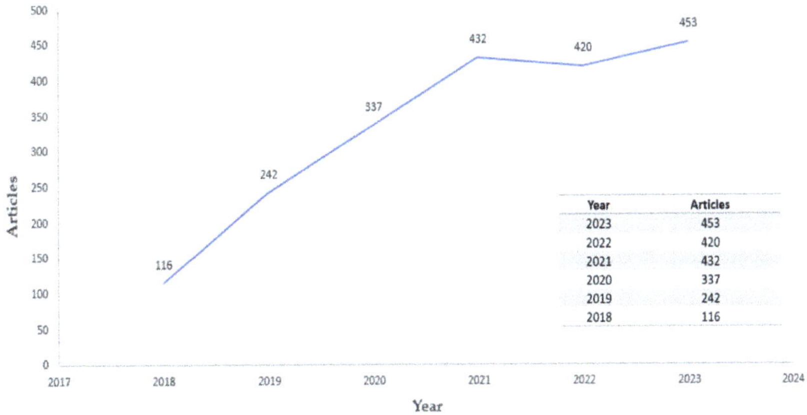

Figure 2. Distribution of urban transportation studies used in the research by year.

According to Figure 2, there has been growing interest in urban transportation analysis by scientists worldwide. Although the number of publications fluctuated over time, it showed an overall upward trend. There were 453 publications in 2023, 420 in 2022, 432 in 2021, 337 in 2020, 242 in 2019, and 116 in 2018. Notably, 2023 recorded the highest increase in research on urban transportation.

4.1.3. Average Annual Citations

Figure 3 shows the status of the annual citations in the bibliometric analysis, focusing on urban transportation research. In Figure 3, the column called "MeanTCperArt" refers to "Mean Total Citations per Article", which represents the average number of citations received by each article in the dataset over the entire study period. Essentially, this is the total number of citations divided by the total number of articles. Furthermore, the column called "MeanTCperYear" in Figure 3 refers to the "Mean Total Citations per Year". This metric calculates the average number of citations per article per year, normalized for the varying lengths of time articles have been available to accumulate citations, providing a more time-sensitive perspective on citation impact.

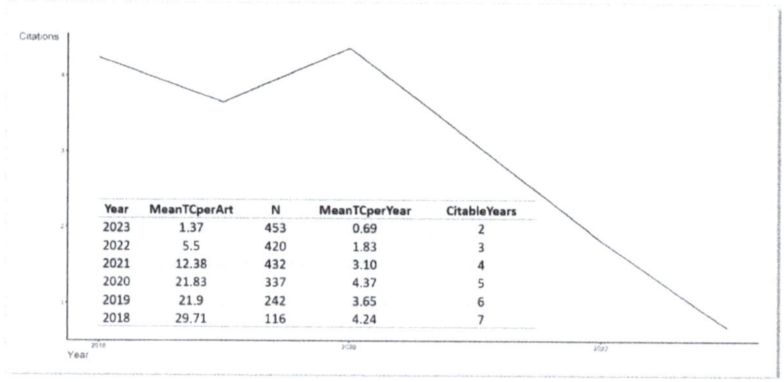

Year	MeanTCperArt	N	MeanTCperYear	CitableYears
2023	1.37	453	0.69	2
2022	5.5	420	1.83	3
2021	12.38	432	3.10	4
2020	21.83	337	4.37	5
2019	21.9	242	3.65	6
2018	29.71	116	4.24	7

Figure 3. Annual average citation trends of studies in the field of urban transportation.

4.1.4. Sankey Diagram

The three parameters to be associated (country, author, and keyword) are specified by the software Rstudio-Bibliometrix-Biblioshiny (version 4.1.3) in this "Three Field Plot" layout; Figure 4 highlights the most important ones for each parameter.

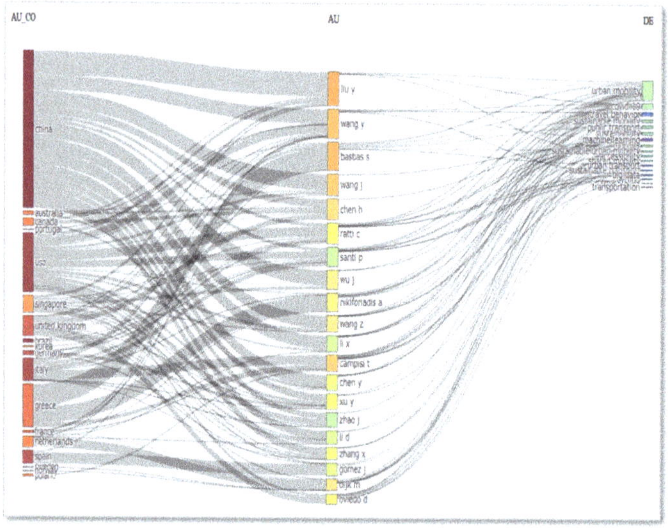

Figure 4. Three field drawings (country, author, and keywords).

The box sizes in Figure 4 show the strength of the relationship between the components. The size of the boxes in this graph corresponds to the influential aspects of the literature. The leading country is "China", the leading author is "Liu Y.", and the leading keyword is "urban mobility".

Specifically, the large box representing China indicates not only its dominant publication volume, but also its central role in international collaborations within urban transportation research. Similarly, the prominence of Liu Y. highlights his influential contributions, which have shaped many of the emerging trends in the field. The substantial size of the 'urban mobility' keyword box underscores a growing research focus on sustainable and integrated transportation systems, reflecting shifts in policy and technological innovation. These visual insights help to pinpoint key drivers of the literature and suggest potential directions for future research in urban mobility.

4.1.5. Most Cited and Most Published Journals

Table 1 presents the ranking of urban transportation publication sources based on the number of publications.

Table 1. Publication status of journals on urban transportation.

Sources	Articles
SUSTAINABILITY (SWITZERLAND)	200
IEEE TRANSACTIONS ON INTELLIGENT TRANSPORTATION SYSTEMS	48
JOURNAL OF TRANSPORT GEOGRAPHY	45
CASE STUDIES ON TRANSPORT POLICY	42
TRANSPORT POLICY	38
IEEE ACCESS	37
SUSTAINABLE CITIES AND SOCIETY	37
TRANSPORTATION RESEARCH PART D: TRANSPORT AND ENVIRONMENT	37
TRANSPORTATION RESEARCH PART A: POLICY AND PRACTICE	36
CITIES	35

According to Table 1, "Sustainability (Switzerland)" ranks first with 200 articles, followed by "IEEE Transactions on Intelligent Transportation Systems" with 48 articles and "Journal of Transport Geography" with 45 articles.

Table 2 highlights the ranking of urban transportation publication sources based on citations to publications.

Table 2. Citation status of journals on urban transportation.

Sources	Articles
SUSTAINABILITY (SWITZERLAND)	1347
JOURNAL OF TRANSPORT GEOGRAPHY	865
TRANSPORT POLICY	844
TRANSPORTATION	747
JOURNAL OF TRANSPORT GEOGRAPHY	721
TRANSPORT POLICY	666
TRANSPORTATION RESEARCH PART A: POLICY AND PRACTICE	638
IEEE TRANSACTIONS ON INTELLIGENT TRANSPORTATION SYSTEMS	606
TRANSP RES REC	602
CITIES	579

According to Table 2, "Sustainability (Switzerland)" ranks first with 1347 citations. "Journal of Transport Geography" follows with 865 citations, and "Transport Policy" ranks third with 844 citations.

4.1.6. Most Important Connections

Table 3 shows the studies published in the field of urban transportation by the institutions or affiliated organizations of the authors participating in the study.

Table 3. Relevant organizations contributing to research in the field of urban transport.

Affiliation	Articles
SOUTHEAST UNIVERSITY	85
ARISTOTLE UNIVERSITY OF THESSALONIKI	62
MASSACHUSETTS INSTITUTE OF TECHNOLOGY	60
UNIVERSITY OF SÃO PAULO	47
UNIVERSITY COLLEGE LONDON	46
UNIVERSIDAD POLITÉCNICA DE MADRID	45
TSINGHUA UNIVERSITY	39
DELFT UNIVERSITY OF TECHNOLOGY	35
UNIVERSITY OF LEEDS	35
TONGJI UNIVERSITY	33

"Southeast University" ranked first with 85 publications between 2018 and 2023, while "Aristotle University of Thessaloniki" ranked second with 62 publications and "Massachusetts Institute of Technology" ranked third with 60 publications.

4.1.7. Frequently Cited Authors by Number of Publications

A total of 2000 research papers on urban transportation were published by 5609 authors in different publications. Figure 5 highlights the most influential authors in terms of their publications worldwide.

Several authors have made significant contributions to urban mobility and transportation research. Yaqi Liu, associated with ORCID ID 0000-0002-2296-165X, has authored 16 articles, focusing on areas such as vehicle group dynamics in urban road networks. Sotirios Basbas, affiliated with the Aristotle University of Thessaloniki, Greece, has published 14 articles, with research interests including the effects of the COVID-19 pandemic on bike-sharing systems. Yongxin Liu, holding ORCID ID 0000-0003-4527-8623, serves as an Assistant Professor of Mathematics at Embry–Riddle Aeronautical University in Daytona Beach, Florida, and has contributed 14 articles related to urban mobility. Jian Wang, with 13 publications, has explored topics such as reverse reconstruction methods of vehicle group situations in urban road networks. Extensive research by these authors has significantly advanced our understanding of urban transportation systems. Therefore, Liu Y ranks top with 16 articles, Basbas S and Wang Y rank second with 14 articles, and Wang J ranks third with 13 articles. The prominence of authors like Liu Y, who leads with 16 publications, reflects the concentrated research efforts in regions or institutions with strong academic contributions. By highlighting these leading authors, bibliometric analysis emphasizes key figures that drive the conversation in sustainable urban mobility. This information is not only valuable for researchers seeking to identify potential collaborators, but also for those looking to stay at the forefront of emerging trends and groundbreaking studies in the field.

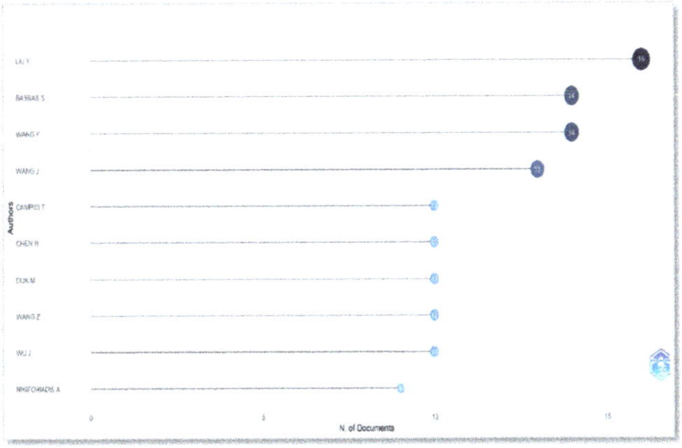

Figure 5. Publication numbers of the most prolific authors.

4.1.8. Most Productive Countries

The countries with the highest numbers of publications are given in Figure 6.

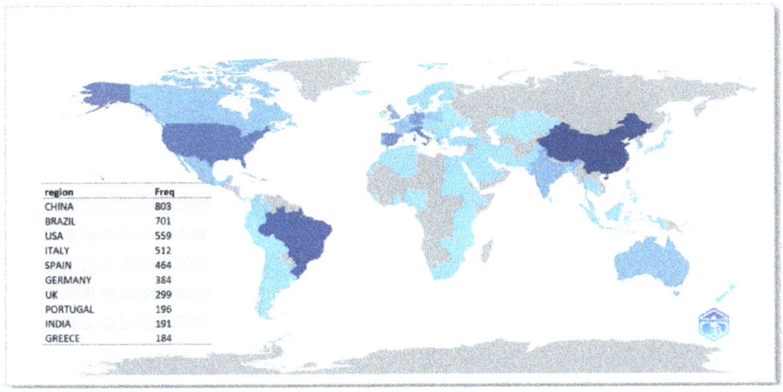

Figure 6. Countries by their number of publications.

On the map (Figure 6), dark blue represents China, blue represents countries like India, Serbia, Iran, Turkey, the USA, and the UK, and gray represents countries with no publications. As shown in Figure 6, China has the highest number of publications (803). China was followed by Brazil (701), the USA (559), and Italy (512).

The most cited countries are listed in Table 4.

As shown in Table 4, the USA has the highest number of citations (2697), followed by China (2285), Spain (1886), and Italy (1829).

Figure 7 shows the global cooperation map.

Countries with strong connections are more likely to collaborate. In the field of urban transportation, China cooperates with the USA, the United Kingdom, Hong Kong, Singapore, and Australia. The USA collaborates with the United Kingdom, Italy, and Singapore, while Brazil cooperates with the USA, and Germany collaborates with the Netherlands. It can be inferred that China is the country with the most international collaborations.

Table 4. Countries by number of citations.

Country	Total Citations	Average Article Citations
USA	2697	23.70
China	2285	12.40
Spain	1886	17.50
Italy	1829	16.80
United Kingdom	1094	15.00
Germany	1044	11.50
Netherlands	908	20.60
Brazil	892	7.80
Portugal	779	17.70
Greece	741	17.20

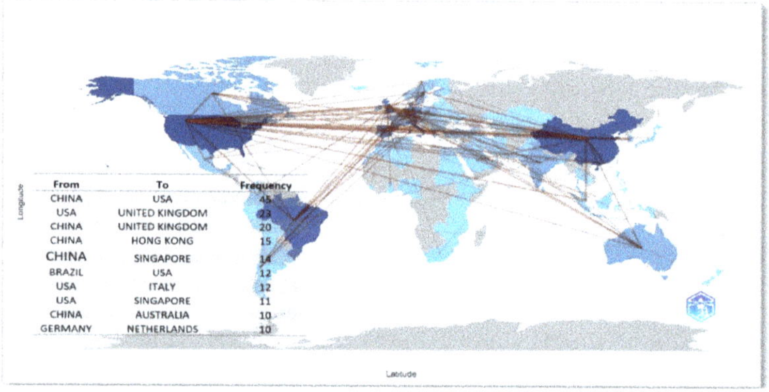

Figure 7. Country cooperation map.

4.1.9. Most Cited Document

The most productive paper is the most cited paper. Table 5 lists the 10 most productive articles.

Table 5. The most effective articles.

Paper	DOI	Total Citations (TC)	TC per Year	Normalized TC
Aloi, A., 2020, Sustainability	10.3390/su12093870	329	65.80	15.07
Liang, X., 2019, IEEE Trans Veh Technol	10.1109/TVT.2018.2890726	305	50.83	13.93
Jenelius, E., 2020, Transp Res Interdiscip Perspect	10.1016/j.trip.2020.100242	280	56.00	12.83
Teixeira, J.F., 2020, Transp Res Interdiscip Perspect	10.1016/j.trip.2020.100166	278	55.60	12.73
Dembski, F., 2020, Sustainability	10.3390/su12062307	201	40.20	9.21
Faisal, A., 2019, J Transp Land Use	10.5198/jtlu.2019.1405	190	31.67	8.68
Lyons, G., 2018, Transp Res Part A Policy Pract	10.1016/j.tra.2016.12.001	171	24.43	5.76
Campisi, T., 2020, Sustainability	10.3390/su12218829	168	33.60	7.70
Ma, Y., 2018, J Clean Prod	10.1016/j.jclepro.2018.06.213	162	23.14	5.45
Mouratidis, K., 2021, Sustainable Cities Soc	10.1016/j.scs.2021.103182	143	35.75	11.55

The most cited article is "Effects of the COVID-19 Lockdown on Urban Mobility: Empirical Evidence from the City of Santander (Spain)" by Aloi et al. (2020), published in Sustainability with 329 citations. This article is followed by "A Deep Reinforcement Learning Network for Traffic Light Cycle Control", published in IEEE Transactions on Vehicular Technology by Liang et al. (2019) with 305 citations. The third most cited article is "Process Safety and Environmental Impacts of COVID-19 on Public Transport Ridership in Sweden: Analysis of Ticket Validations, Sales, and Passenger Counts" by Jenelius and Cebecauer (2020), published in the journal Transportation Research Interdisciplinary Perspectives with 280 citations. In this way, Teixeira and Lopes (2020) ranks 4th with 278 citations, Dembski et al. (2020) 5th with 201 citations, Faisal et al. (2019) 6th with 190 citations, Lyons (2018) 7th with 171 citations, Campisi et al. (2020) 8th with 168 citations, Ma et al. (2018) 9th with 162 citations and Mouratidis and Papagiannakis (2021) 10th with 143 citations.

4.2. Analysis of Science Mapping

Scientific mapping employs computer tools to visualize, analyze, and simulate various scientific and technological processes.

4.2.1. Conceptual Structure Map

This section shows the conceptual structure map, the network map based on the author's keywords, and the theme map. The conceptual structure map is shown in Figure 8.

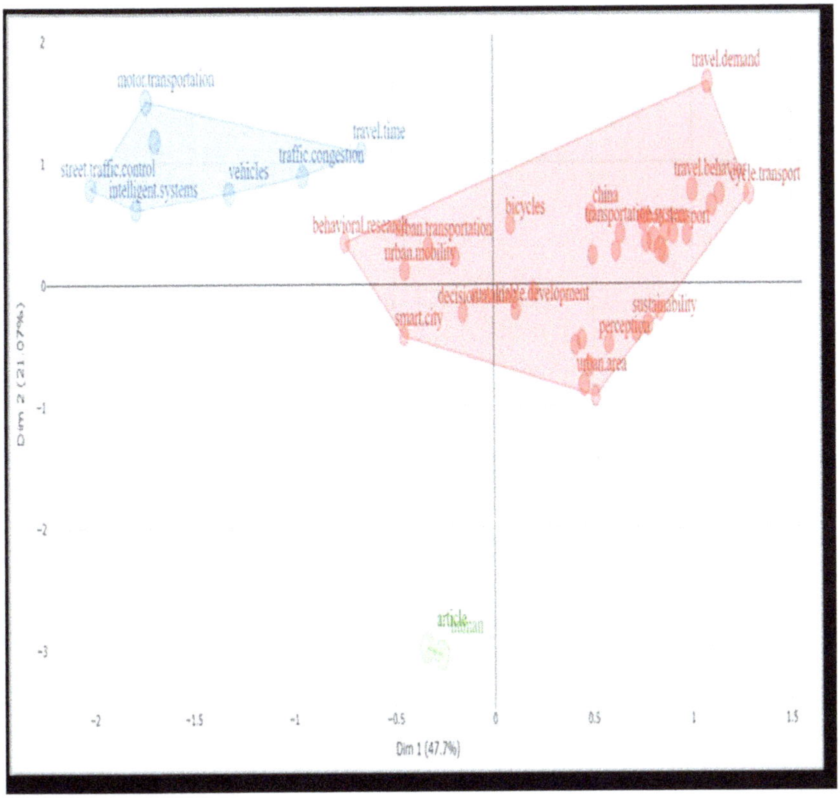

Figure 8. Conceptual structure map.

The red and blue clusters in Figure 8 illustrate the results of a conceptual structural analysis based on a factor analysis of keywords in urban transportation research. Each cluster corresponded to a distinct thematic grouping of related concepts and ideas, as identified by their significant high factor loads along two primary dimensions. The red cluster focuses on keywords associated with sustainability and innovative urban mobility solutions, such as "urban mobility", "smart city", and "bicycles", emphasizing themes like eco-friendly transportation, technological integration, and active travel modes. In contrast, the blue cluster group keywords focused on the efficiency and functionality of motorized transportation systems, including "travel time", "vehicles", and "intelligent systems". These clusters reflect the dual focus of urban transportation research on advancing sustainability (red cluster) and optimizing motorized transportation efficiency (blue cluster). This approach offers a holistic understanding of the field's conceptual structure. Lastly, the green cluster represents a weaker thematic grouping involving vague terms, such as "article" and "human".

When the keywords of publications in the field of urban transportation were analyzed using factor analysis, certain ideas emerged within the red cluster with high factor loads in the first dimension. Keywords such as "urban mobility", "smart city", "urban area", "bicycles", and "travel behavior" are collected here. Keywords such as "intuitive motor transportation", "travel time", "vehicles", and "intelligent systems" are gathered in the blue cluster.

In this sense, let us delve into the significance of each keyword, especially within the context of urban planning and development.

To begin with, in the context of the red cluster, it is essential to mention that "urban mobility" refers to the efficient and effective movement of people and goods within urban areas. It is crucial for the functioning of cities, affecting economic productivity, quality of life, and environmental sustainability. Efficient urban mobility systems reduce congestion, lower emissions, and improve accessibility to various urban services [27]. On the other hand, a "smart city" utilizes digital technologies and data to improve urban services and infrastructure. This includes various applications, from traffic management to energy distribution, aimed at enhancing the quality of life for residents while promoting sustainability [28]. Smart cities integrate various ICT (Information and Communication Technology) tools to optimize urban mobility, making transportation systems more efficient and responsive to real-time conditions. Additionally, an "urban area" is a region characterized by high population density and vast human-built features compared to surrounding areas. Understanding the dynamics of urban areas is essential for planning and implementing effective mobility solutions. Urban areas are the primary focus for developing smart city initiatives and enhancing urban mobility [29]. Furthermore, "bicycles" represent a sustainable and efficient mode of urban transport. Promoting bicycle use in urban areas can reduce traffic congestion, lower greenhouse gas emissions, and improve public health. Bicycle infrastructure is often a key component of smart city strategies to promote eco-friendly and active modes of transport [30]. In addition, "travel behavior" refers to the study of how individuals make decisions about their travel, including modes of transport, travel times, and routes. Understanding travel behavior is critical for designing effective urban mobility systems and policies. Analyzing travel behavior helps in forecasting demand, planning infrastructure, and implementing policies that encourage sustainable travel choices [17].

Secondly, in the context of the blue cluster, it is crucial to indicate that "motor transportation" encompasses the use of motor vehicles such as cars, trucks, buses, and motorcycles for the movement of people and goods. It is the backbone of modern urban mobility, essential for economic activities and daily commuting. However, it also contributes significantly to urban challenges like congestion, air pollution, and greenhouse gas emissions [31]. Moreover, "travel time" refers to the duration spent moving from one location to another. It is a critical factor in transportation planning and policymaking because it affects productivity, quality of life, and economic efficiency. Reducing travel time through efficient transportation systems can significantly enhance urban living and economic performance [32]. Also, "vehicles" are the primary means of motor transportation, playing a crucial role in urban mobility by facilitating the movement of people and goods. However, high reliance on personal vehicles can lead to traffic congestion, pollution, and increased demand for parking spaces [9]. Moreover, "intelligent systems" in transportation refer to the application of advanced technologies and data analytics to improve the efficiency, safety, and sustainability of transportation networks. This includes Intelligent Transportation Systems (ITSs), which utilize real-time data, automated control systems, and communication technologies to enhance traffic management, reduce congestion, and optimize travel times [33].

4.2.2. Thematic Map

The clusters in the motor theme section of the diagram represent the clusters belonging to the most basic fields of study of the keyword under study. The clusters in the core theme section of the diagram indicate clusters where the keyword of interest is fundamental and in a state of transformation, i.e., continuing to evolve. The emerging or declining theme section of the diagram illustrates newly emerging or declining keywords, while the niche theme represents keywords that are the focus of significant work but remain isolated from other studies. Therefore, to understand which scientific studies belong to which themes and which ones are at the forefront of the keyword under consideration, it is necessary to consider the basic theme and motor theme sections of the diagram.

A thematic review of urban transportation articles was conducted using bibliometrics and author's keywords to identify the main review topics of the field. Figure 9 shows that studies in the field of urban transportation are grouped under four themes, albeit with different intensities

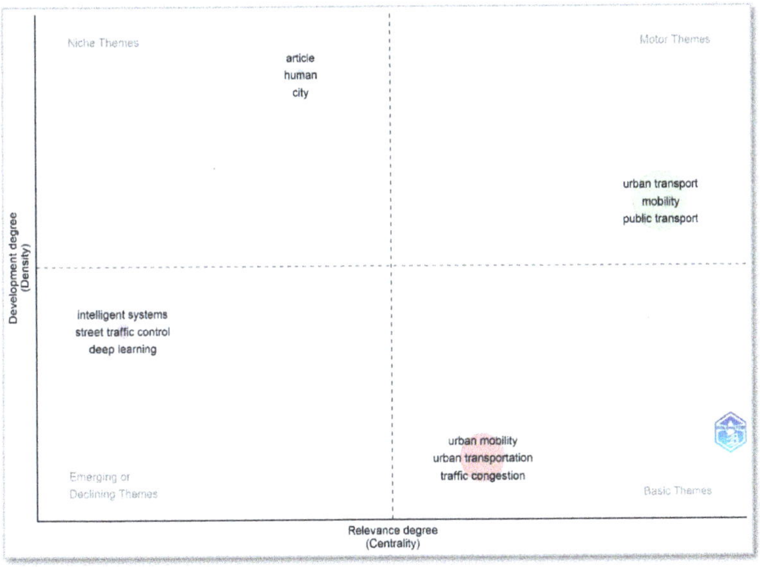

Figure 9. Thematic map based on keywords.

The four categories in Figure 9—"Niche Themes", "Motor Themes", "Emerging or Declining Themes", and "Basic Themes"—represent distinct thematic classifications of research topics in urban transportation based on their centrality and density. "Niche Themes" are highly specialized topics with limited connections to the broader field, often representing advanced focused research areas that contribute to specific aspects of urban transportation. "Motor Themes" are well-developed and influential topics that are both highly central and dense, forming the core of the field's research and driving significant scientific progress. "Emerging or Declining Themes" are topics with low centrality and density, indicating areas that may be in the early stages of development or experiencing a decline in scholarly interest. Lastly, "Basic Themes" are foundational topics with high centrality but lower density, serving as the bedrock of the field and encompassing broad general concepts that support various subfields. This categorization provides valuable insights into the current state and evolution of urban transportation research.

Considering the author's keywords, keywords such as "urban transport", "mobility, public transport", "urban mobility", and "traffic congestion" come to the fore in scientific studies. Let us explore the importance of each of these keywords within the context of urban transportation and mobility. From this point of view, "urban transport" encompasses all modes of transportation used within urban areas, including public transit, private vehicles, cycling, walking, and emerging modes like shared mobility services. It is essential for facilitating the movement of people and goods within cities, supporting economic activities, and ensuring access to services and opportunities. Effective urban transportation systems are critical for reducing traffic congestion, minimizing environmental impact, and enhancing the quality of urban life [31]. Additionally, "mobility" refers to the ability of individuals and goods to move seamlessly and efficiently from one place to another. In urban contexts, mobility is a critical factor influencing economic productivity, social inclusion, and overall quality of life. Enhancing mobility entails optimizing transport infrastructure, services, and policies to ensure people can access jobs, education, healthcare, and other essential services with ease and minimal delay [17]. Moreover, "public transport" includes systems such as buses, trams, trains, and subways that provide shared transportation services to the public. It is a crucial component of urban transportation systems, offering an efficient and sustainable alternative to private car use. Public transportation can help reduce traffic congestion, lower emissions, and provide affordable mobility options for all urban residents, including those who cannot afford or choose not to use private vehicles [10]. Moreover, "urban mobility" is the broader concept encompassing all forms of movement within urban areas, including public transport, private vehicles, cycling, and walking. It involves the interplay of various modes of transport and infrastructure to ensure efficient, sustainable, and inclusive movement of people and goods. Enhancing urban mobility requires integrated planning and policies that address the diverse needs of urban populations and promote sustainable transport options [34]. Furthermore, "traffic congestion" occurs when there is an excess demand for road space, leading to slower speeds, longer trip times, and increased vehicular queuing. It is a significant challenge in urban areas, contributing to air pollution, economic losses, and reduced quality of life. The effective management of traffic congestion involves strategies such as improving public transport, implementing traffic management systems, and encouraging alternative modes of transport like cycling and walking [35].

4.2.3. Trending Topics

The trending topics in the literature in the field of urban transportation from 2018 to 2023 are given in Table 6.

According to Table 6, "Urban transport" was the top trending topic in the first quarter of 2020, the second quarter of 2021, and the third quarter of 2022. "Mobility" was also a hot topic in the same periods. Currently, "COVID-19, vehicles, autonomous vehicles, and reinforcement learning" are the top trending topics in this area.

Table 6. Trending topics in the literature of urban transportation by year (2018–2023).

Item	Freq	Year_q1	Year_Med	Year_q3
urban transport	401	2020	2021	2022
mobility	306	2020	2021	2022
urban mobility	295	2020	2021	2022
COVID-19	106	2021	2022	2023
vehicles	102	2020	2022	2023
autonomous vehicles	54	2021	2022	2023
surveys	38	2020	2020	2022
urban policy	37	2019	2020	2022
taxicabs	33	2019	2020	2022
reinforcement learning	33	2022	2023	2023

4.2.4. Keyword Co-Occurrence Analysis

Keywords, serving as descriptive terms or phrases for an article, significantly influence the prominence of its statements. To visualize the interconnectedness of research findings, we employed VOSviewer to analyze the co-occurrence of keywords within the study and its broader knowledge base. This analysis yielded a comprehensive list of 5796 keywords extracted from the indexed terms associated with the article. As shown in Figure 10, 235 keywords were considered for review after increasing the threshold setting in the VOSviewer application to three.

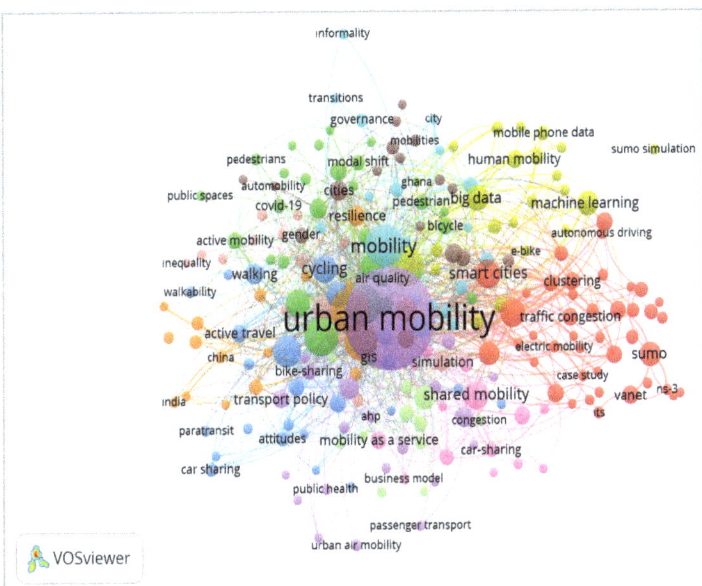

Figure 10. Keywords for co-occurrence.

A different set of keywords is created with each color. In Figure 10, each circle represents the presence of a specific term and sub-domain of urban transportation in the transportation domain. A circle with a similar color represents the distribution in a comparison region. The largest cluster, by the number of elements, is called "urban mobility". This cluster contains keywords such as the AHP, simulation, and a GIS. The second-largest cluster is named "smart cities". It contains keywords such as traffic congestion, VANET, and SUMO. The third-largest cluster is called "mobility" and includes keywords such as informality, transitions, governance, and city.

Leveraging AHP, Simulation, and GIS for Enhanced Urban Mobility Planning

Here is an explanation of the importance of the AHP, simulation, and a GIS in the context of urban mobility according to the largest cluster. Within this scope, firstly, the Analytic Hierarchy Process (AHP) is a decision-making framework that helps prioritize and evaluate various aspects of urban mobility. It involves breaking down complex decisions into a hierarchy of simpler problems and comparing them pairwise to derive priority scales. The AHP helps in making informed decisions about urban mobility infrastructure by weighing factors such as cost, environmental impact, and social benefits [36]. For example, it can prioritize where to invest in new public transit routes versus road expansions. Policymakers use the AHP to evaluate and compare different transportation policies' potential impacts, helping ensure that the chosen policies align with broader urban planning goals [37]. The process involves various stakeholders, ensuring that multiple perspectives are considered, leading to more balanced and accepted decisions [38].

Secondly, simulation techniques enable the creation of virtual urban mobility systems, providing a platform to assess and forecast the consequences of various scenarios. By employing simulation models, researchers can gain valuable insights into traffic flow dynamics, identify potential congestion hotspots, and evaluate the impact of proposed interventions, such as adjustments to traffic signal timings or modifications to road networks [39]. For example, cities can simulate the impact of a new traffic signal system before implementation. They are critical for planning emergency evacuations by modeling various disaster scenarios and optimizing evacuation routes [40]. Simulations can optimize public transportation schedules, routes, and capacities, ensuring efficient service and reducing wait times [41].

Thirdly, Geographic Information Systems (GISs) play a pivotal role in collecting, managing, and analyzing spatial data pertinent to urban mobility. A GIS is essential for optimizing route planning across different transportation modes by incorporating real-time traffic data and geographic insights. [42]. For instance, it helps in designing bus routes that minimize travel time and maximize coverage. A GIS aids in the maintenance and management of urban infrastructure by providing precise geographical data on roads, bridges, and public transit networks [43]. A GIS enables the evaluation of environmental impacts by analyzing spatial data on air quality, noise pollution, and green spaces. This capability is vital for promoting sustainable urban mobility planning [44].

Temporal Dynamics and Emerging Trends in Urban Mobility Keywords

In addition to these studies, the keywords "overlay visualization" were colored differently based on the year of publication using the VOSviewer program, which also allowed for the determination of the time intervals in which they appeared in the literature [45]. In our illustration, 2020 is the typical (yellow) year when newly added terminology was published. The amount of time that had passed since the items' publication determined their colors. Figure 11 displays the blue–green–yellow color range, covering the period from May 2020 to early 2022.

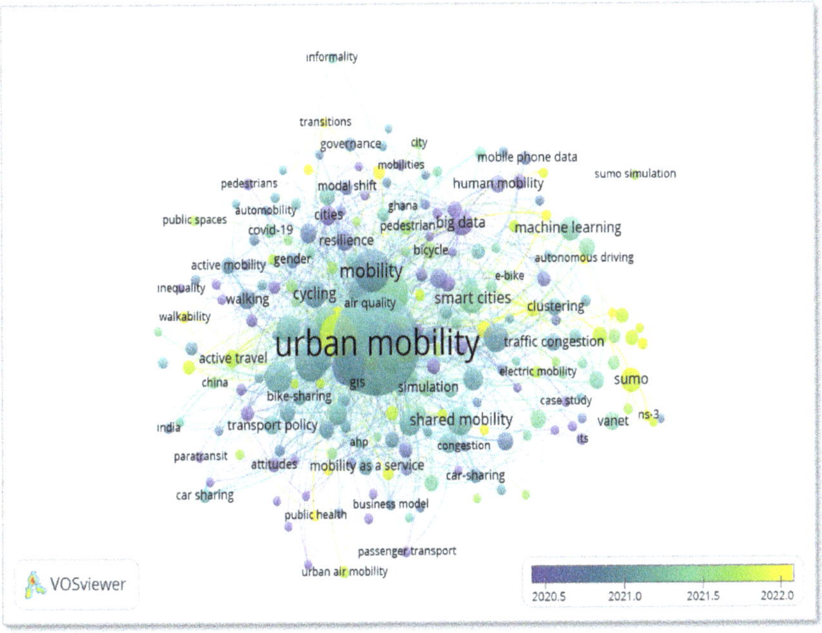

Figure 11. Overlay network of keywords.

While previously used keywords such as "passenger transport", "walking", and "big data" were the keywords intensively studied in the period from May 2020 to early 2021, it can be said that keywords such as "public health", "walkability", "SUMO", and "ns-3" have just started to be used in the literature.

In this context, it is essential to briefly explain the popularity of the keywords "passenger transport", "walking", and "big data" in urban mobility. To begin with, passenger transportation systems—including public transit, taxis, and ride-sharing services—are essential for moving large numbers of people efficiently within urban areas. With growing urban populations, the need for effective passenger transport solutions has increased significantly [41]. Public transportation reduces traffic congestion and lowers carbon emissions, contributing to environmental sustainability [9]. Secondly, walking is a fundamental mode of transportation that offers significant health and environmental benefits. Walking improves physical health and reduces the risk of various diseases [46]. It is also the most eco-friendly mode of transport, helping reduce reliance on motor vehicles and lowering pollution levels [47]. Thirdly, big data involves the analysis of vast datasets to improve urban mobility systems. Big data analytics provides insights into traffic patterns and commuter behavior, aiding in better decision making [48]. It allows for the real-time monitoring and management of transportation networks, enhancing operational efficiency [49]. In summary, "passenger transport", "walking", and "big data" are popular in urban mobility due to their contributions to efficiency, sustainability, health, and informed decision making, which are crucial for developing modern and resilient urban transportation systems.

New Dimensions in Urban Mobility: Integrating Health, Walkability, and Advanced Simulations

In addition to the definition of the important terms mentioned earlier, here is a short explanation of the newly emerging concepts in the literature known as "public health", "walkability", "SUMO", and "ns-3", particularly within the context of urban mobility. To begin with, public health refers to the science and practice of protecting and improving the health of communities through education, policymaking, and research for disease and injury prevention. Effective urban mobility strategies can significantly enhance public health by reducing pollution, promoting physical activity, and decreasing traffic-related injuries [50]. Secondly, walkability refers to how friendly an area is to walking, which depends on factors like the presence of sidewalks, safety, and the proximity of amenities. High walkability is associated with better health outcomes, reduced traffic congestion, and increased social interaction [51]. Thirdly, SUMO (Simulation of Urban Mobility) is an open-source, highly portable, microscopic, and continuous road traffic simulation package designed to handle large road networks. It is used to simulate and analyze traffic patterns, optimize traffic light settings, and test new mobility concepts [52]. SUMO is widely recognized in urban transportation research for enabling the evaluation of transportation system performance under varying conditions, such as changes in infrastructure, traffic demand, or policy interventions. In the context of urban transportation research, SUMO is used in the following ways.

Optimize traffic flow: By simulating different traffic scenarios and optimizing signal timings, urban planners can reduce congestion and improve the efficiency of transportation networks.

Assess sustainability: SUMO allows for the evaluation of environmental impacts, such as vehicle emissions, which is essential for studying sustainable urban mobility.

Test policy measures: SUMO provides a platform to model the effects of different policies, such as congestion pricing, on travel behavior and overall system efficiency.

The adoption of SUMO in academic and professional studies highlights the increasing reliance on simulation-driven methods in urban transportation planning. These approaches offer a versatile and economical means to evaluate intricate urban settings before applying practical solutions. Consequently, SUMO is integral to facilitating informed decisions aimed at achieving intelligent and sustainable urban mobility systems.

New Horizons in Urban Mobility: Enhancing Community Health and Digital Innovation

Lastly, ns-3 (Network Simulator 3) is a discrete-event network simulator for Internet systems, targeted primarily for research and educational use. It is widely used to simulate and study the performance of network protocols and Internet systems, including those related to vehicular communications and Intelligent Transportation Systems (ITSs) [53]. However, its application has expanded into urban transportation research, particularly in studies that explore the interaction between transportation systems and communication networks, such as ITSs. The significance of ns-3 in urban mobility research lies in its ability to achieve the following.

Simulate communication networks in ITSs: ns-3 can model the exchange of data between vehicles (vehicle-to-vehicle, V2V) and between vehicles and infrastructure (vehicle-to-infrastructure, V2I). This capability is essential for the development of connected vehicle technologies and autonomous driving systems, which are central to the future of urban mobility.

Evaluate traffic management systems: ns-3 is a valuable tool for simulating and optimizing traffic management protocols that depend on real-time communication to enhance safety and efficiency. For instance, it can model dynamic traffic signal controls or real-time traffic monitoring systems to test and refine their performance under various scenarios.

Assess the impact of communication delays: With communication networks being integral to modern transportation systems, ns-3 allows researchers to investigate how network delays or failures affect urban mobility solutions. This is especially critical for evaluating system performance during congestion or emergencies.

By emphasizing the increasing synergy between transportation systems and communication technologies, ns-3 empowers researchers to tackle the complexities of incorporating digital innovations into transportation infrastructure. Its application becomes ever more pertinent as cities aim to develop smarter, more interconnected, and more efficient urban mobility systems.

In summary, public health, walkability, SUMO, and ns-3 are important terms in urban mobility due to their focus on improving community health, enhancing pedestrian-friendly environments, simulating traffic patterns, and analyzing network performance, respectively.

4.3. From Descriptive Statistics to Meaningful Insights

The descriptive statistics provided in this section highlight significant trends and emerging themes in urban transportation research. The observed increase in keywords such as "autonomous vehicles", "SUMO", and "reinforcement learning" aligns with the technological evolution of the transportation sector, emphasizing the growing importance of simulation and machine learning tools. These advancements enable researchers and practitioners to model complex systems, optimize traffic flow, and evaluate sustainable transportation strategies with greater precision.

Furthermore, the prominence of keywords like "public health" and "walkability" indicates a paradigm shift in urban planning toward more people-centric and health-conscious design approaches. This trend reflects the increasing recognition of mobility as a determinant of public health, urging policymakers to prioritize pedestrian infrastructure, reduce vehicle dependency, and promote active travel modes.

These findings offer practical insights for urban planners and policymakers. For instance, the prominence of the keyword cluster "urban mobility", encompassing terms such as "AHP", "GIS", and "simulation", highlights the critical role of multi-criteria decision-making tools and spatial data analysis in designing efficient, sustainable, and inclusive transportation systems. Policymakers are urged to integrate these tools into their evaluation of various transport policies and investment decisions.

Additionally, the analysis underscores the continued influence of topics like "COVID-19" on urban mobility research, demonstrating how global disruptions can reshape transportation dynamics. These insights lay the groundwork for further investigation into the interplay between technology, public health, and sustainability in defining the future of urban mobility.

5. Discussion

5.1. Bibliometric Analysis of Urban Transportation Publications

This study utilized Biblioshiny, a web-based platform for bibliometric analysis, to gather and analyze a large volume of academic publications in the urban transportation field. Biblioshiny enables data collection, assessment, and visualization, offering insights into the chronological progression of urban transportation research, prominent authors, journals, and keywords. Additionally, VOSviewer, a sophisticated network visualization tool, was employed to construct network maps illustrating keyword trends and their evolution over time, unveiling emerging research themes and trends in urban transport.

The comprehensive bibliometric analysis aimed to offer readers an extensive overview of the academic advancement of urban transportation studies. In this context, between 2018 and 2023, this study evaluated 2000 articles on urban transportation from 617 journals in the Scopus database. The number of publications increased by an average of 31.32% annually. The average age of the publications was 2.92 years, and each article received an average of 12.19 citations. Out of 5609 authors, 204 published single-authored papers. Additionally, 30.15% of the total documents had international co-authorship.

5.2. *Publication Trends and Leading Contributors in Urban Transportation Research*

In 2023, urban transportation research reached its peak with 453 publications, surpassing the output of previous years, including 2018, 2019, 2020, 2021, and 2022. However, 2018 stands out for achieving the highest citation rate, with an average of 29.71 citations per study in this domain.

Furthermore, the primary country is "China", the main author is "Liu Y.", and the primary keyword is "urban mobility" concerning the Sankey diagram analyzing the significant features of the literature on urban transportation.

In addition, the analysis reveals that Sustainability (Switzerland) leads in the number of publications on urban transportation, contributing 200 articles, followed by IEEE Transactions on Intelligent Transportation Systems with 48 articles and Journal of Transport Geography with 45 articles. In terms of citations, Sustainability (Switzerland) also ranks first with 1347 citations, while the Journal of Transport Geography and Transport Policy follow closely with 865 and 844 citations, respectively.

Moreover, in the area of urban transportation, "Southeast University" emerged at the top with 85 publications from 2018 to 2023, followed by "Aristotle University of Thessaloniki" with 62 publications and "Massachusetts Institute of Technology" with 60 publications, according to studies from the institutions or affiliated organizations of the authors.

5.3. *Key Contributors and Global Collaboration in Urban Transportation Research*

A total of 2000 research papers on urban transportation were authored by 5609 individuals across diverse publications. Among the most influential contributors globally, Liu Y ranks first with 16 articles, followed by Basbas S and Wang Y, each with 14 articles, and Wang J in third place with 13 articles.

Concerning the discussions about urban transportation, China has the highest number of broadcasts (803), followed by Brazil (701), the USA (559), and Italy (512). The most frequently mentioned country is the USA (2697), followed by China (2285), Spain (1886), and Italy (1829). In the field of urban transportation, China collaborates with the USA, the United Kingdom, Hong Kong, Singapore, and Australia. The USA collaborates with the United Kingdom, Italy, and Singapore. Brazil collaborates with the USA, and Germany with the Netherlands. It can be concluded that China is the most active in international cooperation in this area.

The most cited papers tend to reflect the productivity of their authors. For example, the article titled "Effects of the COVID-19 Lockdown on Urban Mobility: Empirical Evidence from the City of Santander (Spain)" by Aloi et al. (2020) leads with 329 citations. In second place is "A Deep Reinforcement Learning Network for Traffic Light Cycle Control" by Liang et al. (2019), published in IEEE Transactions on Vehicular Technology, has 305 citations. The third-most-cited paper, "Process Safety and Environmental Impacts of COVID-19 on Public Transport Ridership in Sweden: Analysis of Ticket Validations, Sales, and Passenger Counts" by Jenelius and Cebecauer (2020), published in Transportation Research Interdisciplinary Perspectives, has been cited 280 times.

5.4. Keyword Factor Analysis: Major Themes in Urban Transportation

Keyword factor analysis in the urban transportation literature reveals two major clusters. The first cluster includes terms such as urban mobility, smart city, urban area, bicycles, and travel behavior, reflecting a focus on sustainable and behavioral aspects of urban mobility. The second cluster comprises keywords like intuitive motor transportation, travel time, vehicles, and intelligent systems, highlighting technological advancements and efficiency in transportation systems.

The relationship between these keywords collected in the first cluster is intrinsic to developing sustainable, efficient, and livable urban environments. In this sense, smart city technologies enhance urban mobility by providing real-time data and analytics. For example, smart traffic management systems can optimize traffic flow and reduce congestion, directly improving urban mobility [54]. Also, effective urban mobility solutions are tailored to the specific needs of urban areas, which vary widely in terms of population density, infrastructure, and economic activities. Urban mobility strategies must consider the unique characteristics of each urban area to be effective [55]. Bicycles are a cornerstone of sustainable urban mobility strategies. Incorporating cycling infrastructure into urban planning allows cities to reduce dependence on motor vehicles, resulting in decreased emissions and better public health outcomes [30]. Moreover, smart city initiatives collect and analyze data on travel behavior to inform transportation planning and policy. Understanding how people move through the city allows for the development of more efficient and user-friendly transport systems [27]. Furthermore, travel behavior in urban areas is influenced by factors such as the availability of public transport, walkability, and cycling infrastructure. Policies aimed at modifying travel behavior, such as congestion pricing or incentives for public transportation use, can lead to more sustainable urban mobility patterns [17].

Recognizing the connections among these keywords enables urban planners, policymakers, and researchers to craft holistic strategies. These strategies aim to improve urban living by advancing mobility, fostering sustainability, and leveraging technological innovations.

On the other hand, the relationship between the keywords collected in the second cluster is integral to developing efficient and sustainable urban transportation systems. In this context, the efficiency of motor transportation directly impacts travel time. Factors such as traffic congestion, road infrastructure, and traffic management practices significantly influence how long it takes to travel from one point to another. Reducing travel time can improve economic productivity and quality of life [56]. Also, vehicles are the core components of motor transportation. The type, number, and efficiency of vehicles in use affect traffic flow, road safety, and environmental impact. Policies promoting fuel-efficient and electric vehicles can mitigate negative environmental impacts [31]. Moreover, intelligent systems are designed to optimize travel by reducing delays, improving traffic flow, and efficiently managing traffic incidents through real-time data utilization. For instance, adaptive traffic signal control systems dynamically adjust signal timings based on live traffic conditions, significantly minimizing delays and enhancing travel times [33]. Moreover, intelligent systems can enhance the efficiency and safety of vehicle operations. Features such as automated driving, vehicle-to-vehicle communication, and real-time navigation contribute to reducing accidents, improving fuel efficiency, and easing traffic congestion. The integration of ITSs with vehicles is a crucial aspect of developing smart and sustainable urban transportation systems [57].

5.5. Enhancing Urban Mobility Through Intelligent Systems and Integrated Planning

In addition, the application of intelligent systems in motor transportation can transform urban mobility by reducing congestion, improving safety, and lowering emissions. Smart traffic management, dynamic toll pricing, and real-time public transit information are examples of how intelligent systems optimize motor transportation [58]. Understanding these interconnections empowers urban planners, policymakers, and engineers to design transportation systems that are not only efficient and sustainable, but also tailored to the demands of modern urban populations.

Additionally, a comprehensive analysis of scholarly articles on urban transportation was carried out using bibliometrics and author-provided keywords to determine the primary focus areas in the field. The findings revealed that terms like urban transport, mobility, public transport, urban mobility, and traffic congestion are prominently featured in academic research.

The connection among these keywords is crucial for creating effective and long-lasting urban transportation systems. In this context, firstly, urban transportation systems are designed to enhance mobility within cities. Efficient urban transportation options, such as well-connected public transportation networks and safe cycling infrastructure, improve overall mobility by providing reliable and diverse options for movement [31]. Secondly, public transportation is a cornerstone of urban transportation systems. Expanding and improving public transportation services can significantly enhance urban mobility, reduce reliance on private cars, and alleviate traffic congestion [10]. Thirdly, high-quality public transportation services improve mobility by offering affordable and efficient travel options. Public transportation can connect various parts of a city, ensuring that all residents have access to essential services and opportunities, thus promoting social inclusion and reducing inequality [17]. Fourthly, effective urban mobility strategies are essential for managing traffic congestion. By promoting public transport, cycling, and walking, cities can reduce the number of private vehicles on the road, thus easing congestion and improving travel times for all modes of transport [35]. Fifthly, improving public transportation remains a vital strategy for mitigating traffic congestion. Efficient, reliable, and convenient public transportation options provide a compelling alternative to private car usage, thereby reducing the volume of vehicles on the road and alleviating congestion [10]. Lastly, comprehensive urban transportation planning must address congestion through a multifaceted approach, combining infrastructure enhancements, demand management strategies, and policies that promote sustainable transport modes. This integrated approach ensures that urban transportation systems can effectively support expanding populations and economic activities without succumbing to overwhelming congestion [35]. By grasping these interrelationships, urban planners, policymakers, and researchers can develop holistic strategies that improve mobility, advance sustainability, and integrate technological innovations to enhance urban living.

5.6. Integrative Frameworks and Emerging Trends in Urban Mobility

In addition, based on the keyword co-occurrence analysis carried out using VOSviewer to emphasize the common findings of the study with keywords and the study's knowledge base, the most extensive cluster, known as "urban mobility", includes keywords like AHP, simulation, and GIS. The integration of the AHP, simulation, and a GIS offers a holistic framework for tackling urban mobility challenges. Specifically, the AHP helps prioritize projects, simulation tests these priorities under various scenarios, and a GIS maps out and visualizes impacts and feasibility. This integration ensures that decisions are underpinned by rigorous data analysis, predictive modeling, and comprehensive stakeholder input, fostering the development of effective and sustainable mobility solutions. It allows cities to create adaptive systems that can respond to changing conditions and emerging challenges in urban mobility. By combining these tools, urban planners and policymakers can develop smarter, more efficient, and more sustainable urban transportation networks. In summary, the AHP, simulation, and a GIS each serve essential functions in urban mobility planning by providing the analytical, predictive, and spatial capabilities required for sound decision making and efficient management. Their combined use results in more informed, streamlined, and sustainable urban mobility strategies.

In addition to these analyses, the VOSviewer software version 1.6.20 was used to analyze the keywords "overlay visualization" and color them based on the year of publication. This helped identify the time intervals in which these keywords appeared in the literature. The analysis revealed that, from May 2020 to early 2021, keywords like passenger transport, walking, and big data were heavily studied, while newer keywords such as public health, walkability, SUMO, and NS-3 have recently started to emerge in the literature. It seems that emerging terms such as public health, walkability, SUMO, and NS-3 are gaining traction in the literature, signaling a shift toward new research frontiers that will shape the future of urban mobility planning.

6. Conclusions

This bibliometric exploration into urban transportation research underscores the dynamic and rapidly evolving nature of this field from 2018 to 2023. Analyzing 2000 articles from 617 journals within the Scopus database reveals critical trends and emerging themes shaping the discourse on urban transportation.

Key findings include a substantial Annual Growth Rate of 31.32% in publications, with 2023 marking the peak year for research output. China has emerged as a leading contributor to urban transportation research, both in terms of publication volume and international collaborations. Prominent authors such as Liu Y. and influential journals like "Sustainability (Switzerland)" and "IEEE Transactions on Intelligent Transportation Systems" underscore the global and interdisciplinary efforts in advancing this field.

The analysis identified two primary keyword clusters. The first emphasizes sustainable urban mobility, incorporating terms like urban mobility, smart cities, and travel behavior. The second focuses on motor transportation efficiency, with keywords such as travel time, vehicles, and intelligent systems. These clusters reflect the dual priorities of advancing sustainability and improving transportation efficiency through technological innovation.

Emerging keywords such as public health, walkability, SUMO (Simulation of Urban Mobility), and ns-3 (a discrete-event network simulator) indicate new directions in urban transportation research. These terms suggest a growing interest in the intersection of transportation with broader societal impacts and technological advancements. The inclusion of public health emphasizes the increasing recognition of the health implications of urban mobility, while walkability highlights a shift towards more pedestrian-friendly urban environments. Advanced simulation tools like SUMO and ns-3 demonstrate the increasing reliance on sophisticated modeling and analysis techniques to tackle complex urban transportation challenges, reflecting the field's commitment to integrating societal impacts with technological advancements.

The integration of tools such as the AHP (Analytic Hierarchy Process), simulation, and GISs (Geographic Information Systems) demonstrates a comprehensive approach to urban mobility planning. These methodologies enable urban planners to prioritize projects, test various scenarios, and visualize their impacts, leading to more informed and sustainable decisions.

In conclusion, this bibliometric analysis provides valuable insights into the current state and future directions of urban transportation research. The findings underscore the importance of sustainability, technological innovation, and interdisciplinary collaboration in addressing the challenges of urban mobility. As new themes like public health and walkability gain prominence, they will likely shape the development of more holistic and inclusive urban transportation systems. These insights provide valuable guidance for researchers, policymakers, and urban planners in crafting strategies that elevate urban living by advancing mobility, promoting sustainability, and leveraging technological innovations.

6.1. Novel Insights and Contributions

It is believed that the insights of this paper offer valuable contributions to the field by emphasizing the evolving trends in urban mobility and the novel applications of simulation tools and interdisciplinary collaborations. In this context, three main subheadings are identified.

Emerging Themes and Keywords: The analysis of this study identifies emerging keywords such as public health, walkability, SUMO, and NS-3, which reflect growing interdisciplinary concerns in urban transportation. These themes are gaining prominence in recent years and signal a shift toward more holistic and integrated approaches to urban mobility, incorporating public health and advanced simulation tools, areas not extensively discussed in the previous literature.

Growth of International Collaborations: The bibliometric analysis reveals China's dominant role in urban transportation research, especially in international collaborations, suggesting a shift in global research leadership and influence. This insight has implications for future research priorities and collaborations across regions.

Technological Integration: This study's focus on the integration of the AHP, simulation, and a GIS provides a novel framework for decision making in urban mobility. The integration of these tools provides a holistic framework for assessing and planning sustainable urban transportation systems, addressing gaps that have been underexplored in previous research.

6.2. Future Directions and Practical Implications

Building upon the key findings of this study, several future directions and practical implications emerge that could guide urban transportation research and practice.

Advancing Public Health and Walkability Research

The growing emphasis on themes like public health and walkability underscores the importance of exploring the connections between urban mobility and health outcomes, paving the way for interdisciplinary research in this area. Future studies could explore how urban transportation systems can actively promote healthier lifestyles through policies that prioritize pedestrian-friendly infrastructure, reduced vehicular emissions, and improved access to green spaces.

Leveraging Simulation Tools for Policy Development

The adoption of advanced simulation tools such as SUMO and NS-3 offers immense potential for urban planners to test and refine transportation policies in virtual environments. Future research could prioritize the practical implementation of these tools to assess the effects of mobility solutions, such as congestion pricing, shared mobility services, and multimodal integration strategies.

Enhancing Sustainability through Technological Innovations

Intelligent Transportation Systems (ITSs), electric and driverless cars, and smart city platforms are examples of emerging technologies that present chances to improve transportation effectiveness and lessen environmental impact. Taking into account the social and environmental ramifications of these technologies, researchers and policymakers should give top priority to incorporating them into long-term urban development plans.

Strengthening International Collaboration

The dominant role of China in urban transportation research underscores the importance of fostering international partnerships. Future efforts should aim to facilitate global knowledge exchange, particularly between countries with advanced technological capabilities and those facing unique transportation challenges. This collaboration has the potential to accelerate the adoption of innovative and context-specific solutions.

Policy Integration and Cross-Sectoral Approaches

Urban mobility is closely linked to other domains, such as housing, land use, and environmental planning. In order to ensure that infrastructure expenditures result in both mobility and environmental benefits, policymakers should take into account integrated methods that address transportation concerns alongside more general urban development issues.

By addressing these future directions, researchers and practitioners can build on the insights from this study to develop urban transportation systems that are not only sustainable, but also equitable and adaptive to emerging challenges. These practical implications aim to bridge the gap between academic research and real-world applications, ensuring that urban mobility solutions effectively meet the needs of diverse stakeholders.

This research serves as a foundation for policymakers, urban planners, transportation authorities, and researchers aiming to improve sustainable mobility. By highlighting global trends, key themes, and innovations in urban transportation, this study offers valuable insights for developing strategies that enhance urban mobility, promote sustainability, and address challenges such as congestion, pollution, and social equity.

6.3. Limitations of This Study

While this bibliometric analysis provides valuable insights into recent trends and innovations in urban transportation research, several limitations should be noted. Firstly, this study relies solely on data from the Scopus database, which, while comprehensive, may exclude relevant studies indexed in other databases such as Web of Science or IEEE Xplore. Secondly, the analysis focuses on a relatively short timeframe (2018–2023), which, while capturing emerging trends, might overlook longer-term developments in the field. Thirdly, this study's reliance on keyword- and citation-based metrics may not fully capture the qualitative aspects of research impact or the nuances of interdisciplinary collaboration. Addressing these limitations in future studies could provide us with a more comprehensive understanding of the field.

Funding: This research received no external funding.

Institutional Review Board Statement: Not applicable.

Informed Consent Statement: Not applicable.

Data Availability Statement: The data gathered during the survey (2000 articles from 617 journals in the Scopus database) are available from the corresponding author upon request.

Acknowledgments: The author sincerely appreciate the technical support of Gülay DEMİR as an advisor on this article, who read the whole article and gave some essential feedback.

Conflicts of Interest: The author declares no conflicts of interest.

Abbreviations

The following abbreviations are used in this manuscript:

AGR	Annual Growth Rate
AHP	Analytic Hierarchy Process
AI	Artificial Intelligence
AVs	Autonomous Vehicles
DOI	Digital Object Identifier
EVs	Electric vehicles
GIS	Geographic Information System
ICT	Information and Communication Technology
IEEE	Institute of Electrical and Electronics Engineers
IoT	Internet of Things
ITS	Intelligent Transportation System
MCDM	Multi-Criteria Decision Making
MeanTCperArt	Mean Total Citations per Article
MeanTCperYear	Mean Total Citations per Year
ns-3	Network Simulator 3
SDGs	Sustainable Development Goals
SOTA	State of the Art
SUMO	Simulation of Urban Mobility
TC	Total Citations
TOD	Transit-Oriented Development
USA	United States of America
VANET	Vehicular ad hoc Network
VMT	Vehicle Miles Traveled
VOS	Visualization of Similarities

References

1. Geels, F.W. A Socio-Technical Analysis of Low-Carbon Transitions: Introducing the Multi-Level Perspective into Transport Studies. *J. Transp. Geogr.* **2012**, *24*, 471–482. [CrossRef]
2. Aria, M.; Cuccurullo, C. bibliometrix: An R-Tool for Comprehensive Science Mapping Analysis. *J. Informetr.* **2017**, *11*, 959–975. [CrossRef]
3. Lucas, K. Transport and Social Exclusion: Where Are We Now? *Transp. Policy* **2012**, *20*, 105–113. [CrossRef]
4. Uteng, T.P.; Turner, J. Addressing the Linkages between Gender and Transport in Low- and Middle-Income Countries. *Sustainability* **2019**, *11*, 4555. [CrossRef]
5. Demir, G.; Chatterjee, P.; Pamučar, D. Sensitivity Analysis in Multi-Criteria Decision Making: A State-of-the-Art Research Perspective Using Bibliometric Analysis. *Expert Syst. Appl.* **2024**, *237*, 121660. [CrossRef]
6. Demir, G.; Chatterjee, P.; Zakeri, S.; Pamučar, D. Mapping the Evolution of Multi-Attributive Border Approximation Area Comparison Method: A Bibliometric Analysis. *Decis. Mak. Appl. Manag. Eng.* **2024**, *7*, 290–314. [CrossRef]
7. Demir, G.; Chatterjee, P.; Kadry, S.; Abdelhadi, A.; Pamučar, D. Measurement of Alternatives and Ranking According to Compromise Solution (MARCOS) Method: A Comprehensive Bibliometric Analysis. *Decis. Mak. Appl. Manag. Eng.* **2024**, *7*, 313–336. [CrossRef]
8. Newman, P.; Kenworthy, J. *Sustainability and Cities: Overcoming Automobile Dependence*; Island Press: Washington, DC, USA, 1999.
9. Litman, T. Evaluating Public Transit Benefits and Costs. Victoria Transport Policy Institute. Available online: https://www.vtpi.org/tranben.pdf (accessed on 1 January 2024).
10. Cervero, R. *The Transit Metropolis: A Global Inquiry*; Island Press: Washington, DC, USA, 1998.
11. Gonzalez, R.; Mollins, J.; Vinot, M. Electric Vehicles in Urban Transportation: Environmental Benefits and Costs. *Renew. Sustain. Energy Rev.* **2019**, *101*, 138–145. [CrossRef]

12. Xie, D.; Gou, Z.; Gui, X. How Electric Vehicles Benefit Urban Air Quality Improvement: A Study in Wuhan. *Sci. Total Environ.* **2024**, *906*, 167584. [CrossRef]
13. Tao, M. Dynamics between Electric Vehicle Uptake and Green Development: Understanding the Role of Local Government Competition. *Transp. Policy* **2024**, *146*, 227–240. [CrossRef]
14. Litman, T. Autonomous Vehicle Implementation Predictions: Implications for Transport Planning. Victoria Transport Policy Institute. Available online: https://www.vtpi.org/avip.pdf (accessed on 1 January 2024).
15. Olayode, I.O.; Du, B.; Severino, A.; Campisi, T.; Alex, F.J. Systematic Literature Review on the Applications, Impacts, and Public Perceptions of Autonomous Vehicles in Road Transportation Systems. *J. Traffic Transp. Eng. (Engl. Ed.)* **2023**, *10*, 123–136. [CrossRef]
16. Shaheen, S.; Cohen, A.; Zohdy, I. Shared Mobility: Current Practices and Guiding Principles. U.S. Dep't Transp. Available online: https://rosap.ntl.bts.gov/view/dot/31164 (accessed on 1 January 2024).
17. Banister, D. The Sustainable Mobility Paradigm. *Transp. Policy* **2008**, *15*, 73–80. [CrossRef]
18. Grengs, J. Nonwork Accessibility as a Social Equity Indicator. *Int. J. Sustain. Transp.* **2015**, *9*, 1–14. [CrossRef]
19. Docherty, I.; Marsden, G.; Anable, J. The Governance of Smart Mobility. *Transp. Res. Part A Policy Pract.* **2018**, *115*, 114–125. [CrossRef]
20. Li, X.; Wang, Y.; Zhao, F. End-to-End Heterogeneous Graph Neural Networks for Traffic Assignment. *J. Transp. Res.* **2021**, *38*, 512–523.
21. Zhang, Z.; Li, Y.; Liu, H. Nation-wide Human Mobility Prediction Based on Graph Neural Networks. *IEEE Trans. Intell. Transp. Syst.* **2022**, *23*, 1672–1685.
22. Liu, T.; Meidani, H. End-to-End Heterogeneous Graph Neural Networks for Traffic Assignment. *Transp. Res. Part C Emerg. Technol.* **2024**, *165*, 104695. [CrossRef]
23. Twinomurinzi, H.; Mageto, J. Smart Mobility in Urban Areas: A Bibliometric Review and Research Agenda. *Sustainability* **2023**, *15*, 6754. [CrossRef]
24. Ceccato, V.; Gaudelet, N.; Graf, G. Crime and Safety in Transit Environments: A Systematic Review of the English and the French Literature, 1970–2020. *Public Transp.* **2022**, *14*, 105–153. [CrossRef]
25. Mirabi, N.; Noorid, M. Urban Mobility and Transportation Planning: A Bibliometric Analysis. *Sustainability* **2022**, *14*, 5544. [CrossRef]
26. Long, Z. Bibliometric Insights into the Implications of Urban Built Environment on Travel Behavior. *ISPRS Int. J. Geo-Inf.* **2023**, *12*, 453. [CrossRef]
27. Papa, E.; Lauwers, D. Schrenk, M., Popovich, V.V., Zeile, P., Elisei, P., Beyer, C., Eds.; Smart Mobility: Opportunity or Threat to Innovate Places and Cities. In *Proceedings of the 20th International Conference on Urban Planning and Regional Development in the Information Society*; CORP—Competence Center of Urban and Regional Planning: Ghent, Belgium, 2015; pp. 543–550.
28. Albino, V.; Berardi, U.; Dangelico, R.M. Smart Cities: Definitions, Dimensions, Performance, and Initiatives. *J. Urban Technol.* **2015**, *22*, 3–21. [CrossRef]
29. Batty, M. *The New Science of Cities*; MIT Press: Cambridge, MA, USA, 2013; ISBN 9780262019521.
30. Pucher, J.; Buehler, R. *City Cycling*; MIT Press: Cambridge, MA, USA, 2012; ISBN 9780262517812.
31. Rodrigue, J.-P. *The Geography of Transport Systems*, 5th ed.; Routledge: New York, NY, USA, 2020; ISBN 9780367364601.
32. Mokhtarian, P.L.; Chen, C. TTB or Not TTB, That Is the Question: A Review and Analysis of the Empirical Literature on Travel Time (and Money) Budgets. *Transp. Res. Part A Policy Pract.* **2004**, *38*, 643–675. [CrossRef]
33. Mahmood, A.; Siddiqui, S.A.; Sheng, Q.Z.; Zhang, W.E.; Suzuki, H. Trust on Wheels: Towards Secure and Resource Efficient IoV Networks. *Computing.* **2022**, *114*, 1461–1480. [CrossRef]
34. Gehl, J. *Cities for People*; Island Press: Washington, DC, USA, 2010; ISBN 9781597265737.
35. Downs, A. *Still Stuck in Traffic: Coping with Peak-Hour Traffic Congestion*; Brookings Institution Press: Washington, DC, USA, 2004; ISBN 9780815701307.
36. Saaty, T.L. *The Analytic Hierarchy Process*; McGraw-Hill: New York, NY, USA, 1980; ISBN 9780070543713.
37. Vaidya, O.S.; Kumar, S. Analytic Hierarchy Process: An Overview of Applications. *Eur. J. Oper. Res.* **2006**, *169*, 1–29. [CrossRef]
38. Ishizaka, A.; Labib, A. Review of the Main Developments in the Analytic Hierarchy Process. *Expert Syst. Appl.* **2011**, *38*, 14336–14345. [CrossRef]
39. Law, A.M.; Kelton, W.D. *Simulation Modeling and Analysis*; McGraw-Hill: New York, NY, USA, 2000; ISBN 9780070592926.
40. Timmermans, H. The Saga of Integrated Land Use-Transport Modeling: How Many More Dreams Before We Wake Up? In *Moving through Nets: The Physical and Social Dimensions of Travel*; Elsevier: Amsterdam, The Netherlands, 2003; ISBN 9780080440323.
41. Ceder, A. *Public Transit Planning and Operation: Theory, Modeling, and Practice*; Elsevier: Amsterdam, The Netherlands, 2007; ISBN 9780750661667.

42. Longley, P.A.; Goodchild, M.F.; Maguire, D.J.; Rhind, D.W. *Geographic Information Systems and Science*; Wiley: Hoboken, NJ, USA, 2005; ISBN 9780470870013.
43. Clarke, K.C. *Getting Started with Geographic Information Systems*; Prentice Hall: Upper Saddle River, NJ, USA, 2003; ISBN 9780131438891.
44. Goodchild, M.F. Citizens as Sensors: The World of Volunteered Geography. *GeoJournal* **2007**, *69*, 211–221. [CrossRef]
45. Demir, G.; Moslem, S.; Duleba, S. Artificial Intelligence in Aviation Safety: Systematic Review and Biometric Analysis. *Int. J. Comput. Intell. Syst.* **2024**, *17*, 279. [CrossRef]
46. Morris, J.N.; Hardman, A.E. Walking to Health. *Sports Med.* **1997**, *23*, 306–332. [CrossRef] [PubMed]
47. Frank, L.D.; Engelke, P.O.; Schmid, T.L. *Health and Community Design: The Impact of the Built Environment on Physical Activity*; Island Press: Washington, DC, USA, 2004; ISBN 9781559639170.
48. Chen, C.; Pan, S.; Zhang, J.; McHugh, J. Using Big Data to Understand and Enhance Urban Mobility. *Transp. Res. Part C Emerg. Technol.* **2014**, *37*, 145–162. [CrossRef]
49. Wang, H. Big Data in Urban Transportation: A Case Study on Big Data Applications in Mobility as a Service. *IET Intell. Transp. Syst.* **2018**, *12*, 1010–1016. [CrossRef]
50. Giles-Corti, B.; Vernez-Moudon, A.; Reis, R.; Turrell, G.; Dannenberg, A.L.; Badland, H.; Foster, S.; Lowe, M.; Sallis, J.F.; Stevenson, M. City Planning and Population Health: A Global Challenge. *Lancet* **2016**, *388*, 2912–2924. [CrossRef] [PubMed]
51. Frank, L.D.; Sallis, J.F.; Conway, T.L.; Chapman, J.E.; Saelens, B.E.; Bachman, W. Many Pathways from Land Use to Health: Associations Between Neighborhood Walkability and Active Transportation, Body Mass Index, and Air Quality. *J. Am. Plan. Assoc.* **2005**, *72*, 75–87. [CrossRef]
52. Behrisch, M.; Bieker, L.; Erdmann, J.; Krajzewicz, D. SUMO–Simulation of Urban Mobility: An Overview. In Proceedings of the SIMUL 2011, the Third International Conference on Advances in System Simulation, Barcelona, Spain, 23–28 October 2011; ISBN 9781612081540.
53. Henderson, T.R.; Roy, S.; Floyd, S.; Riley, G.F. ns-3 Project Goals. In *Proceedings of the 2008 ACM Workshop on ns-2: The IP Network Simulator*; ACM: New York, NY, USA, 2008; pp. 13–18. [CrossRef]
54. Neirotti, P.; De Marco, A.; Cagliano, A.C.; Mangano, G.; Scorrano, F. Current Trends in Smart City Initiatives: Some Stylised Facts. *Cities* **2014**, *38*, 25–36. [CrossRef]
55. Gakenheimer, R. Urban Mobility in the Developing World. *Transp. Res. Part A Policy Pract.* **1999**, *33*, 671–689. [CrossRef]
56. Levinson, D.; Wu, H. Towards a General Theory of Access. *J. Transp. Land Use* **2020**, *13*, 129–149. [CrossRef]
57. Litman, T. Evaluating Transportation Equity: Guidance for Incorporating Distributional Impacts in Transportation Planning. Victoria Transport Policy Institute. Available online: https://www.vtpi.org/equity.pdf (accessed on 1 January 2024).
58. Bishop, R. *Intelligent Vehicle Technology and Trends*; Artech House: Boston, MA, USA, 2005; ISBN 9781580539118.

sustainability MDPI

Review

Bicycle Infrastructure Design Principles in Urban Bikeability Indices: A Systematic Review

Tufail Ahmed [1,*], **Ali Pirdavani** [1,2], **Geert Wets** [1] **and Davy Janssens** [1]

[1] UHasselt, The Transportation Research Institute (IMOB), Martelarenlaan 42, 3500 Hasselt, Belgium; ali.pirdavani@uhasselt.be (A.P.)
[2] UHasselt, Faculty of Engineering Technology, Agoralaan, 3590 Diepenbeek, Belgium
* Correspondence: tufail.ahmed@uhasselt.be

Abstract: Bicycling is a sustainable form of micromobility and offers numerous health and environmental benefits. Scientific studies investigating bikeability have grown substantially, especially over the past decade. This paper presents a systematic literature review of the developed urban bikeability indices (BIs). The paper provides insight into the scientific literature on bikeability as a tool to measure bicycle environment friendliness; more importantly, the paper seeks to know if the BIs consider bicycle infrastructure design principles. Data extraction included identifying the geographical location, essential indicators, sample size and distribution, data source, the unit of analysis, measurement scale, methods used to weigh indicators, and identification of studies using bicycle design principles in BIs. The database search yielded 1649 research articles using different keywords and combinations, while 15 studies satisfied the inclusion criteria. The studies were found to be conducted in various geographical locations. The unit of analysis for developing the index varied across studies, from street segments or bicycle lanes to zones within the city or even the entire city. The most commonly utilized method in developing urban BIs was a scoring and weighting system to weigh the indicators. The weighting methods include an equal weight system, survey-based and literature review-based methods, expert surveys, the analytic hierarchy process, and a weighted linear combination model. The essential criterion is bicycle infrastructure, such as bike lanes, routes, and bicycle paths as 14 studies considered it for the construction of the BIs. The review findings suggest a lack of consideration of all five bicycle infrastructure design principles, as only three studies considered them all, while others only included a subset. Safety and comfort are the most commonly considered principles, while coherence is the least considered principles in the BIs. It is crucial to consider all five bicycle infrastructure design principles to create a bicycle-friendly environment and attract more people to this sustainable mode of transportation.

Keywords: sustainable mobility; micromobility; bicycling; bikeability; bicycle friendliness

1. Introduction

Cycling and walking are considered healthy and sustainable modes of transportation and are recognized and endorsed by governments worldwide [1,2]. Bicycles take up less road space and have zero carbon emissions compared to motorized modes of transportation, so their use in cities is primarily viewed as advantageous to the environment and air quality [3–5]. In the past, road officials emphasized motorized vehicles' safe movement, hence giving less attention to the green modes of transportation [6]. However, policymakers' paradigm has been shifted to provide and improve cycling quality to increase the share of this form of micromobility [6,7]. This necessity to shift travel choices from motor vehicles to eco-friendly bicycles is driven by traffic congestion, air pollution, and other transportation problems [8,9].

Cycling requires appropriate infrastructure, which is essential to its attractiveness [10]. For example, bike lanes are crucial for bicyclist safety and comfort [11]. Dedicated and protected bike lanes reduce the risk of crashes and injuries, as they provide separation from the roadway by means of physical barriers [12,13]. Similarly, bicycles need safe parking facilities, which are crucial for protecting bikes from theft and ensuring convenience for cyclists [14]. Studies show that current and prospective cyclists are willing to pay for better parking facilities to enhance their personal safety and protect their bikes from theft and vandalism [15]. Furthermore, bicycle signals may be required at junctions for cyclists to cross safely [16]. The absence of traffic signals increases the risk of accidents, as it leads to confusion and errors in judgment, contributing to a higher likelihood of road accidents [17]. At the same time, the lack of bicycling infrastructure and a supportive environment discourages bicycle use [2]. Well-planned bicycle infrastructure has the potential to increase bicycling share, as evidenced by research showing a rise in bicycle activity after implementing new infrastructure [11].

The viability of bicycling as a transport mode depends on the condition, comfort, and safety of the infrastructure [18,19]. Over the years, researchers and practitioners have developed several models to assess bike riders' experience. These bicycle infrastructure assessment methods are objective and subjective. Subjective methods assess perceptions gathered from surveys, interviews, or group discussions [20]. Direct observation using audits and geospatial methods using secondary data, such as geographic information systems and remote sensing, are objective tools that measure the physical characteristics of an environment [21]. Some of the most commonly used metrics are bicycle level of service (BLOS), the bicycle compatibility index (BCI), bicycle safety index rating (BSIR), bicycle suitability rating, CycleRap, and the bikeability index (BI) [6,7,16,21–24].

Bicycle assessment methods date back to the 1980s, when Davis [25] initially proposed aBSIR. Similarly, other methods have been developed to assess bicycle infrastructure using metrics such as level of service, quality of service, level of traffic stress (LTS), and the dynamic comfort index (DCI) [2,26,27]. The concept of bikeability existed before; however, the term bikeability has grown, especially in the last decade, because of the walkability concept [21]. Walkability and bikeability are directly related to the built environment, which affects the accessibility, safety, and comfort of pedestrians and cyclists. Although there are certain similarities between walkability and bikeability, a notable difference is in their evaluation. Bicycling requires equipment and a certain level of expertise to ride, and the significance of infrastructure over land use is more pronounced for cycling than walking [10]. The growth in the concept of bikeability encompasses both the increased use of the term and the evolution and refinement of the underlying concept over time. While planners have been working to improve the conditions for cycling for decades, the term "bikeability" may not have been widely used initially, and its recent prominence reflects a growing awareness and emphasis on creating environments that support and promote cycling. Table 1 summarizes different concepts and their essential considerations when developing the metrics in the literature to assess the bicycle environment.

Table 1. Methods for assessing bicycle environment.

Assessment Category	Relevant Assessment Tools	Important Factors	References
Vibration or Roughness Index	DCI, International Roughness Index, Dynamic Cycling Comfort, Bicycle Environmental Quality Index	Vertical acceleration Bicycle vibration	[2,19,28–30]
Bicycle Level of Service	BLOS, LTS, Quality of service, BSIR, Bicycle Comfort Level Rating	Infrastructure Geometric design Traffic conditions Traffic stress	[1,16,31,32]
Bicycle Safety Index	Bicycle Safety Index	Motorized vehicles (speed, volume, flow, density, and infrastructure)	[7,33]
Bikeability Index	BI, Area-Wide Bikeability Assessment Model (ABAM), Bike Score®	Safety Comfort Attractiveness Directness Coherence	[34–38]

Bikeability can be defined in different ways [39,40]. Bikeability measures how easy, safe, and convenient it is to ride a bike on a particular path or in a particular area [41]. Lowry et al. (2012) defined it as "an assessment of an entire bikeway network for perceived comfort, convenience, and access to important destinations." [22]. According to another definition, it is the extent to which the real and perceived environment are favorable and safe for riding [21]. Some scholars have attempted to explain the difference between bikeability and related concepts, such as bike suitability and friendliness [42]. Bicycle suitability is "an assessment of a linear stretch of a bikeway's perceived comfort and safety" [34]. Hence, bikeability is a superordinate term, both geographically and conceptually [42]. Similarly, bicycle friendliness includes characteristics of bikeability and refers to a community's assessment of many aspects of biking, such as laws and regulations, education initiatives, and cycling acceptance [42,43].

There has been rising interest in bikeability-related studies as the number of publications on the subject has increased dramatically, especially over the past four years. Exponential growth is seen in the number of published papers on "bikeability" since 2010, starting with 10 articles in WOS and 32 in Scopus that year. The number of publications on the topic consistently increased over the years, reaching a peak in 2021, with 72 articles in WOS and 124 in Scopus. The rising interest in bikeability-related studies over the years can be attributed to several factors. The increased awareness about active modes of transportation and the necessity for sustainable cities can be attributed to this uptick [6]. Using bicycles as a means of transport can help reduce traffic congestion, improve air quality, and lower carbon emissions [44]. Also, addressing cycling-related factors, such as safety and accessibility, is essential for promoting active transportation and public health [45]. Bicycle use also has health benefits, reducing the risk of all-cause mortality [46]. Bikeability intends to assess and integrate cycle infrastructure for individual well-being and promote urban environmental sustainability [6]. The growing body of literature in this area can help inform the creation of successful policies to increase cycling in cities. Hence, a comprehensive review of bikeability will be helpful.

Studies and discussions suggest that promoting bicycle use is critical for urban planning and transportation policy [47,48]. Cities and institutes have stressed that bicycle infrastructure design principles are essential elements that play a crucial role in promoting bicycle use [49]. Researchers have developed different tools that can be used to assess a city's or a neighborhood's bikeability—in other words, bicycle friendliness. In addition, diverse criteria, weighing systems, analysis units, and methods have been used to develop BIs. To facilitate a comprehensive review of bikeability tools, synthesizing and critically evaluating the existing literature is imperative.

Additionally, gaining insight into different aspects utilized in BIs is important. Moreover, a vital aspect of this review involves assessing whether urban BIs align with fundamental bicycle infrastructure design principles, such as safety, comfort, attractiveness, directness, and coherence. To inform policymakers about future bicycle planning initiatives and to extract, synthesize, and extend the existing body of knowledge for the scientific community, a comprehensive review of the literature that explores the links between bikeability indices and bicycle infrastructure design principles in these indices is needed yet currently missing from existing the literature to the best of our knowledge. A systematic review paper is presented because it ensures a rigorous and comprehensive synthesis of existing evidence, minimizing bias and providing a reliable foundation for informed decision-making in the targeted subject area [50].

1.1. Overview of the Bicycle Infrastructure Design Principles

People of different ages and cycling abilities should experience and enjoy the built environment [11]. It is understood that people's standards of what is "acceptable" differ, but the concept of "inclusive design" serves as the foundation for all bicycle infrastructure design principles [3,21]. Cycling-friendly infrastructure must meet five internationally recognized criteria: safety, comfort, attractiveness, directness, and coherence [49,51–53].

1.1.1. Safety

The perception of danger could discourage people from taking up cycling. Researchers have found a positive relationship between perceptions of safety and increased cycling [54]. Safety measures for a bicycle include, for example, the type of bicycle infrastructure, motorized traffic speed along a bicycle path, traffic control devices at junctions, street lights for evening and night-time cycling, and buffer space from car parking along the cycle path [35,49].

1.1.2. Comfort

Comfort refers to reduced physical exertion from riding a bicycle on a good network [35]. Bicycling comfortability can be achieved by providing a sufficient width for riders; providing minimal stopping and starting along cycle routes; minimizing steep grades; and, whenever feasible, reducing interaction with high-speed or high-volume motorized traffic. When the mentioned factors are considered, bicycle pathways or cycling routes create an environment that allows cyclists to travel efficiently and comfortably [55,56].

1.1.3. Attractiveness

Cycling is an enjoyable experience partly because of the close connection to the external environment [57]. The visual and aesthetic aspects of the built environment are referred to as attractiveness. This component includes trees and shade, scenery, cleanliness, quality of public open space, aesthetic buildings, and street furniture [58]. Selecting a bicycle as a transport mode depends on the attractiveness of cycling and competing modes such as the bus [59].

1.1.4. Directness

This criterion relates to minimizing traveling distance and time by taking the fastest route between the origin and destination and avoiding intersections or stoppages [60]. Directness is important, as cycling can be an appealing alternative to driving or public transportation, particularly for local journeys [6]. A good cycling route must be direct and eliminate the need for cyclists to undergo diversions [61].

1.1.5. Coherence

Bicycle cohesion (accessibility) is defined as people's ability to reach their primary destinations via direct routes [62]. A bicycle network should connect all primary cycling origins and destination zones/centers. Cycle routes can be made cohesive by the continuity of bicycle routes and proximity to other transport modes for better connectivity [53].

The remainder of this paper is structured as follows. Section 2 describes the systematic methodology adopted for the study. Section 3 discusses the results of the study. Section 4 presents a discussion of the review findings. Lastly, Section 5 presents the study's conclusions and limitations and the future scope of the work.

2. Methodology

We utilized the Preferred Reporting Items for the Systematic Review and Meta-Analyses (PRISMA) procedure for this review [63]. The technique aims to be robust and reproducible by minimizing possible biases in research reviews and transparent in choosing and categorizing papers based on precise eligibility criteria [50,64].

2.1. Search Strategy

The study approach began with identifying the topic, the scope of the work, the research aims, and the objectives. Then, a protocol was developed for the papers to be included in the review following the PRISMA method. We searched Scopus, Web of Science, and ProQuest to find the research papers. The initial search was conducted from February 2023 to March 2023 and updated in December 2023 to identify new studies. Before starting the search of the scientific databases, key concepts were developed to ensure we did not miss any relevant research. Possible synonyms, technical terminology, layperson's terms, acronyms, and abbreviations were considered. Concept 1 included bicycle-related keywords and phrases, such as "bike*", "bicycl*", "bicycl* infrastructure", "cycl* infrastructure", "bikeab*", and "bikeability". Concept 2, on the other hand, included terms like "index*", "assessment tools", "assessment methods", "evaluation criteria", "checklist", "compatibility", and "level of service", focusing on assessment-related terminologies to further narrow the search towards evaluation methods and criteria.

After selecting keywords, the key concepts inside each component were linked using "OR", while the two groups were linked using "AND". Suitable Boolean pairs such as "AND" and "OR" help to drastically reduce the number of results returned, as well as remove undesirable results [50]. The asterisk (*) function in search queries is used to include variations of words, effectively capturing terms like "bicycle" and "bicycling". Additional filters were also applied to narrow down the number of papers. The search queries and filters are mentioned in Table 2. We also performed forward and backward snowballing to identify missed papers while searching scientific databases.

Table 2. Search criteria.

Database	Search Terms	Filters	Articles Found
Scopus	((TITLE-ABS-KEY (bike*)) OR (TITLE-ABS-KEY ("Bicycl* infrastructure")) OR (TITLE-ABS-KEY ("Cycl* infrastructure")) OR (TITLE-ABS-KEY (bikeab*)) OR (TITLE-ABS-KEY (bicycl*)) OR (TITLE-ABS-KEY (bikeability)) OR (TITLE-ABS-KEY (blos))) AND ((TITLE-ABS-KEY (index)) OR (TITLE-ABS-KEY ("Assessment Tool")) OR (TITLE-ABS-KEY ("Assessment methods")) OR (TITLE-ABS-KEY ("Evaluation Criteria")) OR (TITLE-ABS-KEY (checklist)) OR (TITLE-ABS-KEY (compatibility)) OR (TITLE-ABS-KEY ("level of service")))	Language: English Publication period: 2010–2023 Article type: Journal and conference papers Exclude subjects like natural sciences, earth sciences, etc.	1048
Web of Science	((((((TI=(Bike*)) OR TI=(Bicycl*)) OR TI=("Bicycl* infrastructure")) OR TI=("Cycl* infrastructure")) OR TI=(Bikeab*)) OR TI=(Bikeability)) OR TI=(BLOS) AND ((((((((TI=(Index*)) OR TI=("Assessment Tool")) OR TI=("Assessment methods")) OR TI=("Evaluation Criteria")) OR TI=("Evaluation Criteria")) OR TI=(Compatibility)) OR TI=("level of service")) OR TI=(Assess*)) OR TI=(Evaluat*)	Language: English Publication period: 2010–2023 Article type: Journal and conference papers Exclude research areas like ecology, medicine, natural sciences, earth sciences, etc.	576
ProQuest	title(Bike*) OR title(Bicycl*) OR title("Bicycl* infrastructure") OR title("Cycl* infrastructure") OR title(Bikeab*) OR title(Bikeability) OR title(BLOS) AND title(Index*) OR title("Assessment Tool") OR title("Assessment methods") OR title("Evaluation Criteria") OR title(Checklist) OR title(Compatibility)	Language: English Publication period: 2010–2023 Source type: Conference papers and proceedings and scholarly journals Article type: Journal and conference papers	17
Forward and backward snowballing			8

2.2. Eligibility Criteria

The next step was to scrutinize and evaluate the papers to be included in the review. For this purpose, inclusion and exclusion criteria were defined (see Table 3). First, the articles had to be published in peer-reviewed journals or as conference proceedings. All other publications, such as research letters, book chapters, review articles, research notes, editors' comments, reader comments, and book reviews, were excluded. Further requirements for paper selection were that the paper was full-length and published in English after 2010.

Table 3. Inclusion and exclusion criteria adopted for the review.

Inclusion Criteria	Exclusion Criteria
• Research articles and conference papers • Published since 2010 • Full-length paper published in English • Considered only bikeability • The focus area must be an urban area. • Developed a method/tool/application for evaluating/assessing/measuring urban bikeability	• Hybrid methods • Measures only one aspect of bikeability • A method that does not come up with a composition of indicators into an index • Review articles, letters to the editor, opinion articles, book chapters, etc. • Full text not available to authors

The fourth criterion was to see if the paper considered methods for only bikeability. We did not consider hybrid methods that measure urban walkability and bikeability. Also, methods that only considered one aspect of bikeability, for example, bike lanes or bicycle surface quality, were excluded, since they do not provide a holistic picture of the urban bicycle environment. Also, studies with a focus other than urban areas, such as the BI for suburban or rural areas, were excluded, since the attributes that lead to higher or lower bikeability might differ. Lastly, studies should have developed a method, tool, or application for evaluating, assessing, or measuring urban bikeability.

In total, we found 1641 research records by searching three databases. In addition, we also identified 8 research articles based on "snowballing" not found in the initial search. Figure 1 shows the article identification, screening, and selection process. The identified records were imported into Rayyan for duplicate removal. Rayyan is a free web tool designed to assist researchers in managing the literature review process [65]. First, all duplicates (n = 116) found in the three searched databases were removed. After removing duplicates from 1533 articles, 1469 were excluded based on title-and-abstract screening. For the remaining 64 papers, full-text papers (n = 63) were retrieved and assessed against the eligibility criteria for inclusion or exclusion, as the authors failed to retrieve the full text of 1 research article. Forty-eight studies were ineligible, as they did not satisfy the eligibility criteria, i.e., they did not construct an index for the bikeability of an urban area or the scope of the study was outside an urban area. Similarly, some methods were hybrid, considering both walkability and bikeability. The remaining 15 articles were used for this systematic review and comprehensively synthesized to extract the results.

2.3. Data Extraction Process

Selected articles were scrutinized to extract relevant data and comprehensively understand their contents to answer the research questions. Table 4 shows the elements extracted against each category. The extracted information includes the author, year of publication, city, country, research instrument, data source, measurement scale, geographical location of the study, study design, unit of analysis, bikeability variables, and type of measurement. In addition, the retrieved variables were categorized and grouped to identify the studies using bicycle design principles in BIs.

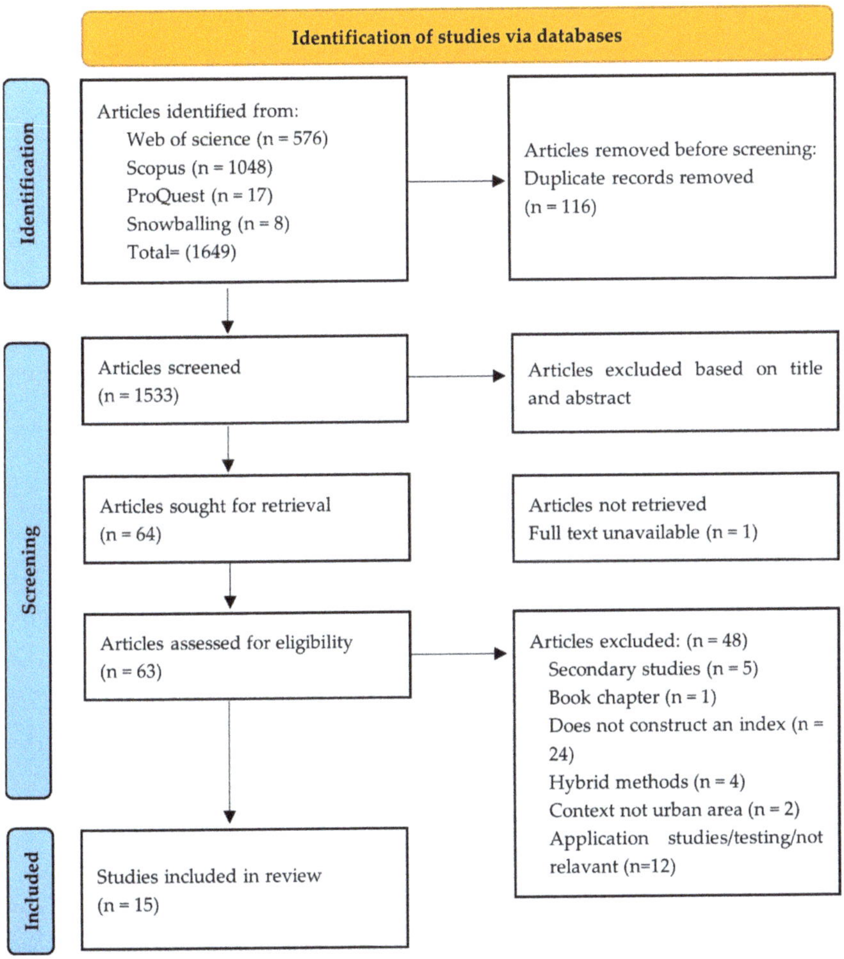

Figure 1. Systematic search process based on the PRISMA framework.

Table 4. Elements extracted during the systematic review.

Category	Extracted Elements
Identifying information	Author's name; title of the research article; publication year
Study setting	The geographical location; description of the study
Research design	Sample size; sample selection; characteristics of the population under study; age
Study methods and unit of analysis	Data source; unit of analysis; methods used; measurement scale
Critical variables	Number of variables considered; important variables
Bicycle design principles in BI	Grouping of indicators; identification of study using bicycle design principles in BI
Finding	Main results specific to BI; important variables that improve BI, other considerations for BI

3. Results

3.1. Geographical Location of the Studies

The studies included in this literature review were conducted in several countries worldwide, including Spain (n = 2) [66,67], Germany (n = 2) [42,68], Singapore (n = 2) [69,70], Greece [57], Japan [69], the United States [34], Colombia [35], Taiwan [36], Canada [71],

Russia [22], Austria [72], and China [73]. Two studies were conducted across two countries (the United States and Canada, Singapore and Japan) [38,69].

3.2. Formulation of BI

The BIs offer a valuable tool for assessing how suitable an environment is for cycling. Authors have used various nomenclatures for BIs, such as the ABAM, BikeDNA, etc. [36,38]. The formulation of an index for urban bikeability is a multi-step process [35]. The first step is identifying criteria, such as traffic, safety, comfort, or connectivity [66,69]. For each criterion, a list of indicators is identified. These indicators can include but are not limited to the presence and condition of bike lanes, pavement quality, traffic volume and speed, connectivity, land use, topography, bike parking facilities, and bicycle-sharing systems [6]. The selection of indicators depends on the context of the BI; for example, micro-level indicators are selected to develop BIs for street-level assessment. Studies then assigned a score or values to these indicators. For example, a scoring system was used in developing Munich's BI [42].

The next step is assigning each indicator a weight based on its perceived importance in facilitating or hindering bikeability. Assigning weights involves population surveys, stakeholder consultations, expert judgments, or empirical research to reflect each factor's relative importance in an area's overall bikeability [35,36,68]. By combining these weighted indicators and the scores of indicators, the BI generates a single score that reflects the overall bikeability of a specific location. This final BI score allows for easy comparison between different areas and can be a valuable asset for urban planners seeking to promote cycling within their cities. Equation (1) shows one example of a BI [66]. In the equation, each criterion has a list of indicators that compute the relevant score, i.e., Ti, Ii, or Ci.

$$Bi = (0.4 \times Ti) + (0.15 \times Ii) + (0.15 \times Ci) + (0.1 \times Pi) + (0.2 \times Si), \quad (1)$$

whereas

Bi = bikeability index
Ti = Traffic indicators
Ii = Infrastructure indicators
Ci = Connectivity indicators
Pi = Parking space indicators
Si = Topography indicators.

Equation (2) is another example of a BI developed to measure the mobility of biking in Mediterranean cities [67].

$$BI = \alpha P + \beta C + \gamma L, \quad (2)$$

whereas:

P: average of the parameters of the segment
C: number of cyclists in the segment
L: length of the studied segment.
α, β, γ: coefficients associated with the variables.

3.3. Study Demographics and Sample Size

The sample characteristics varied, with some studies using small sample sizes while others used larger sample sizes. The minimum sample size was 10 respondents [36,42], while the maximum number of respondents to a bikeability survey was 1402 [71]. One study did not disclose the sample size and characteristics [67]. The data were collected from the urban population and tourists. In contrast, three studies did not collect data from respondents, and they were either based on other methods, such as BLOS and bike suitability, or objective methods [22,38,73].

A few studies also reported the gender distribution. In the studies that disclosed the respondents' genders, men were predominant; only one study had more female respondents than men [72]. Only three studies reported the age distribution of the sample population, with one study considering respondents only under 45 years of age [57], while the other study considered respondents in the range of 18–65 years of age [34]. In another study, 40% of the participants were younger than 35 [72].

Only one study consulted experts to weigh bikeability indicators, with a majority of experts from Germany (58%), followed by other European countries (23%) and America (19%) [68]. In addition, most of the participants were researchers (77%), while the remaining 23% worked in practice. Two studies used focus group discussion and opinion surveys [34,71]. Also, two studies reported using actual travel behavior data [71,72].

3.4. Methods Used to Develop BIs

Table 5 provides information on the unit of analysis and method used for the development of the BI in selected studies. A scoring and weighting system was the most common method to assess urban bikeability. Of the 15 studies, 7 used a scoring and weighting system [35,42,66–68,71,72]. In the scoring method, the indicators of the BIs are given points against a well-defined point score system. The studies usually included complete guidelines based on the standard for each indicator. An audit tool is usually used to collect field data for each bikeability indicator, which are then compared with the guidelines, based on which a score is assigned. The studies used scores from 0 to 1; a score of 0 represents a bikeability indicator that does not exist at all, while a score of 1 shows that the indicator is present according to the standard [35]. Some studies used scores to define if a bikeability component is bicycle-friendly or unfriendly [72]. The second essential component of point-scoring BIs is weighing individual indicators. Usually, each indicator's weight is assigned based on user opinion surveys or experts [35,68]. In addition, the system applies to both individual street segments and grid cells [35,71]. The overall bikeability result is also measured in terms of points, i.e., 0–1 or 0–100, with lower points meaning less bikeable, while the higher points mean more bikeable [35,38,73].

Table 5. General characteristics of the selected studies.

Paper ID	Authors	Country	Data Source	Unit of Analysis	Method	Sample Size	Sample Characteristics
1	Codina et al. (2022) [66]	Spain	Local bike-user self-reported survey	100 × 100 m scale City level	Scoring and weighting	290	DNM *
2	Karolemeas et al. (2022) [57]	Greece	Digital Elevation Model, Open Street Map, existing traffic studies, and General Urban Plan	Street segment	Spatial analysis and AHP	15	12 men 3 women 90% under 45 years old and highly educated
3	Hardinghaus et al. (2021) [68]	Germany	Open geodata, expert surveys	City level	Scoring and weighting	57	37 men 20 women 58% of respondents from Germany 23% from other European countries 19% from America 77% professionals 23% researchers
4	Ito and Biljecki (2021) [69]	Singapore and Japan	SVI, surveys, OpenStreetMap (OSM), land use (LU), Digital Elevation Model (DEM), and Air Quality Index (AQI)	Street segments	Street view imaginary and computer vision	800	DNM

Table 5. *Cont.*

Paper ID	Authors	Country	Data Source	Unit of Analysis	Method	Sample Size	Sample Characteristics
5	Schmid-Querg et al. (2021) [42]	Germany	Field observations and interviews/questionnaires	Road segments and intersections	Scoring and weighting	10	DNM
6	Tran et al. (2020) [70]	Singapore	Land use maps Road network Land use regression Spatial analysis	Road segments	Objective approach	NA	NA **
7	Porter et al. (2020) [34]	United States	Internet-based self-reporting questionnaire, focus group discussion	City level	Exploratory factor analysis	998	520 men 409 Female Mean age: 38 (18–65 considered for data collection) Graduate degree: 33.4% College degree: 43.8% Below college: 22.8%
8	Arellana et al. (2020) [35]	Colombia	Survey questionnaire, secondary data, Google Street view	Road segments	Scoring and weighting	336	208 men 128 women 62.5% belong to socioeconomic strata 1 and 2, 26.5% to strata 3 and 4, and 11% to strata 5 and 6.
9	Ros-McDonnell et al. (2020) [67]	Spain	Secondary data	Bike lanes/roads divided into segments of 100–500 m	Scoring and weighting	DNM	DNM
10	Lin and Wei (2018) [36]	Taiwan	Literature reviews and stakeholder interviews	Zones	Analytic network process	10	DNM
11	Winters et al. (2013) [71]	Canada	Opinion survey Travel behavior Focus groups	10 m grid cells	Scoring and weighting	1402	DNM
12	Lowry et al. (2012) [22]	Russia	Secondary data and primary data on variables, if not maintained previously	Zones	BLOS and bike suitability	NA	NA
13	Krenn et al. (2015) [72]	Austria	Bike trips, questionnaire survey data	100 m × 100 m cells	Scoring and weighting	113	Men: 45% Women: 55% Age <35 years: 40% Age 35–40: 40% Age >51: 20%
14	Winters et al. (2016) [38]	United States and Canada	Secondary data	City level	Weighting and regression	NA	NA
15	Dai et al. (2023) [73]	China	Digital elevation model, Mobile phone signaling data, street view imagery, climate datasets	Road segments	Spatial and temporal analysis	NA	NA

* DNM: Did not mention, ** NA: Not applicable.

One study used the Analytic Hierarchy Process (AHP) [57]. The method involves five steps, as follows: defining indicators, determining parameters, developing scoring rubrics, weighting each parameter using the AHP, and generating a bikeability map for the case study. Another study used street view imagery and computer vision with extracted indicators [69]. The method uses six data sources, i.e., street view imagery, surveys, OpenStreetMap data, land use, a digital elevation model, and air quality index. The indicators are grouped into five categories (connectivity, environment, infrastructure, perception, and vehicle–cyclist interaction). An index called the composite index was proposed, and the critical aspect is that it uses an equal point system for the selected indicators in the index.

An objective approach was also utilized to develop a BI in [70]. The proposed BI was based on four subcriteria, as follows: air quality, accessibility, suitability, and perceptibility. Indicators were identified for all the subcriteria. Interestingly, an objective method was used to measure all the indicators, contrary to other bikeability indices in the literature. In developing a BI, one study used an exploratory factor analysis [34]. To identify and shortlist critical factors that would later be included in the index, an observational, cross-sectional study was conducted to assess multi-level ecological factors and their association with bicycling behavior. A self-reported Internet-based questionnaire assessed the proposed ecological factors of bicycling behavior. The concepts were shortlisted from the literature review, and focus groups were conducted at the two study locations to determine the factors necessary for adopting and maintaining bicycling behavior. The information obtained from literature and focus groups was used to draft an initial survey, which was then rectified after the pilot test. After data collection, the BI creation process involved the following steps: (1) determining the need for domain-specific indices through Spearman rank correlation coefficients; (2) identifying appropriate buffer sizes for environmental variables based on Spearman rank correlation coefficients; (3) conducting a Kaiser–Meyer–Olkin test and exploratory factor analysis to identify essential environmental factors; (4) using factor loadings to create domain-specific indices, ensuring the fit criteria of loadings, the absence of cross-loading, and the presence of at least three variables; (5) evaluating the association between domain-specific bikeability indices and bicycling frequency through correlation coefficients, stratification, and regression analyses, adjusting for clustering by study site and covariates.

One study introduced a BI called the ABAM using an analytic network process (ANP) method [27]. This study adopted the same approach as [34] in shortlisting assessment criteria. However, stakeholders were interviewed instead of cyclists to refine the initial criteria. The ABAM utilizes gray numbers to account for diverse performances within zones, ranking them based on identified interdependent criteria. Another study used the BLOS to determine the bikeability of the bikeway street network [22]. The first step to develop the BI was to calculate the BLOS for the bikeways in the study area. The resultant BLOS score can be used for a set of destinations to assess its bikeability, for example, the bikeability to public parks or commercial destinations. The multinomial logit mode choice model has also been used to determine a BI [68]. BI development was carried out in three steps. A literature review was conducted to identify bikeability indicators, and the expert survey was used to establish a weighting. Finally, an extensive spatial BI was developed by combining the established categories using OpenStreetMap data.

One study used the following four sub-indices to construct a BI: safety, comfort, accessibility, and vitality [73]. It utilized open-source data, advanced deep neural networks, and GIS spatial analysis to eliminate subjective evaluations and provide a more efficient and comprehensive evaluation of bikeability. The weights of each indicator were assessed based on principal component analysis. Another study developed a BI to gain insight into the elements that shape the behavior of residents' cycling activities by using machine learning, deep learning, and trajectory mining algorithms on large, multi-dimensional datasets [73]. The utilized datasets encompass a variety of sources, including bike-sharing trajectory data, digital elevation models, mobile phone signal data, points of interest, street view imagery, air quality monitoring data, and ERA5 climate datasets.

3.5. Unit of Analysis

The unit of analysis varied from the city level to street segments, intersections, and zones. Seven studies developed the BI for street segments or bicycle lanes [35,42,57,67,69,70,73]. The length for which the data were extracted/collected differed in these studies; for example, one study used 500 m aggregation for connectivity indicators and 100 m aggregation for some indicators, i.e., road width and presence of on-street parking [69]. Few studies considered data for the entire segment (road or lane) [35,42,57].

Another study considered bikeability for intersections, and the same indicators were used for intersections and road segments [42]. A bikeability method was also developed for bike lanes/roads [67]. However, the lanes were divided into segments of 100–500 m for better results and a more accurate and detailed assessment of the bikeability of a given segment. Four studies used a scale as the unit of analysis for the bikeability of urban areas [38,66,71,72]. The scale varied from 10 m [71] to 100 × 100 m [66,72], meaning they analyzed bikeability at a very granular level. Two studies considered bikeability in zones within a city [22,36]. In contrast, the remaining two studies considered bikeability at the city level [34,68].

3.6. Summary of the Bikeability Assessment Tools

Table 6 provides a synthesis of indicators affecting bikeability in urban environments, the crucial indicators considered, their assessment methods, and the research findings of each BI. One BI is based on a survey and literature review to identify challenges and hotspots in the built environment [66]. Another BI employs the analytic hierarchy process to evaluate ten indicators, such as slope and junction density, to demonstrate that the road network is the most influential factor in bikeability [57]. Experts were consulted in one study to rank the significance of five indicators [68]. The results indicated that biking facilities along main streets are emerging as the most pivotal element.

Another study used a spatial value index, infrastructure, perception, and vehicle–cyclist interaction to identify 34 indicators that explain more than 65% of the spatiotemporal mobility pattern [69]. Constructing a BI showed that bicycle infrastructure and speed limits are the most critical criteria for bikeability [42]. An index was also developed using a weighted linear combination model, and 12 indicators, including air quality, were combined to calculate the BI [70]. Moreover, exploratory factor analysis was conducted, and it was found that objectively measured environmental variables are more associated with bicycling for transportation and transportation bicycling frequency than with recreation bicycling [34]. Finally, one study used survey data and discrete choice models to rank 20 indicators and found that security is the most critical factor for frequent cyclists whose travel purposes are work and shopping [35].

Table 6 also provides information on different methods of weighting indicators used in BI studies. The weighting methods include an equal weight system, survey-based and literature review-based methods, expert surveys, the analytic hierarchy process, and a weighted linear combination model. These methods determine the relative importance of different indicators in a study or analysis. Some methods, such as exploratory factor analysis and rank survey data using discrete choice models, focus on statistical techniques for the weighting of indicators. Other methods, such as pairwise comparisons through an analytic network process and focus group discussion-based weights, involve subjective input from experts or stakeholders to determine the relative importance of different indicators. Survey-based and equal-weight systems are the most common methods used in BI studies to weigh indicators. Table 6 also provides an overview of the significant findings reported by the studies included in the review.

Table 6. Key indicators, weighting system, and findings of the selected studies.

Paper ID	No. of Indicators	Key Indicators	Weighting System for Indicators	Findings
1	10	Collisions involving bicycles; cyclist volume; nearest cycle path; nearest cyclable lane; intersections of cycle paths; intersections of cyclable lanes; intersections of cyclable paths and cyclable lanes; distance to biking stations; distance to bike racks; percent rise	Survey-based: Findings from the literature review	The proposed index helps show problematic areas. Predicting how often people will cycle. People living in places with more built environment features are more likely to ride.
2	10	Slope; junction density; traffic density; traffic speed; natural environment; built environment; centrality; activity coverage; accessibility to public transport stations; accessibility to bike-sharing stations	Analytic hierarchy process	Two-level hierarchy model. In Level 1, the road network is the most dominant factor. In Level 2, slope and junction density are the most critical factors. Accessibility to bike-sharing stations is the least essential factor.
3	5	Prevalence of neighborhood streets; street connectivity; biking facilities along main streets; green pathways; other cycling facilities	Expert survey-based weights	Biking facilities along main streets are the most crucial component of bikeability. In order of importance, the crucial indicators are street connectivity, the prevalence of neighborhood streets, and green pathways.
4	34	Connectivity No. of intersections with lights; No. of intersections without lights; No. of culs-de-sac Environment Slope; No. of POIs; Shannon land use mix index; air quality index; scenery—greenery; scenery—buildings; scenery—water Infrastructure Type of road; presence of potholes; presence of street lights; presence of bike lanes; No. of transit facilities; type of pavement; presence of street amenities; presence of utility poles; presence of bike parking; road width; presence of sidewalks; presence of crosswalks; presence of curb cuts Perception Attractiveness for cycling; spaciousness; cleanliness; building design attractiveness; safety as a cyclist; beauty; attractiveness for living Vehicle–Cyclist Interaction No. of vehicles; presence of on-street parking; presence of traffic lights/stop signs; No. of speed control devices	Equal weight system	Street view imagery (SVI) can be used to explain more than 65% of the spatiotemporal mobility pattern. The computer visiontechniques and SVI can be used to assess bikeability within and among cities.
5	4	Existence and type of bike path; speed limit; parking facilities for bicycles; quality of intersection infrastructure for bicycles	Survey-based weighting	Bicycle infrastructure is the most fundamental criterion, followed by the speed limit.
6	12	Leisure; transport; commercial; daily route; slope; sinuosity; bike route; greenery; crowdedness; outdoor enclosure; $PM_{2.5}$; BC	Weighted linear combination model	The inclusion of air quality makes a significant difference in calculating bikeability. Air quality, green spaces, and multiple land-use patterns should be improved in low-bikeability areas to enhance cycling mobility.
7	15	Bicycle lanes; separated paths; bicycle sharrows; protected bicycle lanes; bicycle signage; residential density; population density; ozone level; particulate matter; culs-de-sac; intersection density; highway density; distance to transit; parks; tree canopy coverage	Exploratory factor analysis	Environmental variables are not substantially correlated with recreation bicycling. The environmental variables are more significantly associated with bicycling used for transportation.

Table 6. *Cont.*

Paper ID	No. of Indicators	Key Indicators	Weighting System for Indicators	Findings
8	23	Presence of bicycle infrastructure; quality of bike path pavement; obstacles on bike paths; slope of bike paths; width of bike paths; presence of trees; aesthetics of buildings; presence of bicycle infrastructure; presence of traffic control devices; bus traffic flow; vehicle traffic flow; motorcycle traffic flow; pedestrian traffic flow; motorized transport; speed; presence of police officers; presence of security cameras; bike traffic flow; lightning; criminality on roads, directness and coherence; climate; cost of trip	Rank survey data; discrete choice models	Security is the most critical factor for frequent work and shopping cyclists. Bicycle infrastructure is the most crucial factor for sport cyclists. The slope of bike paths is one of the least essential components for comfort.
9	6	Conflicts with other modes of transport; mobility and urban road crossing; obstructions in mobility segments; safety in mobility; signaling and lighting of the bike lane; connection and distribution	Equal-weight system	The BI can identify disparity in situations along the bicycle lane.
10	25	Bikeway density; bikeway width; bikeway exclusiveness; bike parking space density; sidewalk width; sidewalk pavement; parking space for cars/scooters; arcade density; shoulder width, traffic volume; bus route; law enforcement; transit service; public bike service; public bike unavailability; tree shade; green space; air quality; slope; smooth traffic; conflictless traffic; night lighting; intersection density; bikeway ratio; mixed land use	Pairwise comparisons through analytic network process	Hilly terrain negatively affects bikeability. Intra-district biking travel could promote better satisfaction for bikers than inter-district biking travel. Bikeable districts contain large parks and good biking and pedestrian facilities.
11	5	Bicycle route density; bicycle route separation; connectivity of bicycle-friendly streets; topography; destination density	Focus group discussion-based weights	A significant positive correlation exists between the proportion of bicycle work trips and the bikeability score.
12	10	Outside lane width; bike lane width shoulder width; proportion of occupied on-street parking; vehicle traffic volume; vehicle speeds; percentage of heavy vehicles; pavement condition; presence of curbs; number of through lanes	Weighted as adjustment factors	Bikeability increased for the following three scenarios: (1) Adding new bike lanes to the community; (2) Adding new shared-use pathways; (3) Adding both new bike lanes and shared-use pathways.
13	5	Cycling infrastructure; presence of separated bicycle pathways; main roads without parallel bicycle lanes; green and aquatic areas; topography	Equal weight system	Regular cyclists live in more bicycle-friendly neighborhoods than non-cyclists. There is a positive relationship between the BI and cycling behavior. Cycling infrastructure, bicycle pathways, and green areas were positively related, and main roads and topography are negatively related to the used route.
14	4	Bike lanes; hills; destinations and road connectivity; bike commuting mode share	Unequal weight system	Census tracts with the highest bike scores (90 to 100) have mode shares 4.0 higher than the lowest bike score areas (0–25). Bike score correlates moderately with journey-to-work cycling mode share at the city level ($r = 0.52$) and the census tract level ($r = 0.35$).

Table 6. *Cont.*

Paper ID	No. of Indicators	Key Indicators	Weighting System for Indicators	Findings
15	13	Wind speed; slope; precipitation; temperature; sky view index; green view index; sinuosity; $PM_{2.5}$; average speed; public transport; commercial accessibility; number of trajectories; crowdedness	Principal component analysis	Elevated safety, accessibility, and vitality in areas result in higher bikeability scores. Traffic congestion, which lowers cycling speed and actual bikeability, is a potential downside of the higher vitality levels.

3.7. Important Variables Considered in the BIs

Usually, BIs comprise several variables that contribute to the overall score. The systematic review indicated that the essential criterion is bicycle infrastructure, such as bike lanes, routes, and cycle paths, as 14 developed BIs considered it. Only one BI did not consider the presence of bicycle lanes or paths because the BI was based on spatiotemporal bikeability using big data. Topography and trees or greenery along the bicycle path/lane were considered the second most crucial variables in calculating the BI, as mentioned in nine studies. The use of the presence of trees or green areas as an indicator underlines the importance of a pleasant and stimulating environment for cyclists. The city's topography or slope along bicycle paths significantly impacts cyclists' comfort, underscoring the importance of the physical effort needed for biking. Other essential components include traffic density on roads or at intersections (seven studies), vehicular traffic flow (seven studies), availability of street lights (six studies) and access to transit facilities (six studies). Five indicators were used in five BIs. These indicators are bicycle parking facilities; connectivity; traffic speed; safety and security; and density, such as population, residential, or arcade. Bicycle lane width, land use, conflicts, traffic control devices, and aesthetics of the buildings were used in four BIs.

Additionally, ten indicators, i.e., road width, the presence of sidewalks, road signage, pavement condition, parking facilities for vehicles, centrality, particulate matter, road signage, intersections, bike path density, and cyclist volume, were used in at least three BIs. Nine indicators were used at least twice. These indicators include crowdedness [70,73], culs-de-sac [34,69], curbs [22,69], and bicycle path obstacles [34,67]. Other less common indicators in BIs were only considered by one study. Some of these less frequently considered indicators include the ozone layer [34], utility poles [69], activities coverage [57], wind speed [73], and crimes [35].

After identifying the indicators used in developing the BIs, grouping them into five bicycle design principles was necessary. Indicators can be represented by one or more bicycle design principles. Based on the systematic review, all 181 indicators were narrowed down, since some were used with different names although measuring the same feature, such as grade and slope (see Appendix A). The indicators in the BIs were grouped into five bicycle infrastructure design principles. Figure 2 displays the indicators associated with each bicycle design principle.

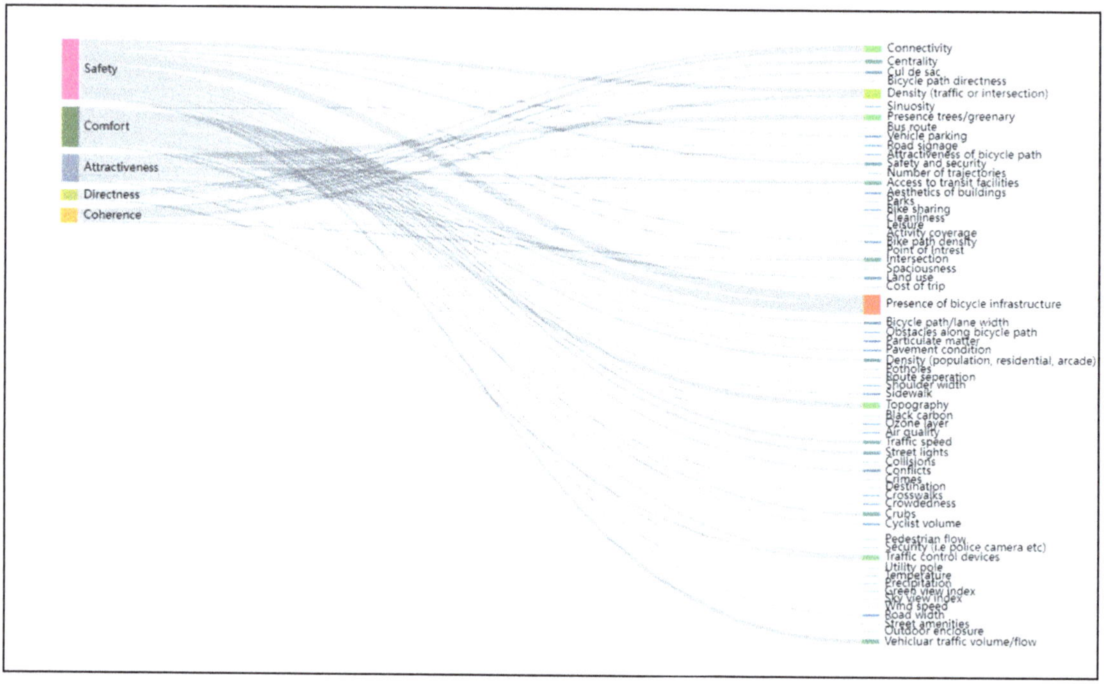

Figure 2. Relevant variables in bikeability design principles.

3.8. Bicycle Infrastructure Design Principles in the BIs

Bikeability assessments are generally based on the following five bicycle infrastructure design principles: safety, comfort, attractiveness, directness, and coherence [74]. These factors can be represented by a collection of components that are properties of each factor [35]. Table 7 shows the BIs developed by various researchers considered in this review. The table provides an overview of the BI studies considering at least one indicator from each of the five bicycle infrastructure design principles. It is evident from Table 7 that only three studies considered all the indicators from the design principles, while others only included a subset. For instance, one BI only considers safety, coherence, and comfort, while some consider all five principles [35,66].

Safety and comfort are the most commonly considered principles, directness and attractiveness are less commonly considered, and coherence is the least considered principle in the studies. Three BIs only considered the following two bicycle infrastructure design principles: safety and comfort [42,67,73]. Based on Table 7, it is clear that the bicycle infrastructure design principles considered in constructing a BI vary across different studies. However, for a comprehensive and holistic assessment of bikeability, it is recommended to consider all five principles, as each contributes to the creation of a safe and attractive cycling environment. Therefore, future studies should develop a BI incorporating all indicators for a more comprehensive bikeability assessment.

Table 7. Synthesis of BIs according to bicycle infrastructure design principles.

Paper ID	BI Index Categories				
	Safety	Comfort	Attractiveness	Directness	Coherence
1	✔	✔			✔
2	✔		✔		✔
3		✔	✔	✔	
4	✔	✔	✔	✔	✔
5	✔	✔			
6		✔	✔	✔	
7	✔	✔	✔	✔	
8	✔	✔	✔	✔	✔
9	✔	✔			
10	✔	✔	✔	✔	✔
11	✔	✔			✔
12	✔	✔			
13	✔	✔	✔		
14	✔	✔	✔		
15	✔	✔	✔	✔	

✔ (Considers at least one indicator).

4. Discussion

The authors comprehensively reviewed various methods and approaches developed for bikeability indices in urban environments. The review findings show that a scoring and weighting system was the most commonly used method for the assessment of urban bikeability. This method is most popular because it offers a systematic and easy-to-follow approach to constructing BIs [35,72]. This approach is utilized in similar research, i.e., BLOS and walkability index research [24]. However, the scoring and weighing system has critics [27]. For example, some BIs used an equal weight system, which is often criticized because the indicators do not affect the index equally [24]. To overcome this problem, studies have used questionnaire surveys to find the weight or importance of indicators [35,71].

For weighing indicators, a sample size that is significant enough is required to ensure the collected data accurately represent the population of interest [74]. Nonetheless, one study used a small sample size to construct an indices and surveyed only ten bicyclists [42]. A small sample size can lead to biased and mostly unreliable conclusions that may not represent the broader population [10]. The importance of variables in BIs based on a smaller size raises concerns about their generalizability, which limits their applicability.

Interestingly, only one study consulted only experts to weigh bikeability indicators, which may suggest a lack of expert involvement [57]. Asadi-Shekari et al. (2019) stated that expert surveys can help researchers ensure that the most important indicators are considered [20]. However, there are also potential drawbacks to relying solely on expert surveys. Experts may have biases that can influence their perceptions, and their opinions may not necessarily reflect the preferences and needs of the users. Therefore, a balanced approach that combines expert surveys with other methods, such as community surveys, can help overcome these limitations [65]. Ahmed et al. (2021) suggested using the mixed approach and argued that this could provide valuable insights in selecting effective indicators [24].

The BIs used various methodologies to collect data, mainly conducted through field surveys or reliant on data provided by government departments or other secondary sources [34,35,69]. This approach is usually time-consuming and requires human and financial resources, while the data may be outdated due to recent developments if relying on secondary sources. Recent BIs have utilized emerging technologies and data sources, including remote sensing images, virtual auditing through SVI, and crowdsourcing, for data collection [36,71–73]. This approach can be more standardized and scalable but comes with technical difficulties in the implementation stage. In addition, remotely sensed imagery cannot capture micro-scale street-level information.

Another crucial finding is that the unit of analysis for the development of the indices varied across studies. Some BIs focused on street segments or bicycle lanes, while others considered intersections, zones within the city, or even the entire city. This variability in the unit of analysis is essential to consider, as it can impact the accuracy of the BI. For example, studies analyzing bikeability at a very granular level, such as 10 m or 100 × 100 m, can provide a more detailed and accurate bikeability assessment for a segment or area [66,72]. In contrast, studies focusing on the city level may not capture the nuances of bikeability in different neighborhoods or streets. Deciding on the BI for a specific context is essential; if street-level bikeability is needed, the BI method developed by Arellana et al. (2020) is very appropriate [35]; however, if macro-scale bikeability of the city is required, the methods developed by Codina et al. and Lin and Wei are beneficial [36,66].

The systematic review results show a lack of uniformity in the number and types of indicators considered for the BIs. The number of indicators considered varied significantly between studies, ranging from 4 [38,68] to 34 [69]. This disparity can make comparing bikeability indicators and scores across different cities challenging, as the indicators' definitions and metrics differ widely. However, the most commonly considered indicators are bicycle infrastructure, greenery along bicycle paths, slopes, vehicular traffic flow/volume, street lights, bicycle path connectivity, and traffic speed. These indicators are related to the sense of comfort and safety along the bicycle pathways, which, when offered, results in a preference of people to choose bicycles over other modes of transport [75]. Past research shows that other indicators, along with cycle infrastructure, such as pavement conditions, road markings, traffic control devices, and crosswalks, play a significant role in getting people to ride bikes [7,76]. However, few studies have considered these essential indicators. Therefore, these indicators need more attention while measuring bikeability, as they can significantly affect rider experience and safety. Furthermore, this review highlights the importance of selecting appropriate indicators for the local context. For instance, indicators like the ozone layer and particulate matter may not be relevant for all cities. In contrast, variables such as motorcycle flow or the presence of police officers may be more crucial for some cities, like cities in the global south [35]. Therefore, it is essential to consider the local context while selecting indicators for BIs.

Studies have developed BIs to assess the quality of cycling infrastructure, but no consensus exists as to which indicators should be included [10,21]. Moreover, the review results suggest a lack of consideration of all five bicycle infrastructure design principles—safety, comfort, attractiveness, directness, and coherence—in developing existing indices. However, these five elements of cycling infrastructure design are universally agreed to promote bicycling [11,53]. Still, the existing indices focus on only a subset of these principles, with safety and comfort being the most commonly considered ones. Zhao et al. (2018) conducted a study in Beijing and Copenhagen to adapt bicycle design solutions and recognized the significance of all five principles [53]. They found that good bicycle infrastructure design always encourages people to cycle.

In this systematic review paper, we found three studies that considered all five design principles of bicycle infrastructure. However, it should be noted that one study was conducted in the global south, and it may not be directly applicable to other regions [35]. The other two are complicated methods and require technical knowledge in their applicability [36,69]. Thus, there is a need to develop a new, easy-to-follow BI that incorporates all indicators from the five design principles for a more comprehensive assessment of bikeability. This new index would provide a more accurate picture of the quality of cycling infrastructure, helping policymakers and urban planners prioritize investments that lead to safer and more attractive cycling environments.

5. Conclusions and Recommendations

Worldwide, the use of micromobility vehicles is significantly increasing in cities. Bicycling is a sustainable micromobility mode. In the past decade, there has been a rising interest in bikeability-related studies, reflecting a growing awareness of sustainable mobility.

Through this systematic review, we wanted to identify the essential indicators covered in bikeability studies. The result indicates that bicycle infrastructure is the most commonly considered indicator in bikeability assessment methods, underscoring its critical role in promoting cycling as a viable and preferred mode of transportation. It is followed by indicators such as trees or greenery along the bicycle path/lane and bicycle comfort factors like slope. Other crucial bikeability indicators include bicycle parking facilities, bicycle path connectivity, vehicular traffic volume, traffic speed, intersection density, and road signage. The critical indicators identified in this review will help urban planners and policymakers to plan well-designed bicycle infrastructure. This will facilitate safer and more efficient travel for cyclists and reduce congestion and pollution from motor vehicles, aligning with broader environmental and public health goals.

BIs are a vital tool in assessing the friendliness of urban settings towards cyclists, encompassing various levels, i.e., street segments to city-level assessment. However, a few issues were identified, such as some BIs using a small sample size for weighing of the index indicators. The second issue found was that there is no consensus on the number of indicators used in BIs. Studies have used from 4 to 34 indicators. Another critical research question was whether urban BIs consider bicycle infrastructure design principles. Since governments and researchers agree that five bicycle infrastructure design principles should be considered to make bicycles an attractive mode for medium and short trips in urban areas, we categorized the indicators used in the studies into safety, comfort, attractiveness, directness, and coherence. The results suggested that the safety and comfort components of bicycle infrastructure were the most commonly considered principles, while coherence was the least considered. However, for a comprehensive and holistic assessment of bikeability, it is recommended to consider all five design principles. Each principle contributes to the creation of a safe, comfortable, and attractive cycling environment, which is crucial for the promotion of cycling and the improvement of infrastructure.

The findings of this literature review emphasize the importance of accurately analyzing bikeability to encourage cycling and enhance infrastructure. This review also highlights the need for a comprehensive and easy-to-follow approach that considers all design principles and emphasizes the importance of a sufficient sample size in data collection. Future studies should aim to develop a BI incorporating all indicators for a more comprehensive assessment and understanding of bikeability in urban environments. One general limitation of this review is that in assessing bicycle infrastructure design principles in BIs, we considered if the BI considered at least one indicator for each bicycle design principle. Future studies should delve deeper, incorporating a broader range of variables to evaluate BIs effectiveness comprehensively.

Author Contributions: Conceptualization, T.A., A.P., D.J. and G.W.; methodology, T.A., A.P., D.J. and G.W.; software, T.A.; validation, A.P. and D.J.; formal analysis, T.A. and A.P.; investigation, T.A.; resources, T.A., A.P., D.J. and G.W.; data curation, T.A., A.P. and D.J.; writing—original draft preparation, T.A. and A.P.; writing—review and editing, T.A., A.P. and D.J.; visualization, T.A., A.P. and D.J.; supervision, A.P. and D.J. All authors have read and agreed to the published version of the manuscript.

Funding: This research received no external funding.

Institutional Review Board Statement: Not applicable.

Informed Consent Statement: Not applicable.

Data Availability Statement: The datasets are available from the corresponding author upon request.

Acknowledgments: We thank the Higher Education Commission (HEC), Pakistan, for funding the Ph.D. research of Tufail Ahmed. We acknowledge Alam Zeb for helping with the PRISMA guidelines for the systematic literature review. We would also like to acknowledge the support and constructive feedback provided by Imran Nawaz.

Conflicts of Interest: The authors declare no conflicts of interest.

Appendix A. Selected Indicators and Their Frequency in the Bis

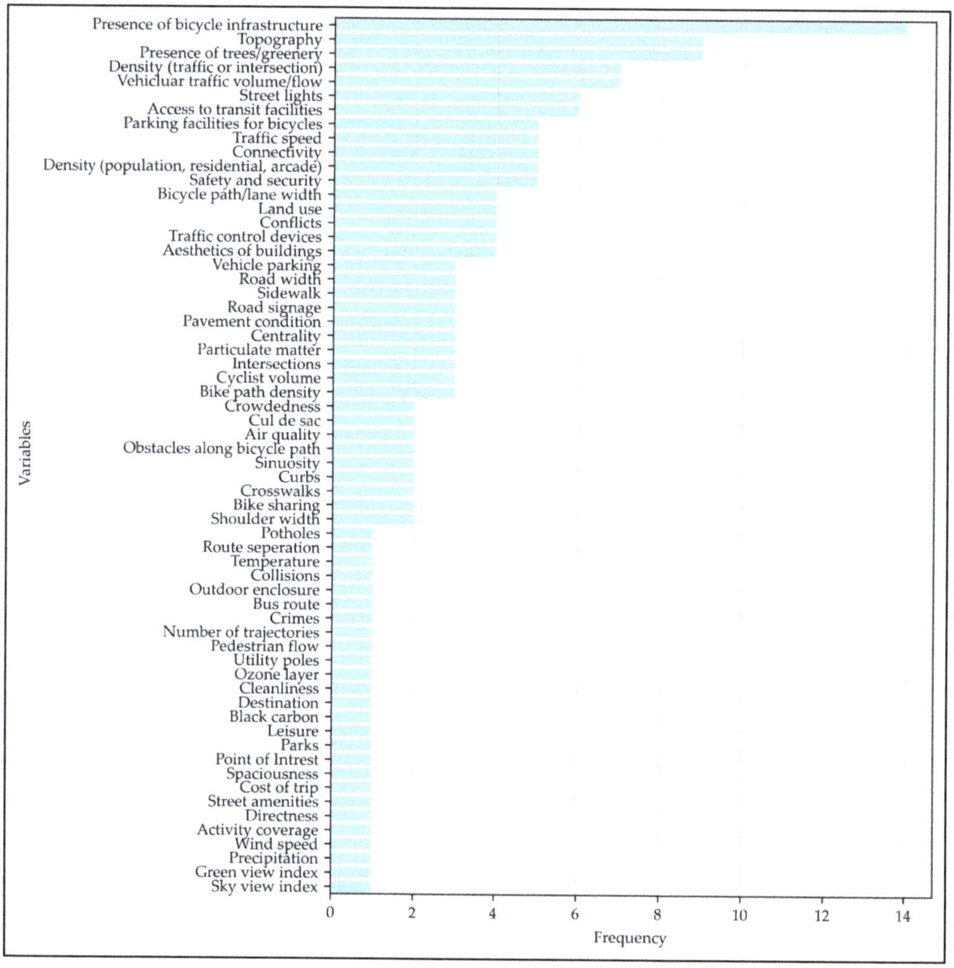

References

1. Kazemzadeh, K.; Laureshyn, A.; Hiselius, L.W.; Ronchi, E. Expanding the Scope of the Bicycle Level-of-Service Concept: A Review of the Literature. *Sustainability* **2020**, *12*, 2944. [CrossRef]
2. Ahmed, T.; Pirdavani, A.; Janssens, D.; Wets, G. Utilizing Intelligent Portable Bicycle Lights to Assess Urban Bicycle Infrastructure Surfaces. *Sustainability* **2023**, *15*, 4495. [CrossRef]
3. Bourne, J.E.; Cooper, A.R.; Kelly, P.; Kinnear, F.J.; England, C.; Leary, S.; Page, A. The Impact of e-Cycling on Travel Behaviour: A Scoping Review. *J. Transp. Health* **2020**, *19*, 100910. [CrossRef]
4. Li, Z.; Wang, W.; Liu, P.; Schneider, R.; Ragland, D.R. Investigating Bicyclists' Perception of Comfort on Physically Separated Bicycle Paths in Nanjing, China. *Transp. Res. Rec.* **2012**, *2317*, 76–84. [CrossRef]
5. Manum, B.; Nordström, T.; Arnesen, P.; Cooper, C.; Gil, J.; Dahl, E.; Chan, R.; Rokseth, L.; Green, S. Using Realistic Travel-Time Thresholds in Accessibility Measures of Bicycle Route Networks. In Proceedings of the 12th International Space Syntax Symposium, Beijing, China, 8–13 July 2019; pp. 238.1–238.14.
6. Castañon, U.N.; Ribeiro, P.J.G. Bikeability and Emerging Phenomena in Cycling: Exploratory Analysis and Review. *Sustainability* **2021**, *13*, 2394. [CrossRef]
7. Asadi-Shekari, Z.; Moeinaddini, M.; Zaly Shah, M. A Bicycle Safety Index for Evaluating Urban Street Facilities. *Traffic Inj. Prev.* **2015**, *16*, 283–288. [CrossRef] [PubMed]

8. Li, W.; Kamargianni, M. Providing Quantified Evidence to Policy Makers for Promoting Bike-Sharing in Heavily Air-Polluted Cities: A Mode Choice Model and Policy Simulation for Taiyuan-China. *Transp. Res. Part A Policy Pract.* **2018**, *111*, 277–291. [CrossRef]
9. Della Mura, M.; Failla, S.; Gori, N.; Micucci, A.; Paganelli, F. E-Scooter Presence in Urban Areas: Are Consistent Rules, Paying Attention and Smooth Infrastructure Enough for Safety? *Sustainability* **2022**, *14*, 14303. [CrossRef]
10. Muhs, C.D.; Clifton, K.J. Do Characteristics of Walkable Environments Support Bicycling? Toward a Definition of Bicycle-Supported Development. *J. Transp. Land Use* **2016**, *9*, 147–188. [CrossRef]
11. Hull, A.; Holleran, C.O. Bicycle Infrastructure: Can Good Design Encourage Cycling? *Urban Plan. Transp. Res.* **2014**, *2*, 369–406. [CrossRef]
12. Cohen, E. Segregated Bike Lanes Are Safest for Cyclists. *CMAJ* **2013**, *185*, E443–E444. [CrossRef]
13. Cicchino, J.B.; McCarthy, M.L.; Newgard, C.D.; Wall, S.P.; DiMaggio, C.J.; Kulie, P.E.; Arnold, B.N.; Zuby, D.S. Not All Protected Bike Lanes Are the Same: Infrastructure and Risk of Cyclist Collisions and Falls Leading to Emergency Department Visits in Three US Cities. *Accid. Anal. Prev.* **2020**, *141*, 105490. [CrossRef]
14. Kellstedt, D.K.; Spengler, J.O.; Maddock, J.E. Comparing Perceived and Objective Measures of Bikeability on a University Campus: A Case Study. *SAGE Open* **2021**, *11*, 1–10. [CrossRef]
15. Heinen, E.; Buehler, R. Bicycle Parking: A Systematic Review of Scientific Literature on Parking Behaviour, Parking Preferences, and Their Influence on Cycling and Travel Behaviour. *Transp. Rev.* **2019**, *39*, 630–656. [CrossRef]
16. Asadi-Shekari, Z.; Moeinaddini, M.; Zaly Shah, M. Non-Motorised Level of Service: Addressing Challenges in Pedestrian and Bicycle Level of Service. *Transp. Rev.* **2013**, *33*, 166–194. [CrossRef]
17. Babić, D.; Babić, D.; Cajner, H.; Sruk, A.; Fiolić, M. Effect of Road Markings and Traffic Signs Presence on Young Driver Stress Level, Eye Movement and Behaviour in Night-Time Conditions: A Driving Simulator Study. *Safety* **2020**, *6*, 24. [CrossRef]
18. McNeil, N.; Monsere, C.M.; Dill, J. Influence of Bike Lane Buffer Types on Perceived Comfort and Safety of Bicyclists and Potential Bicyclists. *Transp. Res. Rec.* **2015**, *2520*, 132–142. [CrossRef]
19. Nuñez, J.Y.M.; Bisconsini, D.R.; Rodrigues da Silva, A.N. Combining Environmental Quality Assessment of Bicycle Infrastructures with Vertical Acceleration Measurements. *Transp. Res. Part A Policy Pract.* **2020**, *137*, 447–458. [CrossRef]
20. Asadi-Shekari, Z.; Moeinaddini, M.; Aghaabbasi, M.; Cools, M.; Zaly Shah, M. Exploring Effective Micro-Level Items for Evaluating Inclusive Walking Facilities on Urban Streets (Applied in Johor Bahru, Malaysia). *Sustain. Cities Soc.* **2019**, *49*, 101563. [CrossRef]
21. Kellstedt, D.K.; Spengler, J.O.; Foster, M.; Lee, C.; Maddock, J.E. A Scoping Review of Bikeability Assessment Methods. *J. Community Health* **2021**, *46*, 211–224. [CrossRef]
22. Lowry, M.; Callister, D.; Gresham, M.; Moore, B. Assessment of Communitywide Bikeability with Bicycle Level of Service. *Transp. Res. Rec.* **2012**, *2314*, 41–48. [CrossRef]
23. Pezzagno, M.; Tira, M. *Town and Infrastructure Planning for Safety and Urban Quality*, 1st ed.; Engineering & Technology, Geography, Politics & International Relations, Urban Studies; CRC Press: London, UK, 2018; ISBN 978-1-351-17336-0.
24. Ahmed, T.; Moeinaddini, M.; Almoshaogeh, M.; Jamal, A.; Nawaz, I.; Alharbi, F. A New Pedestrian Crossing Level of Service (Pclos) Method for Promoting Safe Pedestrian Crossing in Urban Areas. *Int. J. Environ. Res. Public Health* **2021**, *18*, 8813. [CrossRef]
25. Davis, J. *Bicycle Safety Evaluation*; Auburn University, City of Chattanooga, and Chattanooga-Hamilton County Regional Planning Commission: Chattanooga, TN, USA, 1987.
26. Vallejo-Borda, J.A.; Rosas-Satizábal, D.; Rodriguez-Valencia, A. Do Attitudes and Perceptions Help to Explain Cycling Infrastructure Quality of Service? *Transp. Res. Part D Transp. Environ.* **2020**, *87*, 102539. [CrossRef]
27. Asadi-Shekari, Z.; Moeinaddini, M.; Zaly Shah, M. A Pedestrian Level of Service Method for Evaluating and Promoting Walking Facilities on Campus Streets. *Land Use Policy* **2014**, *38*, 175–193. [CrossRef]
28. Calvey, J.C.; Shackleton, J.P.; Taylor, M.D.; Llewellyn, R. Engineering Condition Assessment of Cycling Infrastructure: Cyclists' Perceptions of Satisfaction and Comfort. *Transp. Res. Part A Policy Pract.* **2015**, *78*, 134–143. [CrossRef]
29. Gao, J.; Sha, A.; Huang, Y.; Hu, L.; Tong, Z.; Jiang, W. Evaluating the Cycling Comfort on Urban Roads Based on Cyclists' Perception of Vibration. *J. Clean. Prod.* **2018**, *192*, 531–541. [CrossRef]
30. Micucci, A.; Sangermano, M. A Study on Cyclists Behaviour and Bicycles Kinematic. *Int. J. Transp. Dev. Integr.* **2020**, *4*, 14–28. [CrossRef]
31. Beura, S.K.; Chellapilla, H.; Panda, M.; Bhuyan, P.K. Bicycle Comfort Level Rating (BCLR) Model for Urban Street Segments in Mid-Sized Cities of India. *J. Transp. Health* **2021**, *20*, 100971. [CrossRef]
32. Bai, L.; Liu, P.; Chan, C.-Y.; Li, Z. Estimating Level of Service of Mid-Block Bicycle Lanes Considering Mixed Traffic Flow. *Transp. Res. Part A Policy Pract.* **2017**, *101*, 203–217. [CrossRef]
33. Kamel, M.B.; Sayed, T.; Bigazzi, A. A Composite Zonal Index for Biking Attractiveness and Safety. *Accid. Anal. Prev.* **2020**, *137*, 105439. [CrossRef] [PubMed]
34. Porter, A.K.; Kohl, H.W.; Pérez, A.; Reininger, B.; Pettee Gabriel, K.; Salvo, D. Bikeability: Assessing the Objectively Measured Environment in Relation to Recreation and Transportation Bicycling. *Environ. Behav.* **2020**, *52*, 861–894. [CrossRef]
35. Arellana, J.; Saltarín, M.; Larrañaga, A.M.; González, V.I.; Henao, C.A. Developing an Urban Bikeability Index for Different Types of Cyclists as a Tool to Prioritise Bicycle Infrastructure Investments. *Transp. Res. Part A Policy Pract.* **2020**, *139*, 310–334. [CrossRef]

36. Lin, J.J.; Wei, Y.H. Assessing Area-Wide Bikeability: A Grey Analytic Network Process. *Transp. Res. Part A Policy Pract.* **2018**, *113*, 381–396. [CrossRef]
37. Cain, K.L.; Geremia, C.M.; Conway, T.L.; Frank, L.D.; Chapman, J.E.; Fox, E.H.; Timperio, A.; Veitch, J.; Van Dyck, D.; Verhoeven, H.; et al. Development and Reliability of a Streetscape Observation Instrument for International Use: MAPS-Global. *Int. J. Behav. Nutr. Phys. Act.* **2018**, *15*, 19. [CrossRef] [PubMed]
38. Winters, M.; Teschke, K.; Brauer, M.; Fuller, D. Bike Score®: Associations between Urban Bikeability and Cycling Behavior in 24 Cities. *Int. J. Behav. Nutr. Phys. Act.* **2016**, *13*, 18. [CrossRef]
39. Grigore, E.; Garrick, N.; Fuhrer, R.; Axhausen, I.K.W. Bikeability in Basel. *Transp. Res. Rec.* **2019**, *2673*, 607–617. [CrossRef]
40. Tijana, Đ.; Tomić, N.; Tešić, D. Walkability and Bikeability for Sustainable Spatial Planning in the City of Novi Sad (Serbia). *Sustainability* **2023**, *15*, 3785. [CrossRef]
41. Reggiani, G.; van Oijen, T.; Hamedmoghadam, H.; Daamen, W.; Vu, H.L.; Hoogendoorn, S. Understanding Bikeability: A Methodology to Assess Urban Networks. *Transportation* **2022**, *49*, 897–925. [CrossRef]
42. Schmid-Querg, J.; Keler, A.; Grigoropoulos, G. The Munich Bikeability Index: A Practical Approach for Measuring Urban Bikeability. *Sustainability* **2021**, *13*, 428. [CrossRef]
43. Shaaban, K. Why Don't People Ride Bicycles in High-Income Developing Countries, and Can Bike-Sharing Be the Solution? The Case of Qatar. *Sustainability* **2020**, *12*, 1693. [CrossRef]
44. Buehler, R.; Pucher, J. COVID-19 Impacts on Cycling, 2019–2020. *Transp. Rev.* **2021**, *41*, 393–400. [CrossRef]
45. Ahmed, T.; Pirdavani, A.; Wets, G.; Janssens, D. Evaluating Cyclist Ride Quality on Different Bicycle Streets. *Transp. Res. Procedia* **2024**, *78*, 586–593. [CrossRef]
46. Østergaard, L.; Jensen, M.K.; Overvad, K.; Tjønneland, A.; Grøntved, A. Associations between Changes in Cycling and All-Cause Mortality Risk. *Am. J. Prev. Med.* **2018**, *55*, 615–623. [CrossRef]
47. Qiu, L.-Y.; He, L.-Y. Bike Sharing and the Economy, the Environment, and Health-Related Externalities. *Sustainability* **2018**, *10*, 1145. [CrossRef]
48. Pan, H.X.; Tang, S.; Mai, X.M.; Mou, Y.-J. Overview of Bicycle Transportation Development in Urban Areas. *Urban Transp. China* **2010**, *8*, 40–43.
49. Weikl, S.; Mayer, P. Data-Driven Quality Assessment of Cycling Networks. *Front. Future Transp.* **2023**, *4*, 1127742. [CrossRef]
50. Munir, T.; Dia, H.; Ghaderi, H. A Systematic Review of the Role of Road Network Pricing in Shaping Sustainable Cities: Lessons Learned and Opportunities for a Post-Pandemic World. *Sustainability* **2021**, *13*, 12048. [CrossRef]
51. NSW Government. *Cycleway Design Toolbox Designing for Cycling and Micromobility*; NSW Government: Sydney, NSW, Australia, 2020; p. 96.
52. Transport Scotland. *Cycling by Design 2010*; Transport Scotland: Glasgow, UK, 2011.
53. Zhao, C.; Carstensen, T.A.; Nielsen, T.A.S.; Olafsson, A.S. Bicycle-Friendly Infrastructure Planning in Beijing and Copenhagen—Between Adapting Design Solutions and Learning Local Planning Cultures. *J. Transp. Geogr.* **2018**, *68*, 149–159. [CrossRef]
54. Hong, J.; Philip McArthur, D.; Stewart, J.L. Can Providing Safe Cycling Infrastructure Encourage People to Cycle More When It Rains? The Use of Crowdsourced Cycling Data (Strava). *Transp. Res. Part A Policy Pract.* **2020**, *133*, 109–121. [CrossRef]
55. Sustrans. *Handbook for Cyclefriendly Design*; Sustrans: Bristol, UK, 2004.
56. UK Department for Transport. *Cycle Infrastructure Design Cycle Infrastructure Design*; UK Department for Transport: Norwich, UK, 2020.
57. Karolemeas, C.; Vassi, A.; Tsigdinos, S.; Bakogiannis, D.E. Measure the Ability of Cities to Be Biked via Weighted Parameters, Using GIS Tools. The Case Study of Zografou in Greece. *Transp. Res. Procedia* **2022**, *62*, 59–66. [CrossRef]
58. Manum, B.; Nordström, T.; Gil, J.; Nilsson, L.; Marcus, L. Modelling Bikeability: Space Syntax Based Measures Applied in Examining Speeds and Flows of Bicycling in Gothenburg. In Proceedings of the SSS 2017—11th International Space Syntax Symposium, Lisbon, Portugal, 3–7 July 2017; pp. 89.1–89.16.
59. Gan, Z.; Yang, M.; Zeng, Q.; Timmermans, H.J.P. Associations between Built Environment, Perceived Walkability/Bikeability and Metro Transfer Patterns. *Transp. Res. Part A Policy Pract.* **2021**, *153*, 171–187. [CrossRef]
60. Wahlgren, L.; Schantz, P. Exploring Bikeability in a Suburban Metropolitan Area Using the Active Commuting Route Environment Scale (ACRES). *Int. J. Environ. Res. Public Health* **2014**, *11*, 8276–8300. [CrossRef] [PubMed]
61. Standen, C.; Crane, M.; Collins, A.; Greaves, S.; Rissel, C. Determinants of Mode and Route Change Following the Opening of a New Cycleway in Sydney, Australia. *J. Transp. Health* **2017**, *4*, 255–266. [CrossRef]
62. Nordström, T.; Manum, B. Measuring Bikeability: Space Syntax Based Methods Applied in Planning for Improved Conditions for Bicycling in Oslo. In Proceedings of the SSS 2015—10th International Space Syntax Symposium, London, UK, 13–17 July 2015; pp. 1–14.
63. Pickering, C.; Byrne, J. The Benefits of Publishing Systematic Quantitative Literature Reviews for PhD Candidates and Other Early-Career Researchers. *High. Educ. Res. Dev.* **2014**, *33*, 534–548. [CrossRef]
64. Raad, N.; Burke, M.I. What Are the Most Important Factors for Pedestrian Level-of-Service Estimation? A Systematic Review of the Literature. *Transp. Res. Rec.* **2018**, *2672*, 101–117. [CrossRef]
65. McKeown, S.; Mir, Z.M. Considerations for Conducting Systematic Reviews: Evaluating the Performance of Different Methods for De-Duplicating References. *Syst. Rev.* **2021**, *10*, 38. [CrossRef]

66. Codina, O.; Maciejewska, M.; Nadal, J.; Marquet, O. Built Environment Bikeability as a Predictor of Cycling Frequency: Lessons from Barcelona. *Transp. Res. Interdiscip. Perspect.* **2022**, *16*, 100725. [CrossRef]
67. Ros-McDonnell, L.; de-la-Fuente-Aragon, M.V.; Ros-McDonnell, D.; Carboneras, M.C. Development of a Biking Index for Measuring Mediterranean Cities Mobility. *Int. J. Prod. Manag. Eng.* **2020**, *8*, 21–29. [CrossRef]
68. Hardinghaus, M.; Nieland, S.; Lehne, M.; Weschke, J. More Than Bike Lanes—A Multifactorial Index of Urban Bikeability. *Sustainability* **2021**, *13*, 11584. [CrossRef]
69. Ito, K.; Biljecki, F. Assessing Bikeability with Street View Imagery and Computer Vision. *Transp. Res. Part C Emerg. Technol.* **2021**, *132*, 103371. [CrossRef]
70. Tran, P.T.M.; Zhao, M.; Yamamoto, K.; Minet, L.; Nguyen, T.; Balasubramanian, R. Cyclists' Personal Exposure to Traffic-Related Air Pollution and Its Influence on Bikeability. *Transp. Res. Part D Transp. Environ.* **2020**, *88*, 102563. [CrossRef]
71. Winters, M.; Brauer, M.; Setton, E.M.; Teschke, K. Mapping Bikeability: A Spatial Tool to Support Sustainable Travel. *Environ. Plan. B Plan. Des.* **2013**, *40*, 865–883. [CrossRef]
72. Krenn, P.J.; Oja, P.; Titze, S. Development of a Bikeability Index to Assess the Bicycle-Friendliness of Urban Environments. *Open J. Civ. Eng.* **2015**, *5*, 451–459. [CrossRef]
73. Dai, S.; Zhao, W.; Wang, Y.; Huang, X.; Chen, Z.; Lei, J.; Stein, A.; Jia, P. Assessing Spatiotemporal Bikeability Using Multi-Source Geospatial Big Data: A Case Study of Xiamen, China. *Int. J. Appl. Earth Obs. Geoinf.* **2023**, *125*, 103539. [CrossRef]
74. Bach, B.; van Hal, E.; de Jong, M.I.; de Jong, T.M. Urban Design and Traffic; a Selection form Bach's Toolbox. Stedenbouw en Verkeer; een Selectie uit de Gereedschapskist van Bach. CROW. 2006. Available online: https://research.tudelft.nl/en/publications/urban-design-and-traffic-a-selection-form-bachs-toolbox-stedenbou (accessed on 25 January 2024).
75. Paydar, M.; Kamani Fard, A. The Contribution of Mobile Apps to the Improvement of Walking/Cycling Behavior Considering the Impacts of COVID-19 Pandemic. *Sustainability* **2021**, *13*, 10580. [CrossRef]
76. La Paix, L.; Cherchi, E.; Geurs, K. Role of Perception of Bicycle Infrastructure on the Choice of the Bicycle as a Train Feeder Mode. *Int. J. Sustain. Transp.* **2021**, *15*, 486–499. [CrossRef]

sustainability

MDPI

Review

A Review of Wireless Pavement System Based on the Inductive Power Transfer in Electric Vehicles

Bozhi [1], Mahmoud Mohamed [2,*], Vahid Najafi Moghaddam Gilani [3,*], Ayesha Amjad [4,5], Mohammed Sh. Majid [6], Khalid Yahya [7] and Mohamed Salem [8]

1 Hunan Engineering Research Center for Intelligent Operation and Maintenance of Elevators, Hunan Electrical College of Technology, Xiangtan 411101, China; bozhi_institute_hn@163.com
2 School of Engineering, Cardiff University, Cardiff CF24 3AA, UK
3 Faculty of Medicine and Health Sciences, Université de Sherbrooke, Longueuil, QC J4K 0A8, Canada
4 Faculty of Organization and Management, Silesian University of Technology, 44-100 Gliwice, Poland; ayesha.amjad@polsl.pl
5 Centre for Mechanical Engineering, Materials and Processes (CEMMPRE), University of Coimbra, Polo II, 3030-788 Coimbra, Portugal
6 Computer Techniques Engineering Department, College of Engineering and Technologies, Al-Mustaqbal University, Babil 51001, Iraq; mohammed.shawkat@uomus.edu.iq
7 Department of Electrical and Electronics Engineering, Nisantasi University, Istanbul 34467, Turkey; Khalid.omy@gmail.com
8 School of Electrical and Electronic Engineering, Universiti Sains Malaysia (USM), Nibong Tebal 14300, Penang, Malaysia; salemm@usm.my
* Correspondence: mohamedmt@cardiff.ac.uk (M.M.); vahid.najafi.moghaddam.gilani@usherbrooke.ca (V.N.M.G.)

Abstract: The proliferation of electric vehicles (EVs) hinges upon the availability of robust and efficient charging infrastructure, notably encompassing swift and convenient solutions. Among these, dynamic wireless charging systems have garnered substantial attention for their potential to revolutionize EV charging experiences. Inductive power transfer (IPT) systems, in particular, exhibit a promising avenue, enabling seamless wireless charging through integrated pavements for EVs. This review engages in an in-depth exploration of pertinent parameters that influence the inductivity and conductivity performance of pavements, alongside the assessment of potential damage inflicted by IPT pads. Moreover, the study delves into the realm of additive materials as a strategic approach to augment conductivity and pavement performance. In essence, the review consolidates a diverse array of studies that scrutinize IPT pad materials, coil dimensions, pavement characteristics (both static and dynamic), and adhesive properties. These studies collectively illuminate the intricate dynamics of power transfer to EVs while considering potential repercussions on pavement integrity. Furthermore, the review sheds light on the efficacy of various additive materials, including metal and nanocomposite additives with an SBS base, in amplifying both conductivity and pavement performance. The culmination of these findings underscores the pivotal role of geometry optimization for IPT pads and the strategic adaptation of aggregate and bitumen characteristics to unlock enhanced performance within wireless pavements.

Keywords: electric vehicle; pavement; inductive power transfer; asphalt mixture; concrete

1. Introduction

With the development of urban infrastructure and immigration from rural to urban areas, the usage rate of various transportation modes, particularly passenger cars, has grown [1,2]. This issue causes the world to encounter the main problems, such as air pollution and CO_2 emission [3–5]. Therefore, the necessary infrastructures for electric vehicles (EVs) should be developed [6,7]. One of these critical infrastructures is the charging system of EVs. Researchers worldwide are trying to perform fast charging systems to reduce the time spent on EVs from 8 h up to 35 min [8,9].

In recent years, several studies have been performed on wireless power transfer (WPT) systems to provide sustainable infrastructure for EVs. These systems help a lot to charge EVs faster, leading to a reduction in battery size and EV weight and a decrease in the intensity amount of loading on road pavement. WPT can provide an impressive perspective for the sustainable development of EVs by transferring power in an electromagnetic area [10]. These systems were designed to be both stationary and dynamic [11]. IPT systems are one of the safest and the most efficient of WPT [12]. These systems, which are embedded within the road pavements, have the ability to transfer power in an air gap of 10 to 20 cm through primary coils to the receiving coils with an efficient level of 83 to 92% in EVs.

Therefore, according to the flowchart of Figure 1, in this study, we aimed to comprehensively investigate the impact of IPT pad embedment on pavement damage and explore the role of various pavement layers in facilitating efficient power transfer from the pavement to EVs. Our method involves a thorough analysis of the effects of wireless systems on pavement layers, including their response to repetitive loading and thermal fluctuations, and how these factors influence power transfer efficiency. Additionally, we will carefully examine the damage caused by the embedment of charging units on the pavement surface. Based on our findings, we will propose different methods to enhance pavement properties, ensuring the overall quality and performance of the IPT system.

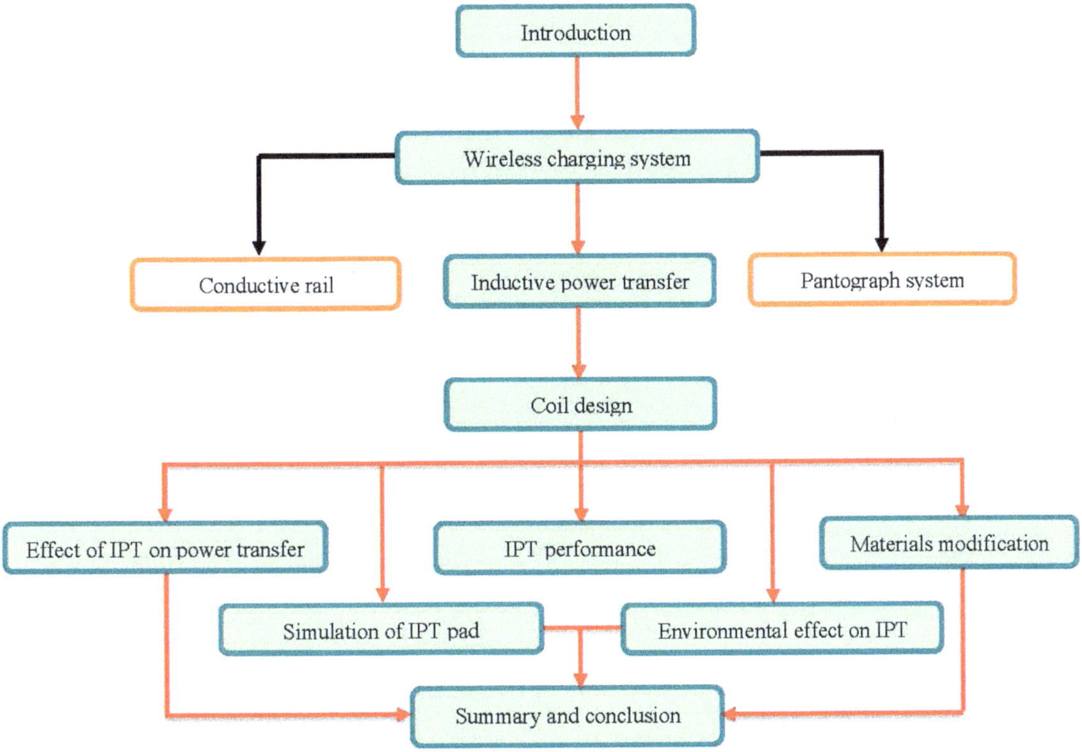

Figure 1. The structure of the present study.

2. Wireless Charging System

The charging of EVs can be operated by various methods. One of these methods is the pantograph shown in Figure 2. The power transfer in this system is fulfilled by physical contact with several weaknesses, such as charging problems in horizontal curves, creating sparks, decreasing the safety of EVs, and also causing traffic congestion due to the low level of speed while they are in charge. Another method for EV charging is a conductive rail, which can be seen in Figure 3. This method has similar problems to the pantograph. However, the third method is the wireless charging system, which has many positive aspects and provides energy without a physical connection [13].

The wireless charging system or WPT is accounted as an appropriate solution to solve EV charging problems [14]. This system provides a safe and efficient condition for EVs and also eliminates the anxiety of lack of charging and time loss of fixed charging stations worked via a plug-in [15]. As can be seen in Figure 4, in this system, unlike the plug-in charging systems, power is transferred by creating an electromagnetic condition using the embedded coils (primary coil) in the pavement to the receiver coils (secondary coil) installed under the electric vehicles [16]. There is an air gap between the embedded coils and receiver coils, which plays an essential role in transferring the amount of power. Fortunately, this system does not require human interference, and it is provided in two forms, stationery and dynamic, which can be described as follows.

In terms of configuration, WPT can be classified as either (1) stationary WPT: charge while the vehicle is not in motion; or (2) dynamic WPT: charge while the vehicle is moving along the roadway [10]. The difference between WPT systems is based on the percentage of efficiency, which is under the effect of the air gap. In other words, the smaller the air gap, the more efficient it is. Coil size is the other significant parameter in the power transfer efficiency. By studying various research worldwide, it is determined that a greater coil can transfer a high amount of power through the inductive wireless system. The system efficiency is lower for dynamic systems than for stationary charging systems, mainly because a certain amount of magnetic flux is generated by the primary coil that is not coupled with the secondary coil. Moreover, the speed of EVs plays an important role in transferring power while EVs are in motion. To put it differently, the lower speed of EVs leads to a higher charging amount of them. Last but not least are the materials of coils and the materials that surround the coil system, like aluminum or ferrite materials, which lead to creating a shield area to increase the magnetic performance.

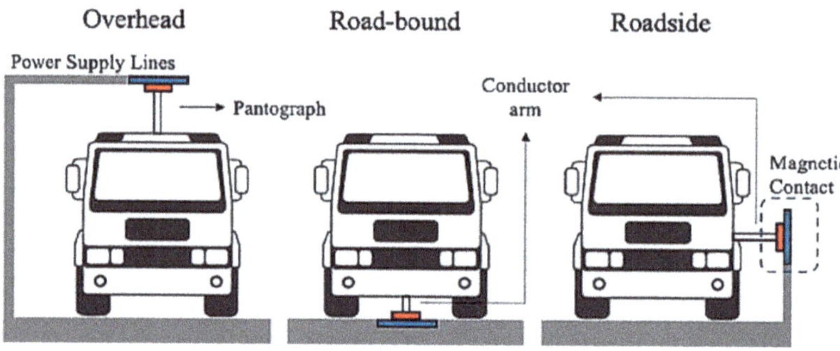

Figure 2. Electrical charging using a pantograph system [17].

One of the most important applications of wireless charging systems is related to passenger cars and heavy vehicles. In addition, they can be used in bus rapid transit (BRT) routes and decrease energy losses. These systems can decrease the size of EV batteries [18] and the delays of stopping buses for charging in stationary places, and consequently, more buses can be in service mode. Another advantage of these systems that can be mentioned is the promotion of the sales market of EVs from 2% in 2020 to 24% in 2050 [19], which not only will decrease CO_2 emissions, but will also decrease the consumption of fossil fuels [20–26]. Despite the mentioned benefits, the large-scale implementation of WPT systems can be one of the challenges for fulfilling such projects, which require a high initial investment and extensive devastating of pavements.

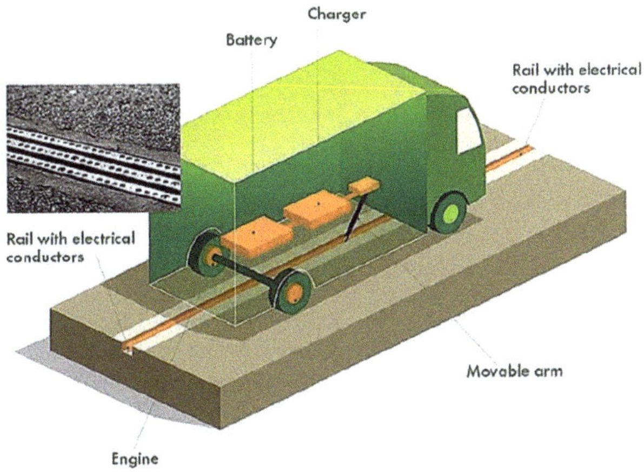

Figure 3. Electrical charging using conductive rail [13].

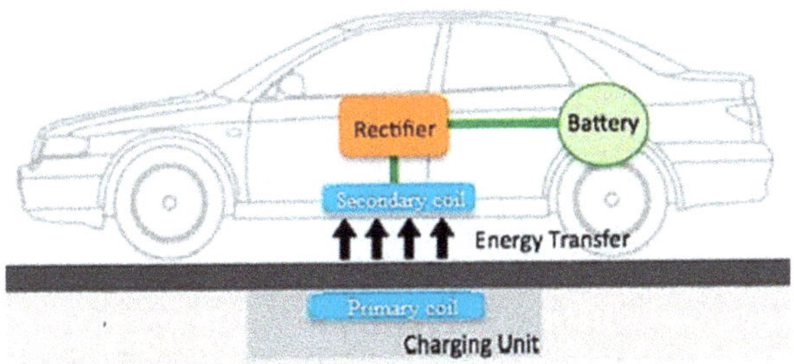

Figure 4. Wireless charging system using inductive power transfer [17].

3. Inductive Power Transfer

For wireless power transfer, various technologies exist, such as magnetic gear, capacitive, microwave, and inductive coupled charging. According to coaxial cable, magnetic gear wireless power transfer comprises two synchronized permanent magnets positioned side by side different from other wireless charging techniques. Capacitive wireless power transfer transfers power through coupled capacitors realized by metal plates. Microwave power transfer comprises a receiver installed in any low-voltage product and a microwave launcher connected to the grid. Inductive power transfer (IPT) transfers power by the use of alternating magnetic fields between primary and secondary coils. The air-core transformer with primary and secondary coils is separated through a small space and transfers power through electromagnetic induction phenomena. IPT systems have superiority and are applicable for e-roads over other mentioned systems for several reasons such as high power transfer (approximately 250 KW), high charging efficiency (71–96%), and the smallest air gap space (7.5–50 cm) [17].

IPT is a technique for coupling electrical power across an air gap without any physical contact, which was introduced over a hundred years ago. Lack of physical contact allows IPT systems to realize advantages over conductive counterparts, including resistance to environmental impacts and comfort. Over the years, various IPT systems have been presented to apply these advantages for multiple applications, such as powering bio-medical devices, powering automatic guided vehicles, and battery charging. By the utilization of IPT systems, EVs can be charged when stationary within a car park, garage, or at traffic signals, or dynamically while in motion. Robust IPT pads must be developed for installation within roadway pavements for both stationary and dynamic IPT EV charging systems to be economically viable, remaining operational for 20–30 years. Figure 5 illustrates an overview of usual IPT systems schematically. IPT systems can be divided into various modular components [27–30]:

- The power is directly derived through the power source, typically from the AC mains grid, to supply either a regulated DC supply to the inverter or a modulated AC frequency derived from the mains frequency.
- The inverter transforms this DC or extremely low frequency (LF) AC input into a greater frequency voltage and current (commonly chosen to be in the LF range of 30–300 kHz) appropriate to drive the output compensation network and magnetics that enhance the power transfer ability of IPT systems. According to standard requirements, a nominal 85 kHz is selected for EV charging systems.
- The primary and secondary pads optimize the coupled magnetic fields generated from both pads. The power is transferred through resonant IPT between the two pads. The primary pad is normally placed on or below the ground, and the secondary one is placed underneath and attached to EVs.
- A secondary controller conditions and regulates the power to the load.
- The load for an EV normally is a battery or an electric motor [30].

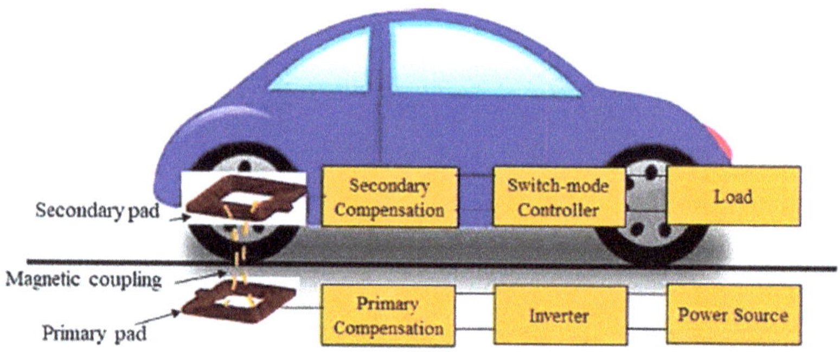

Figure 5. Overview of a typical IPT system [30].

Vehicle speed is a fundamental parameter that can directly impact the charging process in the wireless pavement system. As the speed of the EVs increases, the duration of contact between the vehicle and the IPT pads changes, affecting the power transfer rate. It is essential to investigate the relationship between vehicle speed and charging efficiency to determine the optimal speed range that ensures both effective charging and safe driving conditions [10].

4. Coil Design

As mentioned in the previous section, the coil is one of the important and inseparable components of the IPT pad, which has a role in transferring power between the energy source and receiver. Coils should have a larger dimension than the air gap (100 to 300 mm) distance between vehicles and pavement surface to operate more efficiently. Moreover, a magnetic region should be provided with various materials such as ferrite and aluminum for transferring power efficiently and conducting them to the upper layers [31]. Figure 6 shows an embedded IPT pad within the pavement with four sections: coil, ferrite layer, aluminum plate, and a case in which all details are installed [30].

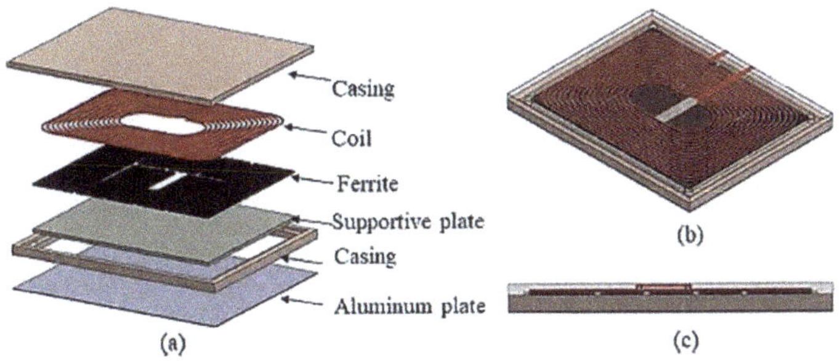

Figure 6. Embedded IPT pad with its components: (**a**) exploded view, (**b**) isometric view, (**c**) cross-section [30].

Several studies have been conducted to promote the level of coil power transfer and decrease the power loss between the pavement and EVs. In most of them, the main structure of coils and usage materials was maintained, and the difference between them is in terms of geometrical aspects. Figure 7 demonstrates different types of coils. For instance, most of the time, circular coils are used in the stationary IPT system, or DD coils can be used for compensating the energy loss, especially for pavements that encounter moisture. The DD-DDQ pad combination can achieve higher coupling and a charging zone more than five times larger in terms of coverage area than that of an equivalent circular pad system [32].

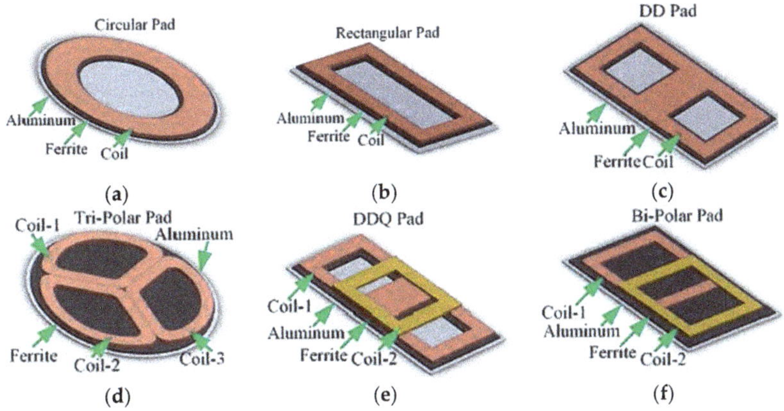

Figure 7. Different types of coil systems [30]: (**a**) Circular pad, (**b**) Rectangular pad, (**c**) DD pad, (**d**) Tri-polar pad, (**e**) DDQ pad, and (**f**) Bi-polar pad.

5. Effect of IPT Pad on Power Transfer Performance

By investigating the different studies, it was determined that various factors such as the depth of the IPT pad, materials of coils, the number of receiver coils, and factors like these have an impact on the power transfer performance of the wireless pavement. The depth of the IPT pad can be considered the thickness of the pavement as well. To put it differently, if the pavements are thicker and the IPTs are embedded in a deeper pavement layer [33], the inductivity performance and dielectric loss will be decreased and increased, respectively. In this case, the upper layer of the IPT pad is concrete, the energy loss was estimated at 3.93 w [17]. However, in the asphalt samples, this energy loss was about 0.01 w. Using materials like rubber for coil coating [17] or increasing materials like copper and ferrite for making IPT pads [34] in the amounts of 12 and 30%, respectively, can cover the weakness of power transfer of the upper layer.

If all of the above are considered, other requirements such as vehicle speed, number of receiving coils, and synchronization of the primary and secondary coils should be provided, and also if the vehicles' speeds on the wireless pavement exceed 80 km/h, the IPT efficiency will be decreased intensively. Therefore, a speed of 40 to 60 km/h can be considered an acceptable speed for efficient performance [35]. Furthermore, the simultaneity of the IPT pad and receiver coil, and the lack of deviation from each other, can decrease the energy loss [17].

Some studies have suggested two coil receivers under electric vehicles in order to solve this problem. Figure 8 shows a sample of two coil receivers. As can be seen, the connection between the primary and secondary coil was never lost, and this issue leads to an increase in the efficiency performance of the IPT pad between 96 and 100% [36].

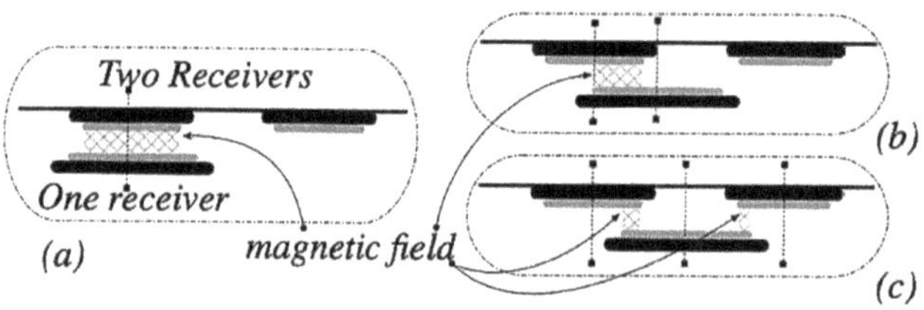

Figure 8. Two installed coils under the electric vehicle and increasing the simultaneity area between primary and secondary coils [36]: (**a**) High magnetic field, (**b**) Medium with one receiver, (**c**) Medium Magnetic field with two receivers.

The wireless pavement system (WPS) based on inductive power transfer (IPT) for electric vehicles (EVs) is undoubtedly a critical infrastructure development in the realm of sustainable transportation. While the current research has focused on various aspects of the system's design and performance, it is essential to acknowledge the potential impact of environmental factors on its stable functioning. In particular, the effects of rain, snow, and sandy soil on the WPS warrant further exploration to ensure its robustness and reliability under diverse weather and terrain conditions.

Rain and snow are common environmental challenges that wireless charging systems may encounter, especially in regions with frequent precipitation. The presence of water on the road surface can affect the efficiency of power transfer and may lead to decreased performance of the wireless pavement system. Investigating the behavior of IPT under wet conditions, analyzing the impact on charging efficiency, and developing appropriate mitigation strategies are crucial steps toward ensuring the system's consistent operation, regardless of weather conditions.

Sandy soil poses another potential challenge for the wireless pavement system. In regions with sandy or loose soil, the structural stability and adhesion of the IPT pads could be compromised. It is essential to study the interaction between the IPT pads and sandy soil, evaluating the risk of displacement or damage during operation. Addressing these concerns will lead to improved design considerations, material selection, and installation techniques, enhancing the long-term sustainability and functionality of the wireless pavement system.

6. Effect of IPT Pad on Pavement Performance

As mentioned in previous sections, IPT pads should be embedded within the pavement layers, and they should not be exposed to traffic loading. Therefore, the pavements play an important role in preventing the IPT system against thermal fluctuation, freeze–thaw cycles [37], and repetitive traffic loading [17]. Different studies have been performed on the effect of charging units (CUs) on pavement damage, and the different conditions of their performance have been evaluated.

Although pavement layers protect the IPT system, the presence of these systems causes a lot of damage to pavement layers. The significant point is that more than twice the amount of damage will be caused than to a conventional pavement layer. Various factors, including the geometry of the IPT pad, embedment depth, and their distance from each other, have an impact on the damage. Figure 9 shows the different types of damage created on the pavement surfaces affected by the embedment of CUs.

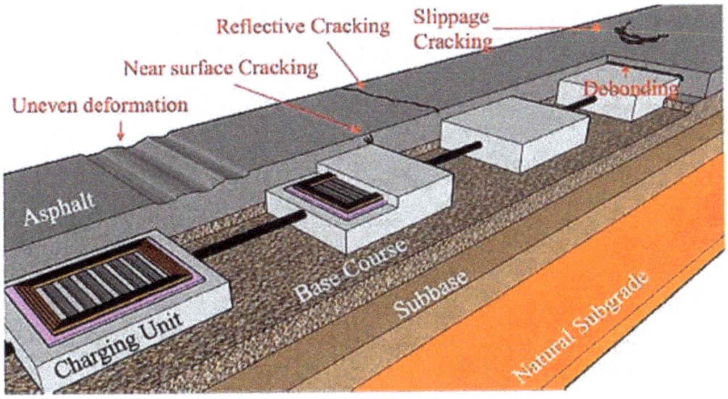

Figure 9. Damage related to the embedment of charging units in pavement [17].

As shown in Figure 9, these charging units can increase the cracks in the pavement surface and cause all kinds of damage such as debonding, slippage, and reflective cracks. These cracks and damage are aggravated at the edge of the CU and the contact surface of the pavement and the CU. Due to the fact that CUs usually are constructed from concrete, lower longitudinal strain and deformation occur exactly on the center and above the IPT surface, and the pavement faces better resistance than the CU shoulder section. In addition, debonding and slippage cracks are among the most common forms of damage in wireless pavements due to horizontal stresses caused by braking and vehicle acceleration. Reflective cracks occur as a result of CUs being closer to each other. To solve this problem, an increase in pavement thickness is suggested [17]. According to the previous section, the distance between IPT pads can be increased from each other, using two coil receivers under EVs instead.

Fatigue and rutting are among the other types of damage to wireless pavements. One of the causes of fatigue is the aluminum plate under the IPT pad, which despite the advantages such as reducing energy loss, has a negative effect on fatigue resistance. To compensate for this weakness and strengthen the electromagnetic performance of wireless pavements, it is better to use a layer of asphalt mixture containing magnetic additives. Despite the various forms of damage to the pavement due to the presence of IPT pads, rutting is one of the failures that appeared with lower intensity than conventional asphalt [8].

During the charging time of EVs on the wireless pavement, the temperature degree of the pavement reaches 87 °C, which can cause softening and thermal stresses, which contributes to other forms of damage. Therefore, bitumen should be used with a higher softening point and viscosity and lower penetration grade. Moreover, if the temperature can be controlled through different additive and bitumen modifications [38], it may be a positive point for the self-healing properties of asphalt mixtures. Otherwise, damage due to fatigue and rutting might develop [17]. The significant point is that the weight of EVs can be reduced by increasing the conductivity of asphalt mixtures and improving the coils' performance. With this measurement, the intensity of loading and damage will be decreased.

The geometry of the IPT system is one of the factors that has an impact on the damage performance of the wireless pavement. The connection areas between the CU and the asphalt experience the weakest performance against damage and the highest stress. Therefore, by rounding the shoulder of the CU with a radius of 3 to 12 cm, the intensity of the stresses on the pavement and damage can be reduced. Moreover, increasing the thickness of the pavement layer above the IPT pad can also reduce the severity of the damage [39]. However, increasing the thickness or the depth of the IPT pad can reduce the power transfer performance [40].

Dielectric properties of materials are one of the factors affecting the energy loss of wireless pavements. In other words, the presence of materials with high dielectric properties reduces power transfer in pavements. Table 1 shows the dielectric characteristic (rε) of different materials.

Table 1. Dielectric characteristics of different materials [41].

Materials	rε
Air	1
Water	81
Snow	6–12
Ice	4
Sand	2–6, 10–30 (wet)
Clay	2–6 (dry), 5–40 (wet)
Limestone	7 (dry), 8 (wet)
Granite	5 (dry), 7 (wet)
Asphalt	2–4 (day), 6–12 (wet)
Concrete	4–10 (day), 10–20 (wet)

As can be seen in Table 1, asphalt pavements have a lower dielectric index than concrete pavements. In other words, by implementing the asphalt pavement, more power can be transferred from the IPT pad to EVs. Among the present materials in Table 1, water has the worst dielectric performance, and in the case of moisture penetration to the pavement texture, the power transfer via the IPT pad encounters several problems [35]. Despite the lower index of asphalt mixtures than concrete, when these pavements are exposed to moisture, they experience a more noticeable decline than concrete pavement [17]. Therefore, it is better to improve the performance of asphalt pavements against moisture damage by using an ant strip additive.

7. Modification of Wireless Pavement Properties to Increase Performance and Conductivity

It is true that things like the deviation between the primary and secondary pads and the lateral movement of vehicles while driving reduce the performance of the IPT system, and by deforming the coils, part of this energy loss can be reduced, but by modifying the properties of asphalt and concrete mixtures, the efficiency of these systems can be improved by creating an electromagnetic conduction zone in the pavement. Various studies have been performed to improve the induction and performance properties of asphalt and concrete mixtures, which have tried to improve the performance of pavements against various failures by modifying the properties of bitumen, concrete, and aggregates in mixtures [42] while increasing the electrical conductivity [43]. Table 2 shows some of the additives used.

As can be seen from Table 2, these materials have mainly metallic properties and their use causes rheological changes in the properties of bitumen or changes in the characteristics of used aggregates. These changes increase the bitumen cohesion and the adhesion of bitumen (concrete) and aggregates and greatly reduce separation. In addition, due to the conductivity of these materials, they can also act as an electromagnetic field in pavements with IPT pads. But in addition to all these advantages, one should also pay attention to the rusting properties of these materials during freeze–thaw cycles and try to use metal additives in small or nano sizes so that they do not corrode, especially in high freeze–thaw cycles, as shown in Figure 10.

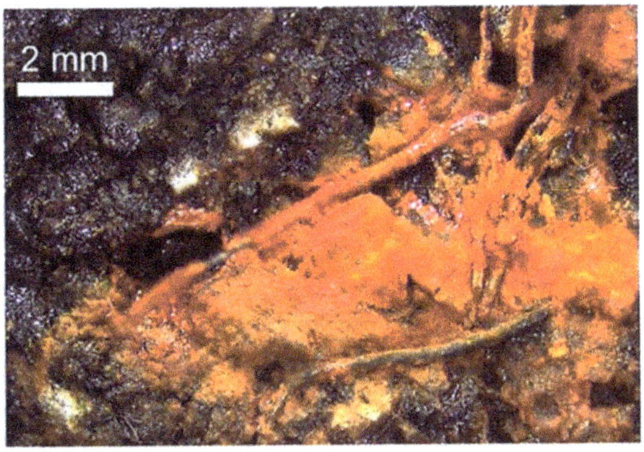

Figure 10. Corrosion of asphalt texture as a result of steel fiber additive [44].

Table 2. Additives used to modify inductive pavements.

Material	Optimal (%)	Test Device	Improvement (%)	Modification with	Reference
Steel wool fiber	6	ITS Electrical resistivity	15.13	Bitumen	[45]
Electric arc furnace	8	ITS, ITSM	69	Aggregate	[46]
Steel slag/steel fiber	6	Cantabro Semi-circle bending fracture Thermal constants	57	Aggregate Bitumen	[47]
Electric arc furnace	3	ITS	50	Aggregate	[48]
Steel wool fiber	N/A	Electromagnetic Induction heating	66	Bitumen	[49]
Metallic waste	4	Electrical resistivity X-ray Thermo physical	✔	Bitumen	[50]
Metallic fiber	1.5	ITS, ITSM	×	Bitumen	[44]
Waste steel shavings	10	Induction heating Electrical resistivity	✔	Aggregate	[51]
Ferrite powder	0.5	ITS Electrical resistivity	17	Limestone filler	[52]
Steel wool fiber	4	Crack-healing X-ray	✔	Bitumen	[53]
Steel fiber	10	Ice-melting	✔	N/A	[54]
Steel fibers and steel wool	10	ITS Electrical resistivity	19	Bitumen	[55]
Steel fibers	6	Induction heating Thermo physical	✔	Bitumen	[56]
Waste steel shavings	8	Heating power	✔	Bitumen	[57]
Steel wool fiber	2	Cantabro fatigue test	✔	Aggregate	[58]

Table 2. *Cont.*

Material	Optimal (%)	Test Device	Improvement (%)	Modification with	Reference
Electric arc furnace Steel slag and copper	N/A	Rutting Creep ITS	47	Aggregate	[59]
Steel fiber	6	Semi-circular bending Induction heating	✔	Bitumen	[60]
Steel wool fiber	1.5	ITS	25	Bitumen	[61]
Micron-scale steel fiber with carbon fiber	0.2	ITS Dynamic modulus	29	Bitumen	[62]
Steel fiber	N/A	Wheel tracking ITS Pull out	✔	Bitumen	[63]
Steel slag	N/A	ITS ITSM Electrical resistivity	34	Aggregate	[64]

✔: The additive materials improve the characteristics of the asphalt mixture, ×: The additive materials have a negative effect.

According to other similar studies, the addition of ferrite materials to concrete mixtures improves specimens by 50% against energy loss [65], and replacing ferrite materials with some aggregates can be suggested as a suitable solution to increase efficiency and conductivity [66]. Other studies have examined the effect of metal materials on the performance of pavements and have concluded that the addition of stainless steel wool and stainless steel fibers can reduce energy losses by up to 50% and improve the performance of pavements [8]. Also, if 1 to 20% of the weight of concrete is replaced with iron and magnetite powders, the pavement efficiency can be improved by up to 86% [67]. In addition, the type of aggregate is one of the influential parameters. Because limestone aggregates have more magnetic properties than basalt aggregates, and if the mixture is more compact, then more power can be transferred [32].

Although many researchers have suggested the construction of charging units in boxes with concrete materials [68], others believe that the creation of an asphalt pavement layer on concrete boxes causes a discontinuity between the materials [69–71]. Therefore, it is suggested that in concrete pavements, IPT pads be placed in the concrete layer, and in asphalt pavements, IPT pads be placed in the asphalt mixture to increase the cohesion and adhesion between different materials. In addition, Chen et al. proposed a method whereby if reinforced pavements such as ultra-thin white-topping pavements as represented in Figure 11 are used for the top layer of IPT pads, the weakness of pavements in energy transfer can be reduced by up to 75%, and in places that are close to the rebars, up to 85% improvement of pavement performance can be observed [72]. In another study, to improve the discontinuity between the IPT pad and its top layer, various materials such as Portland cement, polypropylene (PP), acrylonitrile butadiene styrene (ABS), XPS: polystyrene foam, stretch film, and ferrite were proposed to cover the coils [73]. The results of studies showed that cementitious materials have the highest electrical losses and have negative effects on coils. Among the materials, ABS and PP resins performed best in energy transfer.

In another study, investigators proposed a multilayer asphalt pavement system based on Figure 12, which is mostly comprised of three subsystems from top to bottom [65]: the upper part is constructed with electrically conductive layers with waste steel shavings, the middle part is an asphalt mixture layer with the pre-embedded induction coil, and the lower part is constructed with magnetically absorbing layers with waste ferrites for replacing the conventional waterproof adhesive layer. Based on Figure 13, the role of the lower part was maintaining the electromagnetic power and transferring it to the upper one.

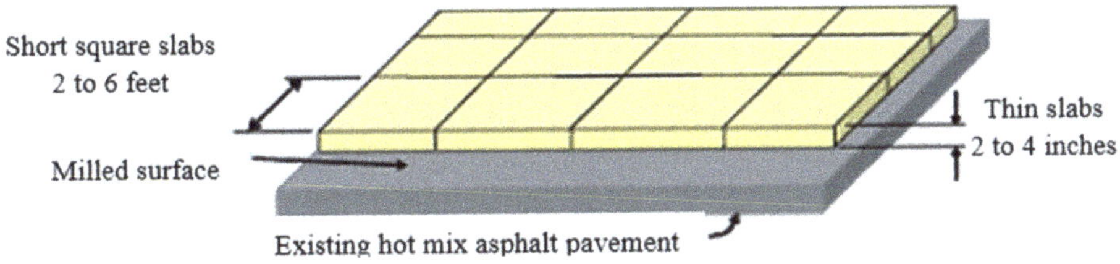

Figure 11. Ultra-thin white-topping method [73].

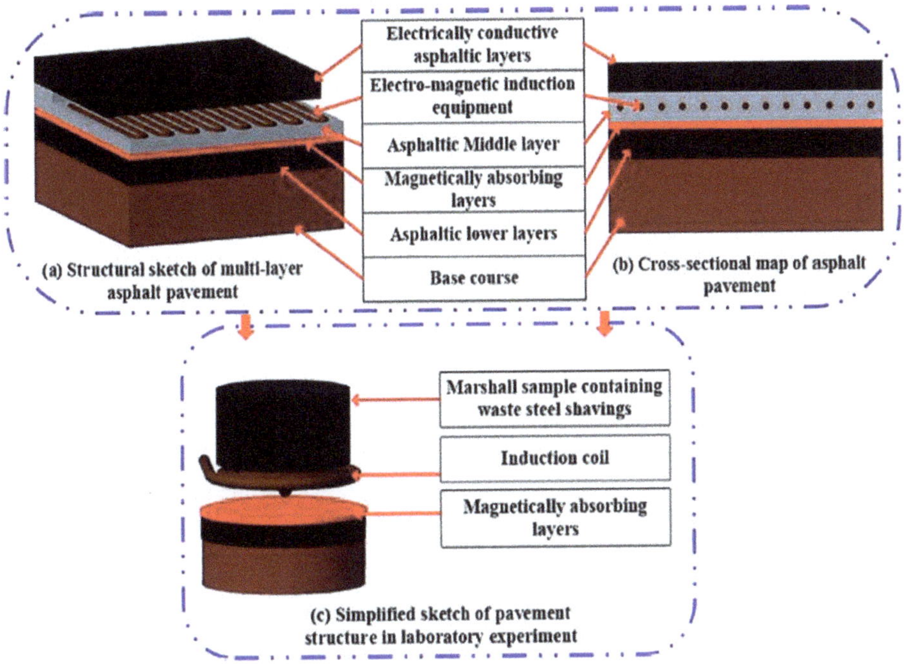

Figure 12. Structural sketch of multilayer asphalt pavements [39].

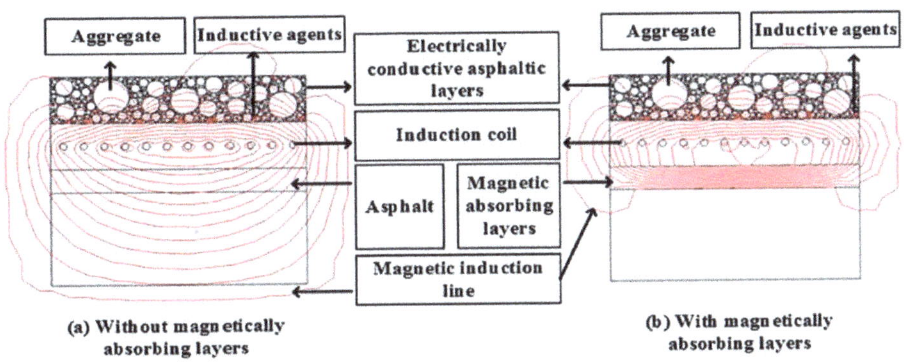

Figure 13. Magnetically absorbing principle of magnetically absorbing layer [39].

The upper layer was made with four various contents of waste steel shavings (2 to 8% by the weight of aggregates) and applied as an inductive agent in asphalts. Moreover, styrene butadiene styrene (SBS) and limestone powder were applied for the modification of bitumen properties to enhance the stability of the multilayer asphalt pavement structure. The result showed that the asphalt magnetically absorbing layers could meet the requirements of conventional waterproof adhesive layers. It was also concluded that 6% was the optimum amount of waste steel shavings. Also, the electrically conductive asphalt layer operated as the magnetic field receiving layer to achieve self-healing and deicing, and snow melting [39]. Based on previous research, nanocomposite bitumens could be applied to enhance the resistance of asphalt mixtures to moisture. These materials contain various types of nano-metal or nano-polymer composites that can simultaneously improve the performance and electrical conductivity of the mixtures.

Most studies investigated the modification of pavement with various additives as a suitable solution to improve the inductive and performance properties. However, in an innovative study, researchers suggested that a partial part of the asphalt in which the IPT pad was embedded be modified with magnetic additive (as shown in Figure 14), and other parts should remain as unmodified asphalt. This design principle creates a pathway that can better connect magnetic fields and guide magnetic flux between receiver coils and transmitter. The enhancement of wireless power transfer was also indicated for charging EVs from the partially magnetized pavement layer over the conventional pavement layer. If the embedment depth transmitter coil is 0.1 m, the wireless charging performance could be increased by 1.5% from 70% to 71.5%. However, once the thickness of the partially magnetized layer is increased to 0.4 m, the improvement rate is remarkably enhanced from 26.6 to 39.9% (by 13.3%) [8].

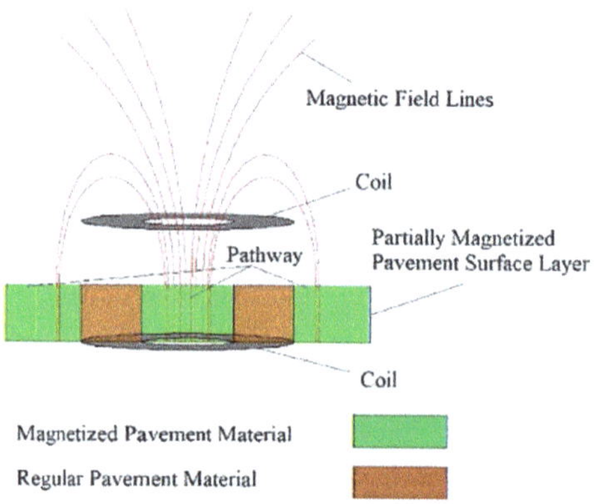

Figure 14. Design concept of partially magnetized pavement layer [8].

8. Simulation and Comparative Analysis of IPT Pad Performance

This section presents an exploration of simulation results concerning the performance of IPT pads within wireless pavement systems. Building upon empirical investigations and comparisons of experimental outcomes, this extension delves into simulation-based insights to comprehensively understand the IPT pad behavior.

Advanced electromagnetic simulation tools were employed to comprehensively assess the IPT pad performance. These simulations modeled the intricate electromagnetic interactions within the pavement structure and IPT pad configuration. Variables encompassed geometric parameters, material properties, and operational conditions for nuanced examination.

Various researchers have studied a collective perspective on IPT pad performance within wireless pavement systems. Sun et al. proposed a framework for magnetically coupled resonant wireless power transmission systems integrated within pavements. Their work highlighted the significance of resonance phenomena and optimized geometries in influencing power transfer efficiency [74]. Amirpour et al. conducted a coupled electromagnetic–thermal analysis of inductive power transfer pads within pavements.

Their research emphasized the interplay between electromagnetic efficiency and thermal effects, contributing to a comprehensive understanding of pad behavior [30]. Aghcheghloo et al. explored the influence of an emulator inductive power transfer pad on asphalt pavement temperature. Their investigation shed light on the role of emulators in simulating real-world scenarios and the resulting thermal implications [75]. Li et al. proposed a wireless power transfer-tuning model using pavement materials as transmission media. Their innovative approach investigated the potential of pavement materials to enhance power transfer efficiency [76]. Guo and Wang introduced a novel design of a partially magnetized pavement for wireless power transfer. Their work showcased the potential for improved efficiency and cost savings [8].

Integration of simulation-based insights alongside experimental results enriches the understanding of IPT pad performance. These outcomes provide a comprehensive perspective on IPT pad behavior within wireless pavement systems. By amalgamating simulation and experimental insights, contribution to wireless pavement system optimization and refinement is advanced. Figure 15 presents a comparative analysis of electrical conductivity across various simulation studies. The discernible outcome underscores the substantial influence of reinforcing the pavement materials above the Inductive Power Transfer (IPT) pad in augmenting the electrical conductivity of wireless road systems. Investigations conducted with the specific aim of enhancing this aspect [8,76] have demonstrated superior electrical conductivity when contrasted with research that primarily addresses the geometric characteristics of IPT pads [30,74].

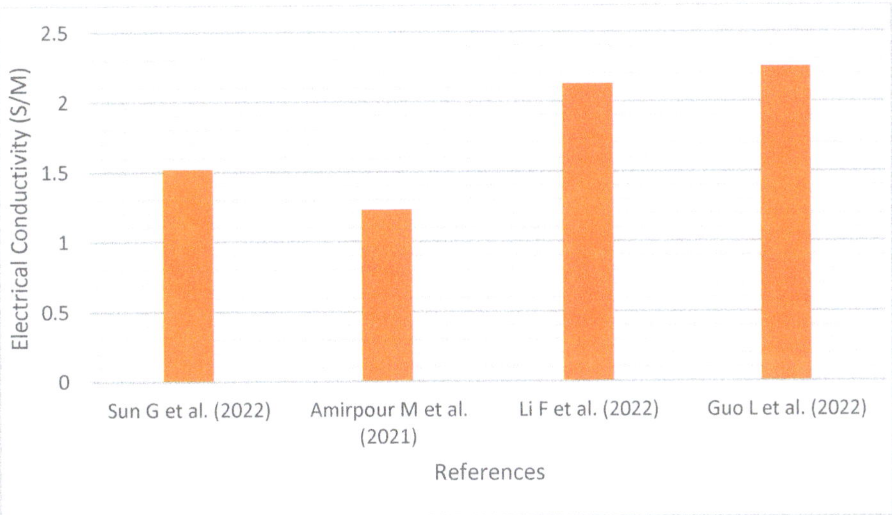

Figure 15. Comparison of simulation results in various studies [8,30,36,74].

9. Effect of Environmental Conditions on the Performance of IPT Pad

The impact of environmental conditions, such as rain and snow, on the performance of IPT systems within the context of wireless pavement charging for EVs, is a crucial aspect warranting thorough investigation. Rain and snow, being common weather conditions encountered in various geographic regions, can potentially influence the charging efficiency and stability of IPT pads integrated into road pavements.

Rainfall on road surfaces introduces moisture, which may affect the dielectric properties of the pavement material and impact the electromagnetic coupling between the IPT pads and the EV's receiving coils [77]. This alteration in electromagnetic characteristics can lead to reduced power transfer efficiency due to increased losses and impedance mismatches. The presence of water on the road surface may also impact the alignment and contact between the charging unit and the EV, affecting the overall system's performance. Addressing these challenges entails a comprehensive understanding of the interactions between water, pavement material, and electromagnetic fields, allowing for the development of strategies to mitigate efficiency loss during rainy conditions [17,78].

Similarly, snow accumulation on road pavements can pose challenges to the proper functioning of IPT systems. Snow buildup can alter the geometry of the road surface and potentially obstruct the alignment between the charging unit and the EV [17]. Additionally, snow's insulating properties can lead to further impedance mismatch and reduced power transfer efficiency [70]. Analyzing the behavior of IPT pads under snow-covered conditions and developing mechanisms for snow removal or adaptation will be instrumental in ensuring the sustained functionality of the wireless pavement charging system in snowy environments [33,76].

Understanding the influence of these environmental conditions on IPT pad performance is essential to developing a robust and reliable wireless pavement system that remains operational under various weather scenarios. By delving into these effects, we can tailor system designs, materials, and operational strategies to ensure optimal performance, thereby enhancing the overall viability and effectiveness of IPT-based wireless charging solutions for electric vehicles.

10. Discussion

The comprehensive analysis of simulation results provides valuable insights into the design, integration, and optimization of IPT pads in pavements for wireless charging of electric vehicles. Each finding holds significant implications for the development of sustainable and effective charging infrastructure. We have discussed the results in detail and their potential impact on the future of wireless power transfer systems and pavement design.

The geometry of the charging unit emerged as a critical factor influencing tension and damage in the pavement. Creating fillet-shaped shoulders with a radius of 3 to 12 cm was found to be an effective strategy to reduce tension and mitigate any potential damage. This highlights the importance of optimizing IPT pad design to ensure minimal adverse effects on pavement integrity and longevity. Further research in this area could explore alternative geometrical configurations to identify the most pavement-friendly charging unit design.

The close application of primary pads within the pavement resulted in the expansion of cracks on the pavement surface. To address this issue, we propose increasing the number of receiving coils under electric vehicles, which significantly enhances charging efficiency by 96 to 100%. This finding underscores the need for precise coil placement and distribution to ensure reliable power transfer while minimizing pavement damage. Future studies could focus on dynamic coil reconfiguration strategies to further improve charging efficiency and reduce potential pavement impacts [79].

Furthermore, the presence of an aluminum plate under IPT pads was found to cause fatigue damage in pavements. To improve pavement induction capability and reduce fatigue damage, we recommend substituting aluminum plates with modified asphalt mixtures containing metal and magnetic additives. This finding highlights the potential of advanced materials to enhance pavement performance under the influence of charging infrastructure. Further research could explore the durability and performance of these modified asphalt mixtures under various loading and environmental conditions.

To address reflective crack formation, we suggest increasing the thickness of the upper layer of IPT pads and modifying aggregate and bitumen with suitable additives. This approach reduces the intensity of reflective cracks and improves the long-term performance of pavements. The importance of pavement composition and construction in preventing reflective cracking is evident, and further investigations could explore innovative materials and construction techniques to mitigate this type of pavement damage [80,81].

Proper integration and alignment of charging infrastructure with the underlying pavement structure is essential to prevent longitudinal strains and pavement deformation. Embedding IPT pads within a case similar to the upper layer is recommended. In concrete pavements, IPT pads should be embedded within the concrete layer, while in asphalt pavements, IPT pads should be embedded within the asphalt layers. This finding highlights the significance of optimal charging unit integration to maintain pavement integrity and durability. Future research could explore dynamic behavior under the presence of embedded charging units to optimize pavement design and performance.

Lastly, the study's suggestion to modify asphalt mixtures with metal additives or nanocomposites based on SBS bitumen opens possibilities for enhancing the self-healing ability and inductivity of pavements. These modifications improve pavement performance against various types of damage such as fatigue, rutting, and moisture susceptibility. Further investigations could delve into the precise mechanisms and long-term behavior of these modified asphalt mixtures to optimize their composition and performance [82–84]. Also, various statistical analyses, machine learning, and optimization methods can be applied for further investigation [85–98].

This comprehensive discussion of simulation results offers valuable insights into the design, integration, and optimization of IPT pads in pavements for wireless charging of electric vehicles. The findings underscore the importance of thoughtful charging unit design, precise coil placement, and pavement materials to achieve efficient power transfer while minimizing pavement damage. The implications of this research have significant relevance for the development of sustainable and effective wireless charging infrastructure, promoting the growth of electric vehicles in the future.

11. Conclusions

Our study focused on investigating the impact of inductive power transfer (IPT) pads on pavement inductivity, conductivity, and durability. Through the exploration of various wireless power transfer systems, IPT pads emerged as a safe and efficient charging solution due to their geometry-sensitive power transfer efficiency. Our investigation revealed several key insights into the effects of IPT pads on pavements. Optimizing the geometry of charging units, particularly by implementing filleted shoulders, mitigated tension and damage in pavements. We also found that embedding more receiving coils under electric vehicles improved charging efficiency and reduced pavement cracks, while modified asphalt mixtures with additives enhanced induction capability and reduced fatigue damage. Furthermore, we underscored the significance of modifying asphalt mixtures with metal additives and nanocomposites based on SBS bitumen to improve pavement performance against various forms of damage. The integration of IPT pads with pavements offers a sustainable charging infrastructure for electric vehicles. However, challenges and considerations remain for its widespread adoption:

1. Scalability and Infrastructure: The scalability of the system to accommodate a growing number of EVs is crucial. A robust and scalable WPS infrastructure is necessary for simultaneous and efficient charging in urban and rural areas.
2. Standardization and Interoperability: Establishing industry-wide guidelines and protocols is essential to ensure compatibility between different EVs and WPS implementations.
3. Environmental Impact: A comprehensive life cycle assessment is needed to evaluate and minimize the ecological footprint of IPT technology.
4. Integration with Smart Grids: Integrating IPT with smart grids can optimize energy management, contribute to grid stability, and support renewable energy integration.
5. Cost-effectiveness: Research efforts should focus on developing cost-efficient materials, processes, and installation techniques to make the technology economically viable.
6. Public Awareness: Collaborative efforts involving stakeholders and the public are essential to promote the adoption of IPT technology.

In summary, our study sheds light on the potential of IPT pads to revolutionize EV charging infrastructure. While challenges exist, addressing them will pave the way for a sustainable and efficient transportation future.

Looking ahead, the integration of IPT into pavements heralds a revolutionary shift in EV charging and urban infrastructure. Autonomous fleets, dynamic charging, grid synergy, and urban planning integration present a horizon of possibilities. By fostering innovation, collaboration, and sustainable practices, IPT-equipped pavements hold the promise of shaping a greener, smarter, and more efficient transportation landscape.

Author Contributions: B., V.N.M.G., M.M. and M.S.M. conceptualized the problem, provided the methodology and analysis, and prepared the original draft; B., A.A., K.Y. and M.S. reviewed and edited the manuscript and provided valuable insights into the overall system. All authors have read and agreed to the published version of the manuscript.

Funding: This work is supported by the Natural Science Foundation of Hunan Province (Project No. 2021JJ60024, 2022JJ60025); Scientific research project of Hunan Provincial Department of Education (Project No. 21C1012).

Institutional Review Board Statement: Not applicable.

Informed Consent Statement: Not applicable.

Data Availability Statement: Data sharing is not applicable.

Conflicts of Interest: The authors declare no conflict of interest.

References

1. Zhang, S.; Zhou, Z.; Luo, R.; Zhao, R.; Xiao, Y.; Xu, Y. A low-carbon, fixed-tour scheduling problem with time windows in a time-dependent traffic environment. *Int. J. Prod. Res.* **2023**, *61*, 6177–6196. [CrossRef]
2. Zhang, X.; Wen, S.; Yan, L.; Feng, J.; Xia, Y. A hybrid-convolution spatial–temporal recurrent network for traffic flow prediction. *Comput. J.* **2022**, bxac171. [CrossRef]
3. Min, H.; Fang, Y.; Wu, X.; Lei, X.; Chen, S.; Teixeira, R.; Zhu, B.; Zhao, X.; Xu, Z. A fault diagnosis framework for autonomous vehicles with sensor self-diagnosis. *Expert Syst. Appl.* **2023**, *224*, 120002. [CrossRef]
4. Zhang, X.; Fang, S.; Shen, Y.; Yuan, X.; Lu, Z. Hierarchical Velocity Optimization for Connected Automated Vehicles With Cellular Vehicle-to-Everything Communication at Continuous Signalized Intersections. *IEEE Trans. Intell. Transp. Syst.* 2023; *Early Access*. [CrossRef]
5. Cao, B.; Zhang, W.; Wang, X.; Zhao, J.; Gu, Y.; Zhang, Y. A memetic algorithm based on two_Arch2 for multi-depot heterogeneous-vehicle capacitated arc routing problem. *Swarm Evol. Comput.* **2021**, *63*, 100864. [CrossRef]
6. Wangsupphaphol, A.; Chaitusaney, S.; Salem, M. A Techno-Economic Assessment of a Second-Life Battery and Photovoltaics Hybrid Power Source for Sustainable Electric Vehicle Home Charging. *Sustainability* **2023**, *15*, 5866. [CrossRef]
7. Alinejad, M.; Rezaei, O.; Habibifar, R.; Azimian, M. A Charge/Discharge Plan for Electric Vehicles in an Intelligent Parking Lot Considering Destructive Random Decisions, and V2G and V2V Energy Transfer Modes. *Sustainability* **2022**, *14*, 12816. [CrossRef]
8. Guo, L.; Wang, H. A novel design of partially magnetized pavement for wireless power transfer to electric vehicles with improved efficiency and cost saving. *Energy Convers. Manag.* **2022**, *252*, 115080. [CrossRef]
9. Abbasi, M.H.; Zhang, J.; Xu, B.; Krovi, V.N. *Reinforcement Learning Based Fast Charging of Electric Vehicle Battery Packs*; SAE Technical Paper; SAE International: Warrendale, PA, USA, 2023; ISSN 0148-7191.
10. Bi, Z.; Kan, T.; Mi, C.C.; Zhang, Y.; Zhao, Z.; Keoleian, G.A. A review of wireless power transfer for electric vehicles: Prospects to enhance sustainable mobility. *Appl. Energy* **2016**, *179*, 413–425. [CrossRef]
11. Bi, Z.; Keoleian, G.A.; Lin, Z.; Moore, M.R.; Chen, K.; Song, L.; Zhao, Z. Life cycle assessment and tempo-spatial optimization of deploying dynamic wireless charging technology for electric cars. *Transp. Res. Part C Emerg. Technol.* **2019**, *100*, 53–67. [CrossRef]
12. Marghani, A.; Wilson, D.; Larkin, T. Performance of Inductive Power Transfer-Based Pavements of Electrified Roads. In Proceedings of the 2019 IEEE PELS Workshop on Emerging Technologies: Wireless Power Transfer (WoW), London, UK, 18–21 June 2019.
13. Balieu, R.; Chen, F.; Kringos, N. Life cycle sustainability assessment of electrified road systems. *Road Mater. Pavement Des.* **2019**, *20* (Suppl. S1), S19–S33. [CrossRef]
14. Liu, G. Data collection in mi-assisted wireless powered underground sensor networks: Directions, recent advances, and challenges. *IEEE Commun. Mag.* **2021**, *59*, 132–138. [CrossRef]
15. Feng, H.; Tavakoli, R.; Onar, O.C.; Pantic, Z. Advances in high-power wireless charging systems: Overview and design considerations. *IEEE Trans. Transp. Electrif.* **2020**, *6*, 886–919. [CrossRef]
16. Chen, Z.; Yin, Y.; Song, Z. A cost-competitiveness analysis of charging infrastructure for electric bus operations. *Transp. Res. Part C Emerg. Technol.* **2018**, *93*, 351–366. [CrossRef]
17. Soares, L.; Wang, H. A study on renewed perspectives of electrified road for wireless power transfer of electric vehicles. *Renew. Sustain. Energy Rev.* **2022**, *158*, 112110. [CrossRef]
18. Jeong, S.; Jang, Y.J.; Kum, D.; Lee, M.S. Charging automation for electric vehicles: Is a smaller battery good for the wireless charging electric vehicles? *IEEE Trans. Autom. Sci. Eng.* **2018**, *16*, 486–497. [CrossRef]

19. Lin, Z.; Li, J.-M.; Dong, J. *Dynamic Wireless Power Transfer: Potential Impact on Plug-In Electric Vehicle Adoption*; SAE Technical Paper; SAE International: Warrendale, PA, USA, 2014. [CrossRef]
20. Pazouki, S.; Olamaei, J. The effect of heterogeneous electric vehicles with different battery capacities in parking lots on peak load of electric power distribution networks. *Int. J. Ambient Energy* **2019**, *40*, 734–738. [CrossRef]
21. Haghifam, M.-R.; Pazouki, S.; Pazouki, S. Renewables and plug in electric vehicles modeling on electricity and gas infrastructures scheduling in presence of responsive demand. In Proceedings of the 2013 3rd International Conference on Electric Power and Energy Conversion Systems, Istanbul, Turkey, 2–4 October 2013.
22. Das, H.S.; Nurunnabi, M.; Salem, M.; Li, S.; Rahman, M.M. Utilization of Electric Vehicle Grid Integration System for Power Grid Ancillary Services. *Energies* **2022**, *15*, 8623. [CrossRef]
23. Habibifar, R.; Aris Lekvan, A.; Ehsan, M. A risk-constrained decision support tool for EV aggregators participating in energy and frequency regulation markets. *Electr. Power Syst. Res.* **2020**, *185*, 106367. [CrossRef]
24. Huang, N.; He, Q.; Qi, J.; Hu, Q.; Wang, R.; Cai, G.; Yang, D. Multinodes interval electric vehicle day-ahead charging load forecasting based on joint adversarial generation. *Int. J. Electr. Power Energy Syst.* **2022**, *143*, 108404. [CrossRef]
25. Abbasi, M.H.; Zhang, J.; Krovi, V. A Lyapunov Optimization Approach to the Quality of Service for Electric Vehicle Fast Charging Stations. In Proceedings of the 2022 IEEE Vehicle Power and Propulsion Conference (VPPC), Merced, CA, USA, 1–4 November 2022.
26. Pazouki, S.; Haghifam, M.-R.; Pazouki, S. Short term economical scheduling in an energy hub by renewable and demand response. In Proceedings of the 2013 3rd International Conference on Electric Power and Energy Conversion Systems, Istanbul, Turkey, 2–4 October 2013.
27. Farhadi, P.; Sedaghat, M.; Sharifi, S.; Taheri, B. Power point tracking in photovoltaic systems by sliding mode control. In Proceedings of the 2017 10th International Symposium on Advanced Topics in Electrical Engineering (ATEE), Bucharest, Romania, 23–25 March 2017.
28. Taheri, B.; Jabari, F.; Foroud, A.A. A green cogeneration microgrid composed of water-source heat pumps, a gravity energy storage, and a bio-fueled gas turbine: Design and techno-economic optimization. *Sustain. Cities Soc.* **2023**, *95*, 104594. [CrossRef]
29. Taheri, B.; Aghajani, G.; Sedaghat, M. Economic dispatch in a power system considering environmental pollution using a multi-objective particle swarm optimization algorithm based on the Pareto criterion and fuzzy logic. *Int. J. Energy Environ. Eng.* **2017**, *8*, 99–107. [CrossRef]
30. Amirpour, M.; Kim, S.; Battley, M.P.; Kelly, P.; Bickerton, S.; Covic, G. Coupled electromagnetic-thermal analysis of roadway inductive power transfer pads within a model pavement. *Appl. Therm. Eng.* **2021**, *189*, 116710. [CrossRef]
31. Covic, G.A.; Boys, J.T. Inductive power transfer. *Proc. IEEE* **2013**, *101*, 1276–1289. [CrossRef]
32. Nagendra, G.R.; Covic, G.A.; Boys, J.T. Sizing of inductive power pads for dynamic charging of EVs on IPT highways. *IEEE Trans. Transp. Electrif.* **2017**, *3*, 405–417. [CrossRef]
33. Chen, F.; Taylor, N.; Balieu, R.; Kringos, N. Dynamic application of the Inductive Power Transfer (IPT) systems in an electrified road: Dielectric power loss due to pavement materials. *Constr. Build. Mater.* **2017**, *147*, 9–16. [CrossRef]
34. Varghese, B.J.; Kamineni, A.; Roberts, N.; Halling, M.; Thrimawithana, D.J.; Zane, R.A. Design considerations for 50 kW dynamic wireless charging with concrete-embedded coils. In Proceedings of the 2020 IEEE PELS Workshop on Emerging Technologies: Wireless Power Transfer (WoW), Seoul, Republic of Korea, 15–19 November 2020.
35. Li, F.; Sun, X.; Zhou, S.; Chen, Y.; Hao, Z.; Yang, Z. Infrastructure material magnetization impact assessment of wireless power transfer pavement based on resonant inductive coupling. *IEEE Trans. Intell. Transp. Syst.* **2021**, *23*, 22400–22408. [CrossRef]
36. Mohamed, N.; Aymen, F.; Issam, Z.; Bajaj, M.; Ghoneim, S.S.; Ahmed, M. The impact of coil position and number on wireless system performance for electric vehicle recharging. *Sensors* **2021**, *21*, 4343. [CrossRef]
37. Throngnumchai, K.; Hanamura, A.; Naruse, Y.; Takeda, K. Design and evaluation of a wireless power transfer system with road embedded transmitter coils for dynamic charging of electric vehicles. *World Electr. Veh. J.* **2013**, *6*, 848–857. [CrossRef]
38. Wang, Z.; Wang, Q.; Jia, C.; Bai, J. Thermal evolution of chemical structure and mechanism of oil sands bitumen. *Energy* **2022**, *244*, 123190. [CrossRef]
39. Liu, K.; Fu, C.; Dai, D.; Jin, C.; Li, W.; Li, S.; Xu, X. Induction heating performance of asphalt pavements incorporating electrically conductive and magnetically absorbing layers. *Constr. Build. Mater.* **2019**, *229*, 116805. [CrossRef]
40. Chen, F.; Balieu, R.; Córdoba, E.; Kringos, N. Towards an understanding of the structural performance of future electrified roads: A finite element simulation study. *Int. J. Pavement Eng.* **2019**, *20*, 204–215. [CrossRef]
41. Chen, F.; Kringos, N. Towards new infrastructure materials for on-the-road charging. In Proceedings of the 2014 IEEE International Electric Vehicle Conference (IEVC), Florence, Italy, 17–19 December 2014.
42. Mahmud, M.H.; Elmahmoud, W.; Barzegaran, M.; Brake, N. Efficient wireless power charging of electric vehicle by modifying the magnetic characteristics of the transmitting medium. *IEEE Trans. Magn.* **2017**, *53*, 1–5. [CrossRef]
43. Liu, K.; Dai, D.; Pan, S.; Wang, F.; Hou, C.; Xu, P. Numerical investigation and thermal predictions of asphalt pavement containing inductive materials under alternating magnetic field. *Int. J. Therm. Sci.* **2020**, *153*, 106353. [CrossRef]
44. González, A.; Norambuena-Contreras, J.; Poulikakos, L.; Varela, M.J.; Valderrama, J.; Flisch, A.; Arraigada, M. Evaluation of asphalt mixtures containing metallic fibers from recycled tires to promote crack-healing. *Materials* **2020**, *13*, 5731. [CrossRef]
45. Hosseinian, S.M.; Najafi Moghaddam Gilani, V.; Mehraban Joobani, P.; Arabani, M. Investigation of moisture sensitivity and conductivity properties of inductive asphalt mixtures containing steel wool fiber. *Adv. Civ. Eng.* **2020**, *2020*, 8890814. [CrossRef]

46. Lizárraga, J.M.; Gallego, J. Self-healing analysis of half-warm asphalt mixes containing electric arc furnace (EAF) slag and reclaimed asphalt pavement (RAP) using a novel thermomechanical healing treatment. *Materials* **2020**, *13*, 2502. [CrossRef]
47. Liu, Q.; Li, B.; Schlangen, E.; Sun, Y.; Wu, S. Research on the mechanical, thermal, induction heating and healing properties of steel slag/steel fibers composite asphalt mixture. *Appl. Sci.* **2017**, *7*, 1088. [CrossRef]
48. Gulisano, F.; Crucho, J.; Gallego, J.; Picado-Santos, L. Microwave healing performance of asphalt mixture containing electric arc furnace (EAF) slag and graphene nanoplatelets (GNPs). *Appl. Sci.* **2020**, *10*, 1428. [CrossRef]
49. Xu, C.; Wang, K.; Li, K.; Zong, Y. Deicing property of asphalt mixture containing steel wool fiber by electromagnetic induction heating. *Coatings* **2021**, *11*, 1276. [CrossRef]
50. Norambuena-Contreras, J.; González, A.; Concha, J.; Gonzalez-Torre, I.; Schlangen, E. Effect of metallic waste addition on the electrical, thermophysical and microwave crack-healing properties of asphalt mixtures. *Constr. Build. Mater.* **2018**, *187*, 1039–1050. [CrossRef]
51. Liu, K.; Dai, D.; Fu, C.; Li, W.; Li, S. Induction heating of asphalt mixtures with waste steel shavings. *Constr. Build. Mater.* **2020**, *234*, 117368. [CrossRef]
52. Peinado, F.; Medel, E.; Silvestre, R.; Garcia, A. Open-grade wearing course of asphalt mixture containing ferrite for use as ferromagnetic pavement. *Compos. Part B Eng.* **2014**, *57*, 262–268. [CrossRef]
53. González, A.; Norambuena-Contreras, J.; Storey, L.; Schlangen, E. Self-healing properties of recycled asphalt mixtures containing metal waste: An approach through microwave radiation heating. *J. Environ. Manag.* **2018**, *214*, 242–251. [CrossRef]
54. Fang, H.; Sun, Y.; Liu, Q.; Li, B.; Wu, S.; Tang, J. Ice melting properties of steel fiber modified asphalt mixtures with induction heating. *IOP Conf. Ser. Mater. Sci. Eng.* **2017**, *182*, 012042. [CrossRef]
55. Liu, Q.; Schlangen, E.; García, Á.; van de Ven, M. Induction heating of electrically conductive porous asphalt concrete. *Constr. Build. Mater.* **2010**, *24*, 1207–1213. [CrossRef]
56. Liu, Q.; Yu, W.; Wu, S.; Schlangen, E.; Pan, P. A comparative study of the induction healing behaviors of hot and warm mix asphalt. *Constr. Build. Mater.* **2017**, *144*, 663–670. [CrossRef]
57. Liu, K.; Fu, C.; Xu, P.; Li, S.; Huang, M. An eco-friendliness inductive asphalt mixture comprising waste steel shavings and waste ferrites. *J. Clean. Prod.* **2021**, *283*, 124639. [CrossRef]
58. Lastra-González, P.; Indacoechea-Vega, I.; Calzada-Pérez, M.A.; Vega-Zamanillo, Á.; Castro-Fresno, D. Assessment of induction heating in the performance of porous asphalt mixtures. *Road Mater. Pavement Des.* **2020**, *21*, 2302–2320. [CrossRef]
59. Oluwasola, E.A.; Hainin, M.R.; Aziz, M.M.A. Comparative evaluation of dense-graded and gap-graded asphalt mix incorporating electric arc furnace steel slag and copper mine tailings. *J. Clean. Prod.* **2016**, *122*, 315–325. [CrossRef]
60. Li, H.; Yu, J.; Wu, S.; Liu, Q.; Wu, Y.; Xu, H.; Li, Y. Effect of moisture conditioning on mechanical and healing properties of inductive asphalt concrete. *Constr. Build. Mater.* **2020**, *241*, 118139. [CrossRef]
61. Karimi, M.M.; Darabi, M.K.; Jahanbakhsh, H.; Jahangiri, B.; Rushing, J.F. Effect of steel wool fibers on mechanical and induction heating response of conductive asphalt concrete. *Int. J. Pavement Eng.* **2020**, *21*, 1755–1768. [CrossRef]
62. Huang, B.; Chen, X.; Shu, X. Effects of electrically conductive additives on laboratory-measured properties of asphalt mixtures. *J. Mater. Civ. Eng.* **2009**, *21*, 612–617. [CrossRef]
63. Wang, H.; Yang, J.; Liao, H.; Chen, X. Electrical and mechanical properties of asphalt concrete containing conductive fibers and fillers. *Constr. Build. Mater.* **2016**, *122*, 184–190. [CrossRef]
64. Ahmedzade, P.; Sengoz, B. Evaluation of steel slag coarse aggregate in hot mix asphalt concrete. *J. Hazard. Mater.* **2009**, *165*, 300–305. [CrossRef] [PubMed]
65. Liu, K.; Fu, C.; Wang, H.; Wang, F.; Xu, P.; Kan, C. Exploring the energy-saving potential of electromagnetic induction pavement via magnetic concentrating technique. *Energy* **2020**, *211*, 118650. [CrossRef]
66. Marinescu, A.; Tudorache, T.; Vintila, A.; Dumbrava, I. A comparative assessment of magnetic concrete versus ferrite for a high power inductive coupler. In Proceedings of the 2021 9th International Conference on Modern Power Systems (MPS), Cluj-Napoca, Romania, 16–17 June 2021.
67. Edwards, K.A.; Al-Abed, S.H.; Hosseini, S.; Brake, N.A. Properties of a magnetic concrete core transformer for application in wireless power transfer systems. *Constr. Build. Mater.* **2019**, *227*, 117041. [CrossRef]
68. Mohamed, N.; Aymen, F.; Alqarni, M.; Turky, R.A.; Alamri, B.; Ali, Z.M.; Aleem, S.H.A. A new wireless charging system for electric vehicles using two receiver coils. *Ain Shams Eng. J.* **2022**, *13*, 101569. [CrossRef]
69. Chabot, A.; Deep, P. 2D Multilayer solution for an electrified road with a built-in charging box. *Road Mater. Pavement Des.* **2019**, *20*, S590–S603. [CrossRef]
70. Chen, F. Inductive power transfer technology for road transport electrification. In *Eco-Efficient Pavement Construction Materials*; Elsevier: Amsterdam, The Netherlands, 2020; pp. 383–399.
71. Chen, F.; Coronado, C.F.; Balieu, R.; Kringos, N. Structural performance of electrified roads: A computational analysis. *J. Clean. Prod.* **2018**, *195*, 1338–1349. [CrossRef]
72. Beeldensa, A.; Hauspiec, P.; Perikd, H. Inductive charging through concrete roads: A Belgian case study and application. In Proceedings of the 1st European Road Infrastructure Congress, Leeds, UK, 18–20 October 2016; 2016.
73. Hanawa, K.; Imura, T.; Abe, N. Basic Evaluation of Electrical Characteristics of Ferrite-less and Capacitor-less Coils by Road Embedment Experiment for Dynamic Wireless Power Transfer. In Proceedings of the 2021 IEEE PELS Workshop on Emerging Technologies: Wireless Power Transfer (WoW), San Diego, CA, USA, 1–4 June 2021.

74. Sun, G.; Yang, Y.; Zhang, J.; Cao, Y.; Tan, X.; Pei, J. Modeling and optimization of pavement scale-model for magnetically coupled resonant in wireless power transmission systems. *Constr. Build. Mater.* **2022**, *319*, 126066. [CrossRef]
75. Aghcheghloo, P.D.; Larkin, T.; Wilson, D.; Holleran, G.; Amirpour, M.; Kim, S.; Bickerton, S.; Covic, G. The effect of an emulator inductive power transfer pad on the temperature of an asphalt pavement. *Constr. Build. Mater.* **2023**, *392*, 131783. [CrossRef]
76. Li, F.; Li, Y.; Zhou, S.; Chen, Y.; Sun, X.; Deng, Y. Wireless power transfer tuning model of electric vehicles with pavement materials as transmission media for energy conservation. *Appl. Energy* **2022**, *323*, 119631. [CrossRef]
77. Feng, Z.; Shimizu, O.; Sumiya, H.; Nagai, S.; Fujimoto, H.; Sato, M. Influence of Contamination Between Receiver Coil and Embedded Transmitter Coil for Dynamic Wireless Power Transfer System. In Proceedings of the 2021 IEEE PELS Workshop on Emerging Technologies: Wireless Power Transfer (WoW), San Diego, CA, USA, 1–4 June 2021.
78. Gil, A.; Taiber, J. A literature review in dynamic wireless power transfer for electric vehicles: Technology and infrastructure integration challenges. In *Sustainable Automotive Technologies 2013: Proceedings of the 5th International Conference ICSAT*; Springer: Berlin/Heidelberg, Germany, 2013.
79. Yuan, C.; Liu, D.-D.; Zhu, Y.-J.; Zeng, T.; Jiang, B.-X.; Tang, C.-X.; Zhou, Y.; He, J.-L.; Li, Q. Effect of charge transport on electrical degradation in polypropylene/organic molecular semiconductor composites for HVDC cable insulation. *Appl. Phys. Lett.* **2023**, *122*, 112904. [CrossRef]
80. Wang, H.; Zhang, X.; Wang, M. Rapid texture depth detection method considering pavement deformation calibration. *Measurement* **2023**, *217*, 113024. [CrossRef]
81. Luo, Z.; Wang, H.; Li, S. Prediction of international roughness index based on stacking fusion model. *Sustainability* **2022**, *14*, 6949. [CrossRef]
82. Khordehbinan, M.; Kaymanesh, M.R. Chemical analysis and middle-low temperature functional of waste polybutadiene rubber polymer modified bitumen. *Pet. Sci. Technol.* **2020**, *38*, 8–17. [CrossRef]
83. Rezaei, S.; Damadi, S.M.; Edrisi, A.; Fakhri, M.; Khordehbinan, M.W. Fatigue analysis of bitumen modified with composite of nano-SiO_2 and styrene butadiene styrene polymer. *Frat. Ed Integrità Strutt.* **2020**, *14*, 202–209. [CrossRef]
84. Shayesteh, A.; Ghasemisalehabadi, E.; Khordehbinan, M.W.; Rostami, T. Finite element method in statistical analysis of flexible pavement. *J. Mar. Sci. Technol.* **2017**, *25*, 15.
85. Wang, H.; Zhang, X.; Jiang, S. A laboratory and field universal estimation method for tire–pavement interaction noise (TPIN) based on 3D image technology. *Sustainability* **2022**, *14*, 12066. [CrossRef]
86. Lin, Z.; Wang, H.; Li, S. Pavement anomaly detection based on transformer and self-supervised learning. *Autom. Constr.* **2022**, *143*, 104544. [CrossRef]
87. Zhu, H.; Xue, M.; Wang, Y.; Yuan, G.; Li, X. Fast visual tracking with siamese oriented region proposal network. *IEEE Signal Process. Lett.* **2022**, *29*, 1437–1441. [CrossRef]
88. Zou, W.; Sun, Y.; Zhou, Y.; Lu, Q.; Nie, Y.; Sun, T.; Peng, L. Limited sensing and deep data mining: A new exploration of developing city-wide parking guidance systems. *IEEE Intell. Transp. Syst. Mag.* **2020**, *14*, 198–215. [CrossRef]
89. Liu, H.; Yue, Y.; Liu, C.; Spencer Jr, B.; Cui, J. Automatic recognition and localization of underground pipelines in GPR B-scans using a deep learning model. *Tunn. Undergr. Space Technol.* **2023**, *134*, 104861. [CrossRef]
90. Xiao, Z.; Shu, J.; Jiang, H.; Min, G.; Chen, H.; Han, Z. Perception task offloading with collaborative computation for autonomous driving. *IEEE J. Sel. Areas Commun.* **2022**, *41*, 457–473. [CrossRef]
91. Yao, Z.; Yoon, H.-S. Hybrid electric vehicle powertrain control based on reinforcement learning. *SAE Int. J. Electrified Veh.* **2021**, *11*, 165–176. [CrossRef]
92. Liu, M.; Gu, Q.; Yang, B.; Yin, Z.; Liu, S.; Yin, L.; Zheng, W. Kinematics model optimization algorithm for six degrees of freedom parallel platform. *Appl. Sci.* **2023**, *13*, 3082. [CrossRef]
93. Luo, P.; Wang, B.; Wang, H.; Ma, F.; Ma, H.; Wang, L. An ultrasmall bolt defect detection method for transmission line inspection. *IEEE Trans. Instrum. Meas.* **2023**, *72*, 1–12. [CrossRef]
94. Zhang, Z.; Li, W.; Yang, J. Analysis of stochastic process to model safety risk in construction industry. *J. Civ. Eng. Manag.* **2021**, *27*, 87–99. [CrossRef]
95. Qu, Z.; Liu, X.; Zheng, M. Temporal-spatial quantum graph convolutional neural network based on Schrödinger approach for traffic congestion prediction. *IEEE Trans. Intell. Transp. Syst.* **2023**, *24*, 8677–8686. [CrossRef]
96. Lv, Z.; Wu, J.; Li, Y.; Song, H. Cross-layer optimization for industrial Internet of Things in real scene digital twins. *IEEE Internet Things J.* **2022**, *9*, 15618–15629. [CrossRef]
97. Lv, Z.; Qiao, L.; Nowak, R. Energy-efficient resource allocation of wireless energy transfer for the internet of everything in digital twins. *IEEE Commun. Mag.* **2022**, *60*, 68–73. [CrossRef]
98. Min, C.; Pan, Y.; Dai, W.; Kawsar, I.; Li, Z.; Wang, G. Trajectory optimization of an electric vehicle with minimum energy consumption using inverse dynamics model and servo constraints. *Mech. Mach. Theory* **2023**, *181*, 105185. [CrossRef]

sustainability

MDPI

Review

A Comprehensive Review of Electric Vehicle Charging Stations with Solar Photovoltaic System Considering Market, Technical Requirements, Network Implications, and Future Challenges

Ali Jawad Alrubaie [1], Mohamed Salem [1,*], Khalid Yahya [2], Mahmoud Mohamed [3,*] and Mohamad Kamarol [1]

[1] School of Electrical and Electronic Engineering, Universiti Sains Malaysia (USM), Nibong Tebal 14300, Penang, Malaysia; ali.jawad@uomus.edu.iq (A.J.A.); eekamarol@usm.my (M.K.)
[2] Department of Electrical and Electronics Engineering, Nisantasi University, 34467 Istanbul, Turkey; khalid.yahya@nisantasi.edu.tr
[3] School of Engineering, Cardiff University, Cardiff CF24 3AA, UK
* Correspondence: salemm@usm.my (M.S.); mohamedmt@cardiff.ac.uk (M.M.)

Abstract: Electric cars (EVs) are getting more and more popular across the globe. While comparing traditional utility grid-based EV charging, photovoltaic (PV) powered EV charging may significantly lessen carbon footprints. However, there are not enough charging stations, which limits the global adoption of EVs. More public places are adding EV charging stations as EV use increases. However, using the current utility grid, which is powered by the fossil fuel basing generating system, to charge EVs has an impact on the distribution system and could not be ecologically beneficial. The current electric vehicle (EV) market, technical requirements including recent studies on various topologies of electric vehicle/photovoltaic systems, charging infrastructure as well as control strategies for Power management of electric vehicle/photovoltaic system., and grid implications including electric vehicle and Plug-in hybrid electric vehicles charging systems, are all examined in depth in this paper. The report gives overview of present EV situation as well as a thorough analysis of significant global EV charging and grid connectivity standards. Finally, the challenges and suggestions for future expansion of the infrastructure of EV charging, grid integration, are evaluated and summarized. It has been determined that PV-grid charging has the ability to create a profit. However, due to the limited capacity of the PV as well as the batteries, the Power system may not be cost effective. Furthermore, since PV is intermittent, it is probable that it will not be able to generate enough electricity to meet consumer demand.

Keywords: EV charging; photovoltaic systems; grid connectivity; standards

1. Introduction

The electrical power and transportation networks are beginning to integrate in a way that was before imaginable thanks to the EV's environmental, technical, and economic potential [1]. The main link between the two is the batteries, which power its EV's traction, control, lights, and air conditioning system. Charging the EV from the power grid, however, places additional load on the utility, especially during high demand hours. Prompting the charge of renewable energy sources is one method to mitigate the grid's negative impact [2]. The use of these clean energy sources is meant to reduce negative environmental consequences while also increasing the overall efficacy of the charging system [3].

Solar energy is becoming widely accepted as a competitive energy source of supplementing the grid due to the ongoing decline in photovoltaic (PV) module prices [4]. In addition, the PV system requires very low maintenance in terms of labor & fuel [5]. The development of energy converting technology, battery management systems, improved installation methods, & design standards have all helped to significantly improve the application for PV to charge EVs (i.e., PVEV charge) [6].

A lot of the time, especially during the day, EV is left lazily sitting in the parking place, exposing to the full sun. It facilitates the expansion of charging options for EVs by making direct use of the "charging-while-parking" concept, it is meant to work in tandem with the standard "charging by halting" method. Installing a photovoltaic system on the parking garage's roof is one easy option for recharging these electric vehicles [7], while the owner of the vehicle is engaged in other activities [8]. The PV powered charging station offers a wide range of advantages, according to the authors in. The savings are particularly significant because charging takes place during the day, while load demand & electricity prices are their highest. Additionally, it has very low CO_2 emissions and small amount of fuel costs. This roof parking facilities are advantageous structural because they offer free shelter from sun & rain, which is important for countries with hot climatic conditions [9].

A PV-power, EV charge station uses PV generation as a secondary power point to recharge EVs, which will cut down on co-emission through fossil fuel-powered plants. In additional words, while the grid is down, EVs may still be charged using PV energy. In addition to reducing peak loads and improving microgrid stability via PV production and V2G, these technologies may also be used to reduce peak loads [10]. However, because to their mobility features, EVs are not the same as an energy storage system. Even if enough EVs are present at the charging station, an V2G may not occur, so would reduce peak power consumption or improve microgrid stability. Even though PV-powered EV charging stations have the potential to increase microgrid stability, there are a number of considerations that must be made [11,12].

The layout of a solar-powered EV charging station is shown in Figure 1. Solar panels, DC/DC converters, EVs, bidirectional EV chargers, as well as bidirectional inverters are the main components of a PV-powered EV charging station. Through a bidirectional inverter, the charging station is connected to the microgrid. The bidirectional inverter allows electricity from the grid to be delivered to the charging station [13,14]. Both bidirectional inverter as well as the microgrid have parallel connections to the local load. The solar array's output may either be utilised to directly serve local customers or added to the utility grid. As a means of monitoring the highest power point of a photovoltaic array, a direct current to direct current (DC/DC) converter is used (MPPT) [15]. The charging and discharging of electric cars are both within the jurisdiction of the bidirectional EV charger. To link EVs to a microgrid, a bidirectional inverter as well as reversible EV charging are required [16].

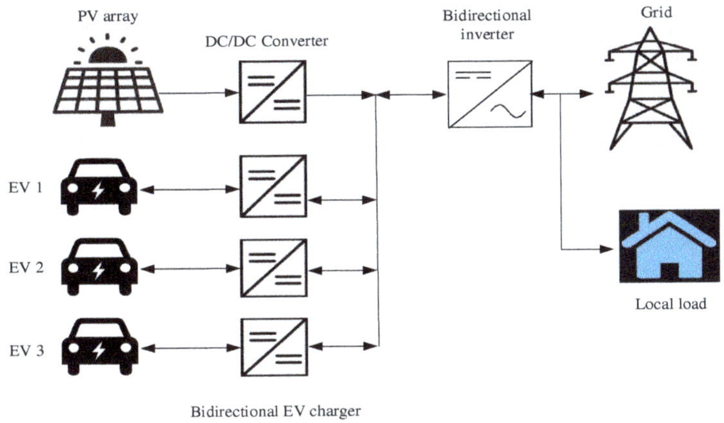

Figure 1. Structure of the investigated PV-powered EV charging station [17].

There are some review papers regarding EVs. Table 1, presents a brief review of some of the existing works along with the most important aspects that investigated in the articles under 7 domains including EV market review and analysis, technical requirements, EV charging infrastructure, grid concepts, overview of the current state of EV and analysis on important global standards, presenting future challenges and suggestions for the development of charging infrastructure, and EV-PV charging system. As it can be seen in the table, most of the mentioned papers are not considered EV-PV stations except Ref. [18] which it is not considered EV's market, and standards.

Table 1. A brief review on some of the related existing review papers carried out on electric vehicle.

Domain	Ref [19] Year 2022 Publisher Elsevier	Ref [20] Year 2021 Publisher Elsevier	Ref [21] Year 2022 Publisher Wiley	Ref [22] Year 2021 Publisher IEEE	Ref [18] Year 2022 Publisher IJRER
EV Market review and analysis	Technical characteristics of 17 EVs are compared. Also, a comprehensive analysis is carried done regarding the countries using EVs as well as introducing the most popular EVs.	The overall view of cost and characteristics of 10 EVs are compared. Also, the number of EVs in some countries are reported as well as the trend of banning of using internal combustion cars	The number of EVs used in some countries and their market's share are shown from 2015 to 2019	There is no analysis regarding EV's market, while in a table a comparison about the technical characteristics of 4 EVs models is done	There is no analysis regarding EV's market
Technical requirements	Investigations of 3 charging techniques considering 5 aspects and the optimum site of charging stations are presented. Analysis of controlling infrastructures including the controlling architecture, centralized and decentralized controlling are presented.	Investigation of fast charging system (DC) considering converters and their instructions is done as well as a comprehensive analysis on different converters.	13 articles are listed regarding the charge standards. Also, different aspects of controlling including harmonics and disharmonic, methods and architecture of charging controls. Discussion regarding the integrating of distribution sources with distribution network is the in the highlights of the study.	Improvement of off -board charger (EVs with IPT) and its infrastructure are investigated, since it addresses the need of a specific protocol for each EV.	Application of passive balancer for energy management of EVs considering various EVs and charge stations are investigated to analyse the output power of PVs. Also, considering: architecture of EV-PV, topology of converters, and 5 optimization methods regarding the improvement of the system
EV Charging infrastructure	Charging methods including BSS, CC, and WPT are discussed along with a comprehensive investigation of various conectors' protocols	On-off board charging systems are discussed along with a comprehensive analysis is carried done on converters as well as cost of charging stations in some countries	Charging infrastructures are discussed	Description of the topology of AC/DC and DC/DC converters. Power factor corrector, two stage onboard chargers and integrated onboard chargers are discussed.	Charge levels and their different moods along with vehicle coil detection system are presented

Table 1. *Cont.*

Domain	Ref [19] Year 2022 Publisher Elsevier	Ref [20] Year 2021 Publisher Elsevier	Ref [21] Year 2022 Publisher Wiley	Ref [22] Year 2021 Publisher IEEE	Ref [18] Year 2022 Publisher IJRER
Grid concepts	Investigation of the positive and negative effects of integration of EVs with grid along with the role of distributors and aggregators on EVGI	An overall description of grid concepts along with the investigation of converter's topologies	Investigation of integration of distributed energy resources with grid considering DER standards and the role of data analysis on DER	A comprehensive description of the effect of charging stations on the grid considering RES, grid stability, demand-supply, assets, and current harmonics	Description of intelligent grid system along with V2G technology and intelligent transportation system
Overview of the current state of EV and analysis on important global standards	25 different standards based on the standards' kind including connector, safety, charging teqniques are presented. Also, presenting the specific standard using by different manufactures	A comprehensive presentation of patents and projects of EV manufactures. Also, a comprehensive study is carried out on the charging stations standards	Investigation of 9 standards implemented in some countries	Charger standards are discussed	There is no description regarding standards
Presenting future challenges and suggestions for the development of charging infrastructure	Discussing the challenges regarding the integrating of grid, V2G technology, and range anxiety. Presenting the challenges and suggestions regarding the integration of grid.	Application of SiC and GaN in converters creates research rooms for fast charging and discharging	DR is considered as a challenge and the solutions are discussed. Challenges and barriers of EV adoption from different aspects including social, policy, and economy are discussed.	Introducing 7 cases that researchers focus on as challenges, such as V2G technology off board charging	Introducing of the different criteria on the charging time as a challenge
EV-PV charging system	There is no description regarding EV-PV	There is no description regarding EV-PV	There is no description regarding EV-PV	There is no description regarding EV-PV	A comprehensive study on EV-PV stations along with their architectures

Considering more aspects of the research regarding EVs would cause to have a comprehensive source for readers. In this review paper, for the first time, all of the mentioned domains in Table 1 are systematically discussed to have an overview regarding other important related factors of EV-PV stations. The contributions of the present work are as follows:

- Electric Vehicle and Plug-in hybrid electric vehicles Charging Systems: global deployment of electric vehicle charging infrastructure, charging systems and their standardization, classification of electric vehicles charging levels, specifications, and standards, standards for electric vehicle charging as well as grid integration, electric vehicle charging standards, electric vehicle grid integration standards, safety standards for electric vehicle, electric vehicle integration in the power grid, modelling of grid-connected electric vehicle-photovoltaic system, electric vehicle smart charging using photovoltaic and grid

- Control Strategies for Power Management of Electric Vehicle/Photovoltaic System: intelligent energy management strategy, energy management strategy for smart home integrated with electric vehicle, and photovoltaic, control strategy for power electronic components
- Recent Studies on Various Topologies of Electric Vehicle/Photovoltaic Systems
- Challenges and Future Work Recommendations: modelling, optimization and control, issue on the integration with smart grid system, challenges and suggestions for electric vehicle charging

Figure 2 shows a flow diagram to see how the bibliometric review and research method and other parts of the current study is carried out.

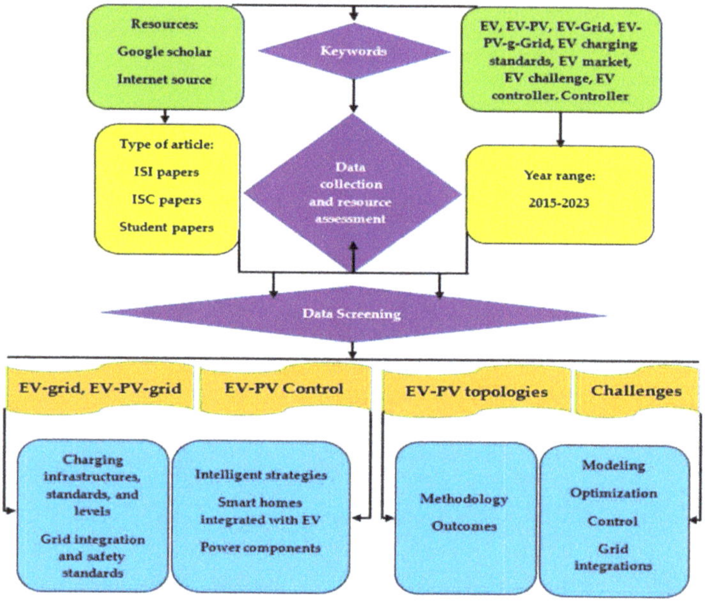

Figure 2. Detailed research method.

2. Electric Vehicle and Plug-in Hybrid Electric Vehicles Charging Systems

2.1. Global Deployment of Electric Vehicle Charging Infrastructure

To that end, the federal government of the United States has mandated that one million electric vehicles be in use by the year 2015, as well as numerous regulations have been put into place to encourage electrification of all spheres of public life. According to the Ontario Ministry of Transportation, using money of its Green Investment Fund, the Canadian state of Ontario plans to construct 500 EV charging stations (EVCSs) in around 250 places throughout the province by the end of 2017. It is estimated that by 2020, Germany will need around 70,000 public on-street charging spaces, according to the country's National Electric Mobility Platform (NPE) [23]. As a solution to the problems caused by China's current approaches to exploiting renewable energy and to keeping up with the ever-increasing energy needs of electric cars, the concept of placing a limited number to solar-powered charging stations to EVs is presented [24]. In May of 2017, the United States, Canada, France, Germany, Japan, the Netherlands, Norway, Sweden, and the United Kingdom formed the Electric Vehicles Initiative (EVI), a multi-govt policy forum with the goal of promoting the global adoption of EVs. South Africa, which joined the EVI in 2016 but remains a member, is actively involved in EVI operations, as do Korea and India [25]. The Indian government as well as major automakers have joined forces to promote e-vehicles as well as other clean fuel options in an effort to lower transportation-related pollution.

Thus, plan (NEMMP) 2020 for the nation's EV mission was initially suggested in 2013, and the following year it was signed into law. Faster Adoption and Manufacturing of (Hybrid &) Electric Vehicles (FAME) is another initiative in this space. The year 2015 also saw the announcement of India's intention to push the use of electric vehicles, and the India Scheme is introduced to support that effort [26]. It subsidises both utilities and their customers. The absence of charging outlets is now the biggest problem with India's infrastructure. There are still not a lot of places to charge your car around the country. Ninety-five percent of the world's electric vehicle stock and registrations are held by EVI members [27].

2.2. Charging Systems and Their Standardization

The classification of EVs charging levels, specifications, and standards, as well as in the next section, we will talk about controls strategies for EV/PV system power management. Finally, a review of the issues and the prospects for the future is provided, with an emphasis on the energy management system.

Electric Vehicle Charging Standards

Full EVs enable total electrification of the transportation industry, in contrast, Electric Vehicle and Plug-in hybrid electric vehicles (PHEVs) are just partially electrified automobiles [28]. Various charging states are shown in Figure 3. In addition to high-power off-board chargers, Plug-in electric vehicles may also be charged continuously at any time of day or night using on-board level 1 or level 2 chargers (PEVs). An on-board integrated charge that can rapidly charge PEVs may combine the best features of conventional on-board and off-board chargers. Different charging levels had been defined by various organization throughout the globe, with an emphasis on the architecture of the vehicle and the kind and size of the battery utilized [29]. As a result, each car charges differently and behaves differently when charging. Therefore, the PV-EV charger must be designed in accordance with the aforementioned specifications.

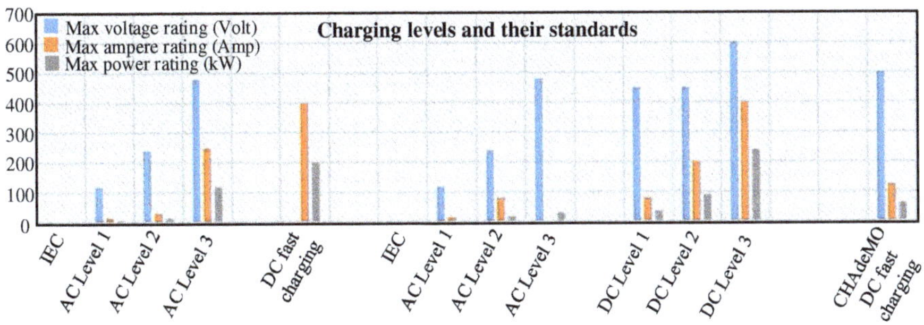

Figure 3. Graphical comparison of different charging levels and related standards [7].

2.3. Classification of Electric Vehicles Charging Levels, Specifications, and Standards

Conductive charging, wireless (or contactless) charging, and battery swapping are the three ways to refill an electric vehicle as classified in Figure 4. The most common and easiest way to charge anything nowadays is via conductivity [30,31]. In conductivity charging, the power source and battery are connected by a cable, whereas in wireless charging (WC), they are not. As opposed to conductive charging, WCs and battery shifting are currently under investigation and development [32,33]. The sections that follow will give further information on these technologies. Figure 4 provides a categorization of several charging methods.

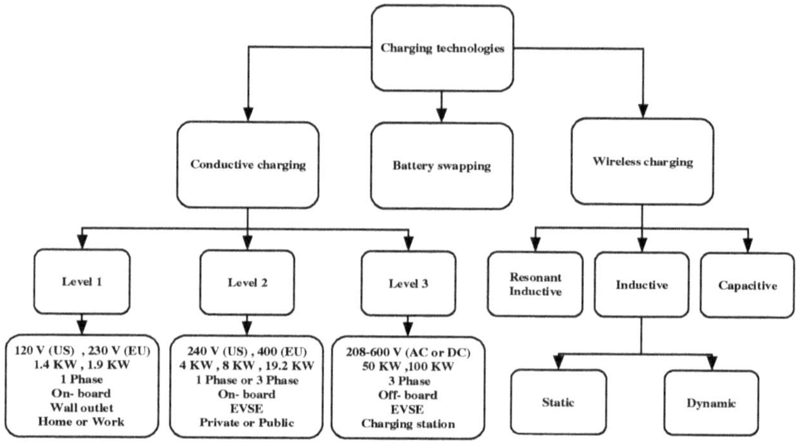

Figure 4. Classification of EVs charging technologies based on conductive charging, wireless charging, and battery swapping [34].

2.3.1. Conductive Charging

EV battery charger play crucial role for development of EVs since the adoption & societal acceptability of EVs is contingent on the ease of access of charging stations and public chargers. Various topologies of single phase & three phase EV charger are discussed [35]. A power factor adjustment unit, a DC-DC converter, and an AC-DC converter are the three components that make up this setup. There are two types of charger systems: on-board (inside the car, for slow charging) as well as off-board (outside the vehicle, for quick charging) (i.e., outside vehicle for fast charging) [36]. In addition, these chargers may be categorized as unidirectional or bidirectional. The gear for unidirectional charging is basic and merely permits electricity to flow from the grid to the electric vehicle. Bidirectional charging permits power to be transferred from the vehicle's charging station to the battery while driving on a public road; also known as "charging" to provide energy to a structure, the grid, or a home [37]. Potentially alleviating some of the stress experienced by EV owners and lowering the amount of energy storage required for onboarding may be achieved via the availability and development the EV charging infrastructure. The Society of Automotive Engineers (SAE) has established three different charging thresholds in its standard J1772 [38]. Level 1 and 2 chargers will be the norm for residential use, level 3 will be used by public charging stations, as reported by Electric Power Research Institute (EPRI) [38].

Level 1 Charging

Due to lack of extra infrastructure and the ability to use any wall outlet, this method of charging is both the slowest and easiest. For Level 1, a typical 120 V/15 A wall outlet is utilized in the US. Only by on-board charger is offered. This EV takes a time to charge completely, although being less expensive than other charging levels. This charging level has the least negative effects on distributing networks because of its low power rating [39].

Level 2 Charging

Level 2 charging utilizes 19.2 kW of charging power and 208 V or 240 V at currents up to 80 A. Because Level 2 requires less time to charge than Level 1, EV owners prefer it. For both private and public charging, specific electric vehicle supply equipment (EVSE) installation may be necessary. Electric vehicles like the Nissan Leaf have this much charging capacity built in [40].

Level 3 Charging

Level 3 is used for quick charging & functions like typical gas station (i.e., less than an hour of charging time) that may be put on major roads and highways. Due to the high charging power, which may surpass 100 kW it is only accessible as an off-board charger and is fed from a 3-phasing circuit with 480 V or greater voltages. It was obvious that level 3 charging is inappropriate for use at home. Its expensive installation raises the possibility of a problem [41]. Level 2 as well as Level 3 charging stations are expected to be widely used in public places including shopping centers, garages, restaurants, hotels, movie theatres, and other entertainment venues to facilitate rapid charging. Expensive charging power is advantageous to the perspective of charging time, but it may also result in peak demand, equipment overloaded on distributing network, and of high installation costs [42,43].

2.3.2. Wireless Charging

With WC, electric vehicles may be charged without a cord or any other direct physical contact here between power source and the battery. To reduce the cost, size, and environmental impact of EVs, advances in WC will reduce the amount of energy needed from the vehicles' onboard batteries. WC could replace conventional conductive charging in the future. The possibility exists to charge the batteries in electric busses using WC. It will function at various voltages (level 1, 2, & 3). 90% is at greatest efficiency for WC ever noted. WCs utilizes the inductive, resonant inductive, and capacitive technologies [34]. We'll go through inductive wireless charging (IWC) to give you an understanding of how the technology works. The AC electricity from the power grid is converted to DC power via the IWC's built-in AC/DC converter. It is then transformed once again into AC power and delivered to the transmitting (or main) coil at a high frequency. These parts are all located below under the roadway. The receiving (secondary) coil in the EV gets electricity from the transmitting coil via the air gap using electromagnetic induction [44–46]. After an AC/DC converter transforms the energy from an incoming power source into a usable direct current, the battery may be charged. Static inductive charging & dynamic inductive charging are two categories for WC. EVs must remain stationary while being charged using static inductive technology. Dynamic inductive charging, however, permits WC while an EV is in motion [47,48].

Existing WC implementations are geared at facilitating one-way flow of power from the grid to a vehicle, but developments in this area will enable electric vehicles to wirelessly withdraw the grid's energy to provide power. These benefits of this technology are user ease, cable-free operation, and electrical safety. The disadvantages of this technique are the poor power transfer efficiency between prices and the expensive infrastructure expenditures compared to conductive charging [49].

2.3.3. Battery Swapping

BSSs are charging station where empty batteries may be swapped out. A fully charged battery will soon replace the EV battery. Electric buses with large batteries that require a long time to charge using conventional conductive charging may employ battery swapping. The BSS or a third party that rents batteries to EV owners is required to maintain a sizable inventory for this technology [50]. The BSS is equipped with a distribution transformer, batteries, battery switching equipment, and AC/DC converters for charging the batteries. According to certain research, BSS might leverage bidirectional charges to provide electricity services using a V2G paradigm. Battery uniformity, significant infrastructure costs, and a huge BSS footprint are obstacles for this technology. A battery switching technology that can change the battery in 90 s was unveiled by Tesla Company in 2013 [51].

2.4. Standards for Electric Vehicle Charging as Well as Grid Integration

As a result of using EVs, the auto and power sectors have more room to expand. Every part of this new technology must be standardized if it is to be used consistently over the world. The standardization of EV charging may be divided into three categories according on the Table 2: EV charging component standards, EVGI standards, & safety requirements. Component-level EV charging standards are developed by the International Organization for Standardization (ISO) and others, whereas ISO focuses on EV standardization as a whole [52,53].

Table 2. Classification table of comparison between different type of EVs charging levels, specifications, and standards.

S.No	Charging Station	Voltage(V)	Power(kW)	Type of Vehicle (Wheels)	Type of Compatible Charger
1	Level 1 (AC)	240	$\leq$35	4, 3, 2	Type 1, Bharat AC-001
2	Level1 (DC)	$\geq$48	$\leq$15	4, 3, 2	Bharat DC-001
3	Level 2 (AC)	380–400	$\leq$22	4, 3, 2	Type 1, Type 2, GB/T, Bharat AD-001
4	Level 3 (AC)	200–1000	22 to 4.3	4	Type 2
5	Level 3 (DC)	200–1000	Up to 400	4	Type 2, CHAdeMO, CCS1, CCS2

The requirements for EV discharging and charging into the grid have been specified. Electric vehicles serve as a DEV during grid charging as well as discharge (DER). As a result, DERs' grid connectivity regulations also apply to EVGI. The IEEE (Institute of Electrical and Electronics Engineers Engineers) as well as Underwriters Laboratories (UL) are two of the most influential organizations in the development of standards for grid connection [54]. The majority of the aforementioned organizations have established its safety requirements for EV charging & grid connectivity. On the other hand, the National Fire Protection Association (NFPA) and National Electrical Code (NEC) are heavily concerned with safety [55]. Subsequent sections elaborate on the norms and regulations established by such groups.

2.5. Electric Vehicle Charging Standards

EV charging infrastructure is included in a few international standards. Whereas IEC is commonly used in Europe, SAE & IEEE are employed by manufacturers headquartered in the United States [56]. The CHAdeMO EV charging protocol was developed in Japan. The Guobiao (GB/T) standard [52] is used in China for both alternating current (AC) and direct current (DC) charging, IEC standards [53] are identical to those of the GB/T for AC charging [55]. The Chinese National Committee for ISO as well as IEC developed this specification. Since the IEC [57] as well as SAE standards [38] for EV charging are the most widely used, we devote a lot of space to them here. From the specifications alone, it is evident that IEC61852 and SAE J1772 [38] are almost similar with the exception of certain language differences. While "level" is used to describe the intensity of an output in SAE, "mode" is the preferred term in the International Electrotechnical Commission (IEC) [57].

International Electrotechnical Commission standards

The IEC is a British organization that establishes norms for many forms of electrical, electronic, as well as related technologies.

- IEC61851 [52]. The IEC 61851 outlines the requirements for charging EVs and PHEVs using on-board and some off equipment utilising 1000 V AC and 1500 V DC supply voltages.
- IEC 61980 [52]. WPT systems up to 1000 V AC or 1500 V DC are covered under the IEC 61980 standard. This requirement applies to the WPT system that may be accessed using local storage facilities [57].
- IEC62196 [58]. IEC62196 specifies the plugs, socket outlets, vehicle connections, as well as vehicle inlets used for electrically conducting electric cars.

- SAE standards. The SAE is professional organization with headquarters in the United States that creates standards for engineering organizations in many sectors.
- SAEJ2293 [59]. For use with either an on-board or external power source, check out SAEJ2293 for detailed specifications. Part one of this standard, J2293-2, covers the information requirements and network architecture of EV charging, Part One of the J2293 standard discusses the energy needs and system design in three different operational scenarios (conductive AC charging, conductive DC charging, as well as inductive charging) [59].
- SAEJ1772 [38]. The current rating of circuit breakers and the voltage rating of chargers are both included in SAEJ1772. Both AC and DC are included in the definition of the standard, and each includes three levels. Almost all modern cars can receive Level 2 AC onboard charging at current that flows of less than 30 A. In terms of DC charging standards, the SAE DC level provides the fastest possible rate. However, a number of other elements, like as infrastructure and battery chemistry, have an impact on the actual charging rate [60].
- SAEJ1773 [38]. This standard outlines the minimal specifications for EV inductively linked charging systems. Inductive charging systems must be manually linked in accordance with SAEJ1773, which also specifies the criteria for the software interface [61].
- SAEJ2847 & SAEJ2836 [62]. These two standards, together with SAEJ1772, specify what electric vehicles and their charging stations must be able to communicate with one another. SAEJ2836 describes the scenario and provides the testing environment; SAEJ2847 details the necessary means of communication [62].
- SAEJ2931 [62]. This standard defines the requirements for digital interaction between electric vehicles, EVSEs, utilities, power utility interfaces, smart meter infrastructure, and home area networks. To make smart grid recharging of electric vehicles possible, a communication network must be constructed in accordance with SAEJ2931 standards.
- SAEJ2954 & SAEJ2954 [62] recommended practice (RP). When it comes to wireless charging, SAEJ2954 only supports level 2 (7.7 kW), while a recently published RP version claims to support level 3. (11 kW). For electric vehicle producers, the updated version will serve as a more uniform testing ground as well as infrastructure companies to evaluate the efficacy and validity of new products. This standard also incorporates autonomous charging, smooth EV parking, and payment setup [63].

2.6. Electric Vehicle Grid Integration Standards

IEEE1547, UL1741, & NFPA70 [32] are three accessible standards and codes. The following list of standards and codes highlights their key features.

- IEEE1547 [32] is a set of standards for integrating decentralized energy sources into centralized electrical grids. Specifically, performance, operation, testing, safety issues, and maintenance needs for interconnecting DERs are discussed, as well as DER installation on both secondary and primary network distribution networks with a PCC aggregate capacity of 10 MVA or less, it is appropriate to all DER technologies [64].
- UL standards [32] UL released a number of standards to address various DER grid interconnection issues. The criteria of power conversion equipment and related protective devices as they pertain to DER grid integration are covered in UL 1741 [32], the most relevant of these standards. Additionally, we adhere to UL62109, UL62109-1, UL62109-2 [32], as well as UL1741 SA [65].

2.7. Safety Standards for Electric Vehicle

This safety safeguard is a necessary component of grid connectivity and EV charging. Even though the majority of standardization bodies include safety requirements, NFPA and NEC place a strong emphasis on safety and security. Below is further information on the codes that these two organizations have created for EV charging and grid interconnection.

2.7.1. National Fire Protection Association Requirements

When it comes to educating the public about fire, electrical, and life safety, NFPA was an early adopter. The EV & its grid integration community has published its NFPA 70 [32] standard, which details wiring and safety procedures for electrical equipment on the customer side of a PCC [66].

They include:

- Electrical equipment and conductors mounted on or inside or outside of public or private buildings & other structure.
- Electrical wires that link the installations to an electrical supply as well as other external conductors & on-site machinery.
- Fiber optic cable.
- Structures utilized by the electric utility but not necessarily a component of power plant, a substation, or a control room [67].

2.7.2. National Electrical Code Standards

NEC [32] is different supplier of standards that focuses on EV safety measures. Additionally, it offers the specifications for EV charging hardware.

- NEC 625 [32], "Electric Vehicle Charging as well as Supply Equipment Systems", details the norms for EV charging infrastructure that is not part of the vehicle itself. Installation guidelines for EV charging station hardware are included. This includes things like conductors, connecting connections, including inductive charging devices, which are all part of the charging infrastructure connected either to feeder or branch circuits [68].
- NEC 626 [32]. The area of truck parking spaces is covered by this standard, which is labeled "Electrified Truck Parking Spaces." It specifies the requirements for the electrical apparatus and conductors used to charge trucks that are located outside of the vehicle. Circuit breakers, grounds, cable diameters, back feed prevention, & other requirements are among them [69].

2.8. Electric Vehicle Integration in the Power Grid

Until recently, there was only tenuous connection between the transportation & electric power industries. The widespread electrification of transportation has significantly altered the established economic strategies of electric utilities [20]. Overall, EVs have given the electrical grid both considerable problems and advantages.

Impacts of Electric Vehicle Integration on the Grid

EVGI has both positive and negative outcomes. The next subsections and Figure 5 illustrate the specifics.

- Negative Impacts: Electric utilities face a huge difficulty as a result of EVs. According to Table 3, excessive EV integration into the distribution network may have an effect on the distribution grid's stability, power losses, voltage and frequency imbalances, load profile, and component capacity [70].
- Positive impacts: It's true that having too many EVs upon that grid might cause issues with power quality degradation, an increase in peak demand, and challenges with power regulation, all of these concerns can be managed by adopting modern power management methods. The benefits of integrating EVs into grid in a strategic manner are summarised in Table 4 [71].

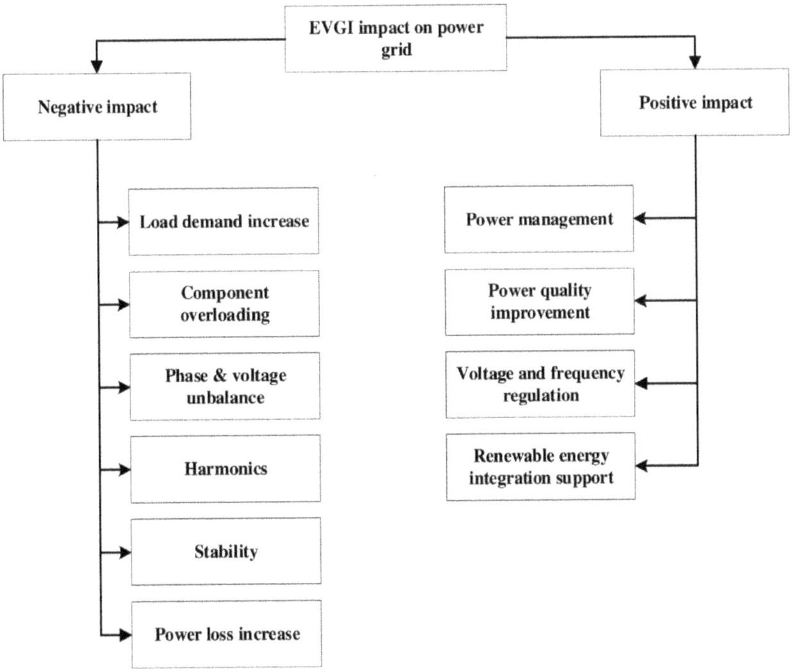

Figure 5. The positive and negative impacts of EV grid integration [32].

Table 3. Negative Impacts of EV grid integration along with the corresponding descriptions [32].

Impacts	Description
Load demand increase	Up to 1000 TWh of extra loads (a 25% increase from present levels) may be added by EVGI. If electric vehicle charging is not controlled, the surge in demand during peak times might be a serious issue for utilities.
Component overloading	Extra-large EVGI values provide an incremental load demand that must be produced and communicated. The new demands are too much for the components of the current power system, which might lead to overloading and reduce transformer lifetime.
Phase and voltage unbalance	Since EV chargers are single phased, charging a lot of EVs at once might lead to phase imbalance.
Harmonics injection	Harmonics are created during the power conversion process by EV chargers since they are power electronic equipment as well as, if their penetration is larger, harmonize the grid. Although the THD level may increase as more chargers are being used, several studies suggest that it is less than 1% due to EV charging.
Power loss	A significant quantity of actual power is used as EVs become more and more integrated into the grid, which results in distribution system power loss. Given that 60% of the cars linked to the distribution system are electric vehicles, During off-peak times, power loss might increase by as much as 40%. Coordinated charging may increase the load factor of the electrical system while minimizing power losses. Power loss in the grid may be reduced by strategically placing charging stations.
Stability	The power system becomes unstable when electric vehicle loads are added since they are nonlinear and consume lot of power rapidly. More electric vehicles on the road makes the power infrastructure more vulnerable to outages and increases the time it takes to get things back to normal. EVGI has the potential to improve the stability of the electricity system if properly handled.

Table 4. Positive Impacts of EV grid integration along with the corresponding descriptions [32].

Power management	The use of planned charging and draining may improve power management. Discharging may be scheduled during peak hours to meet peak load demand.
Power quality improvement	Controlled EVGI may reduce voltage surges brought on by unchecked DER penetration. Flickering voltage may be reduced. When necessary, reactive power may be introduced. Uncontrolled DER harmonic injection may be minimized.
Regulation	Frequency control using grid frequency deviation regulation. Achieving voltage regulation via the production or consumption of reactive power. An equilibrium in power flow is achieved by the use of energy storage. Absorption of power that ramps up. An increase in the stability of isolated electric networks.
Renewable energy support	Using electric vehicles as energy storage might help smooth out the fluctuations in renewable energy production. Using electric vehicles as a buffer for renewable energy might reduce emissions and save money.

2.9. Modeling of Grid-Connected Electric Vehicle-Photovoltaic System

2.9.1. Photovoltaic System

In this study, a diode serves as the *PV* module. As a result, a diode, a current source with a variable output, as well as a series as well as parallel resistance compensate the *PV* cell idea (R_s and R_p). Evidence for the dependability of this paradigm already exists. To determine the output current (I_{pv}) ones from a *PV* system, apply the following [72,73]

$$I_{PV} = I_{Ph} - I_{sat}\left(e^{q(V_{PV}+I_{PV}R_s)/(NkT_{PV})}\right) - (V_{PV} + I_{PV}R_s)/R_P \quad (1)$$

$$I_{sat} = K_1 T^3 e^{(-\frac{qV_g}{kT})} \quad (2)$$

where K represents the Boltzmann constant, I_{sat} represents the saturation current of the diode, Ko and K1 represent constants that are dependent on the features of *PV* cells, in the *PV* model, N represents the input impedance of the an diode, q represents an elementary of an electron, and T represents the temperature at which the *PV* cells are active [74].

A DC/DC converter completes the virtual photovoltaic system. This part controls the PV system's output voltage as well as makes the *PV* voltage compatible with MVDC bus voltage so that the *PV* system can run in maximum power point tracking mode. The DC/DC converter might be switching to a different mode (DC bus voltage sustaining mode) in order to maintain a constant MVDC bus voltage, resulting in a reduction in the quantity of power provided by the *PV* modules [75].

2.9.2. Electric Vehicle as Well as Energy Storage System Batteries

The SimPowerSystems toolbox of Simulink's model has been used to simulate these batteries for both EVs & energy storage system (ESS). One resistance and one variable voltage source are coupled in series to make up this type. As a result, battery voltage may be determined by the steps below [8].

$$V_{bat} = E_{bat} - I_{bat}R_{int} \quad (3)$$

where R_{int} is the internal resistance of the battery, I_{bat} is the current through the battery, and E_{bat} is the open-circuit voltage, that changes with the state of charge or discharge. To prevent the battery from being damaged by excessive charging or discharging, one of the most crucial battery metrics to be monitored is the state-of-charge (*SOC*) [76]. Inferring it from following expression:

$$SOC\ (\%) = SOC\ (\%) - 100\left(\frac{\int I_{bat}dt}{Q}\right) \quad (4)$$

Q = maximum battery capacity.

2.9.3. Direct Current/Direct Current Converters

All of the DC/DC converters' dynamic behavior has been modeled using average-value equivalents. These models are often used to investigate lengthy simulation situations because it may be possible to recreate the converters' dynamic behavior faithfully. This DC/DC converter type consists of a voltage source as well as a current source that may each be adjusted independently [14]. Depending on the converter being utilized, the duty cycle links the input, output voltages & currents. For a complete rundown of a DC/DC converters found in the tested systems, see Table 5.

Table 5. Summary of the dc/dc converters found in the tested systems and functions [77].

Component	Converter	Energy Flow (From–To)	Transfer Function
PV	Unidirectional-boost	(PV–DC bus)–boost	$M_{DC} = \frac{V\alpha}{V_j} = \frac{V_i}{V_o} = \frac{1}{1-D_C}$
EV	Unidirectional-buck	(DC bus–EV)–buck	$M_{DC} = \frac{V_o}{V_i} = \frac{V_i}{V_g} = \frac{D_C}{1-D_C}$
EV	Bidirectional	(ESS–DC bus)–boost	$M_{DC} = \frac{V\alpha}{V_i} = \frac{V_i}{V_o} = \frac{1}{1-D_C}$

2.9.4. Grid Connection

The link to the neighboring 20 kV AC grid is made possible via a delta-wye transformer as well as a three-phase IGBT inverter. This components' representation based on models created. A Space-Vector PWM modulator is used to drive the inverter under control [78].

2.10. Electric Vehicle Smart Charging Using Photovoltaic and Grid

Numerous research has looked at the benefits of PV-based EV charger systems. Reference highlights the benefit of charging the EV with PV and explains how it enables for higher penetration of both PV as well as EV. Additionally, EVs may lessen the consequences of excessive PV production. Reference provides a case study of Columbus, Ohio, to demonstrate that solar-powered EV charging is both cheaper and less polluting than traditional grid-based EV charging [79]. The grid integrated PV system outperforms the other two systems economically, a case study contrasting three different approaches to recharging electric vehicles (grid alone, PV just plus battery storage, as well as grid integrated PV). The authors of reference discuss the use of PV power and EVs as a battery system to reduce peak grid demands. Studies like this show how much better PV-based EV charging is compared with grid-based EV charging. Numerous papers discuss alternative charging algorithms and what role they may play in meeting the financial, technological, and societal goals of PV-based EV charging [80].

The problem formulation determines the kind of optimization model. Convex type problems (linear, mixed-integer, and quadratic) often have cheap processing costs and may reach optimum solutions [81]. Optimization strategies of the meta-heuristic kind, such the Genetic Algorithm and the Particle Swarm Optimization, may be used to solve non-convex problems and provide solutions that are close to optimum while requiring little computing effort. With limited data and processing resource needs, the rule-based algorithm or heuristic type optimization techniques may give adequate solutions for arbitrary instantaneous occurrences (such as the plugging in/unplugging of EVs or the change in PV power, for example) [82].

The reason why PV-based EV charging systems for homes and offices have received more attention than those used in commercial settings is that they need less complex analysis and flexible integration inside the distribution system. Additionally, the majority of research on smart charging focuses on particular ways to improve EV grid integration, for example, variable rates of pricing, market participation, and supplementary offerings A complex system with many different components is needed to simulate the real-life implementation [83].

3. Control Strategies for Power Management of Electric Vehicle/Photovoltaic System

The power flow between the system's four key components must be investigated in order to create the suggested system layout. Electrical grid, PV sources, battery storage, as well as EV charging load are the major components. Power flow management can be used to establish the demand for and size of a bidirectional power flow energy automated device. Consequently, the study introduces their applicability on the examined application in an effort to answer the power flow management issue [83,84].

Heuristic algorithms that take into consideration loaded demand, PV insolation levels, including off utility hours are often used to estimate power flow in energy PV/battery systems. However, the proposed system's solution is made considerably more difficult by a dynamic grid pricing [85,86]. Using the streamlined heuristic concepts to operate the PV/battery system will provide operational cost solutions that differ considerably from of the least cost operation underneath a dynamic grid pricing model. As a result, the pace of research in this field has quickened [87].

The intended architecture for power flow management must allow for non-linear operations. This enables the designed topology to be used to various operational circumstances. To enable the system to function properly under mismatched circumstances and predicting deviations, the topology must feature an online error compensation phase [88,89]. Last but not least, the online optimization step should be built with a short calculation time so that it may be quickly incorporated into real-time controllers. To get over these issues and carry out the suggested optimization, research on smart methods of energy management and flexible power distribution is urgently required [90].

3.1. Intelligent Energy Management Strategy

In its present form, the EV-PV charger should be able to charge an electric vehicle using solar power, but it has no intelligence of its own [91]. According to forecasts, the cost of electricity will be lowest in the morning hours, this being an ideal moment to plug in your electric vehicle to the grid for a recharge. While solar charging is most effective during sunny afternoons. Smart charging algorithms are required for the control of EV-PV systems to be realized [92].

Every car has a predictable period of accessibility as a load, and this condition of charging the automobiles at parking lots has been taken into consideration. While parking the car, the user specifies the billing interval and makes any other relevant price/billing type/etc. selections.

Capacity, chemical composition, open-circuit voltage, battery state of charge, ambient temperature, and so on are all characteristics of a vehicle. Every car will allocate the power to the system based on the user's preferences and the battery's characteristics. So, in order to sum up, dynamic power management of loads takes into account user preferences and load variables to optimize energy consumption [93,94]. Improving demand-side power management, control mechanisms, as well as consumer preferences for utility services are the cornerstones of grid design. A specific system was designed to show how the smart grid works [95,96]. The skeleton of the whole system is shown in Figure 6. All of the advantages outlined and taken into account during the system's design are realized via battery loads, charging, and connection to the power grid all fall within the purview of energy management. Every parking deck is controlled by the Intelligent Energy Management Strategy (IEMS), and each parking deck is made up of different loads [97,98].

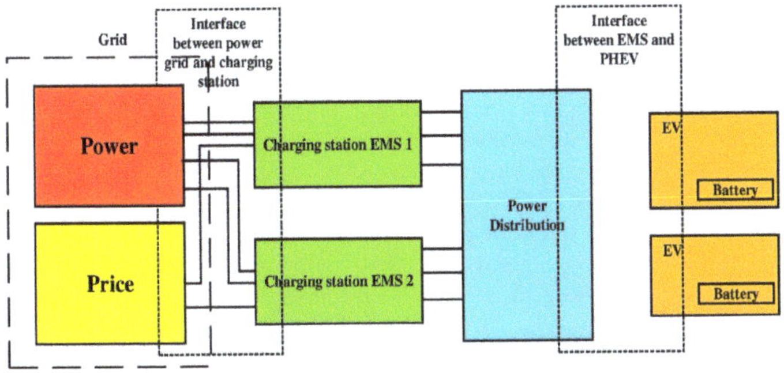

Figure 6. Architecture of the intelligent energy management system considering the performance of a smart grid [99].

3.2. Energy Management Strategy for Smart Home Integrated with Electric Vehicle, and Photovoltaic

To limit load power curve volatility and associated energy costs in homes with PV and EV, energy management measures may be essential. With the proposed method in [99]., electricity may be transferred from the grid to vehicles (G2V), photovoltaics to vehicles (PV2V), as well as vehicles to houses (V2H), therefore reducing daily energy expenses and smoothing out graph of smart home electrical consumption [100]. When PV power generation is unavailable (PV power is equal to 0), stage A of the proposed control technique has three modes of operation, while Stage B has five modes of operation (PV power greater than zero) [101]. The primary objectives of the approach are to manage the energy drawn from the grid and/or PV and the energy given to a load demand by EVs, and to regulate the charging and discharging cycles of EVs. Maintaining the load curve inside the optimal energy consumption range will reduce power bills, the following goals may be attained [102].

- Find out when the price of power is at its lowest and highest.
- Determine when the residence under study uses the most and the least power.
- To save energy costs and increase valley capacity, at moments of low power demand as well as cheap energy costs, charge electric cars from the grid utility.
- Manage the battery's level of charge to avoid overcharging and overdischarging it when charging and discharging, respectively [103].
- Learn when PV production is inversely proportional to the amount of needed power that can be supplied by photovoltaics, and use those periods to charge your electric vehicle from the extra energy you've generated.
- Electric car discharge during peak demand periods has the potential to minimise peak demand [104].

The following data is used as inputs in the proposed method to accomplish the aforementioned aims: the time the EV Tin arrives, the time the EV Tout leaves, SOC of an electric vehicle battery at time t, maximum (SOCmax) and minimum (SOCmin) levels of charge, average power load (Pavg) and average price signal (Cavg), and time-of-use price signal (C) are all variables to consider; the profile of solar power production Ppv (t); and so on [105].

3.3. Intelligent Energy Management Strategy

In its current version, the EV-PV charger can take in solar energy and charge the EV, but it does not have any specialised knowledge on how to do it. The cost of electricity is predicted to remain low throughout the morning; hence, charging an EV from the grid is more profitable in the morning. While bright afternoons are advantageous for solar charging. A smart charging algorithm is required for the control of an EV-PV system [106,107].

As each parked car is available for a certain amount of time as a load, the pricing situation in carparks has been investigated. The user chooses the charging period, price, and kind of charge at the time of parking [108,109].

A vehicle's capacity, chemistry, open-circuit voltage, charging levels, the state of charge, temperature, and similar factors are all additional features. Every car will allocate power to the system based on user selection and battery parameters. To summarise, the system's energy usage is maximised by taking into load factors and customer preferences into consideration while implementing dynamic power management [110]. Control methods, advanced demand-side power management, as well as customer service preferences are some of the primary drivers behind grid layout. A specialised system was designed to demonstrate the benefits of the smart grid. All the advantages outlined and results expected by integrating a power grid, charging infrastructure, and battery loads into an existing energy management system. Each parking structure under the IEMS's jurisdiction holds a broad range of cars and must adhere to its rules [111,112]. The IEMS is automatically updated with information about available power and cost from the grid's perspective at a given time interval. This approach assists IEMS in monitoring several converters at the same time in order to make real-time choices [113–116].

3.4. *Control Strategy for Power Electronic Components*

Power converters are used to link all the parts of the system to a DC power source. The maximum output of a PV array is constantly monitored by the boost converter. When it comes to performance as well as battery life, a bidirectional DC/DC conversion is used for charging and charging the battery. Electric cars need two separate control systems in order to work with a broad variety of battery sizes. We'll look at two types of controllers: (1) the I-V controller, as well as (2) the average sliding mode control [117].

3.4.1. I–V Controller for Electric Vehicle

The EV runs in a current loop mode to enhance an EV's charging capacity. Charging the EV requires just a correct current reference control algorithm [118]. The system's health determines how often this resource is refreshed. Feedback from a comparison of a current drawn from a PI controller is able to maintain a steady-state error of zero by connecting the batteries of a hybrid or electric car to a current reference illustrated in Figure 7 [119,120].

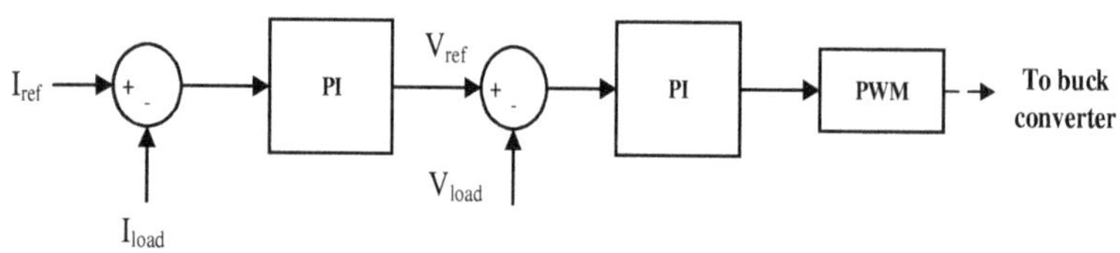

Figure 7. Block diagram of I-V based controller for EV from input to output [100].

Meanwhile, the EV's battery voltage is compared to a value computed by the PI controller. The DC/DC converter is controlled by pulses generated from an external PI block's output. The electric vehicle may be charged to a maximum of 6 kW when overall power generation exceeds consumption [121].

3.4.2. Average Sliding Mode Controller to Electric Vehicle

Nonlinear control is used because conventional linear controllers like proportional-integral-derivative (PI) as well as proportional-integral (PID) are inadequate when dealing with substantial input as well as load voltage fluctuations [122]. Despite variations in the system's features and load, sliding mode management maintains stability and resilience. To improve performance, multi-loop control is used instead of basic loop control, however designing the controller is difficult, especially for higher order converter topologies [123]. Sliding mode control and hysteresis control are two examples of nonlinear control approaches that provide very accurate models. To regulate DC/DC boost converters, several current controllers are employed. In order to fine-tune the current characteristics and attain the targeted dynamic performance, average sliding mode control (SMC) is used, as shown in Figure 8. For this reason, the average SMC reference current is derived from PI current control block's output [19].

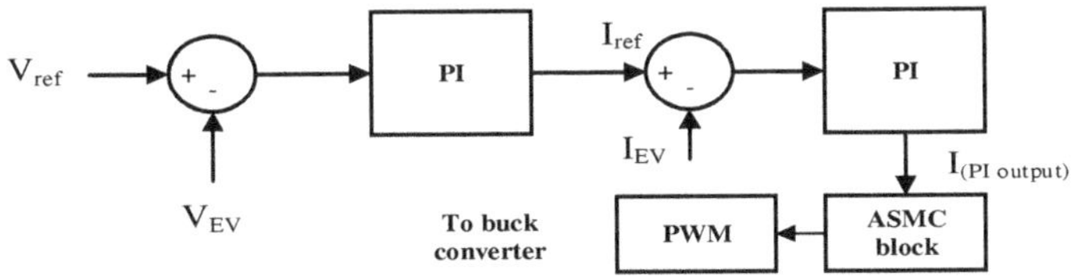

Figure 8. Block diagram of average SMC from input to output [100].

The varying switching frequencies make it very sensitive to overall SMC noise. Few regulated options exist to address the issue of variable switching frequency. After the median SMC has established the allowable signal range, its variable switching frequency is locked in at a fixed value. Figure 9 depicts a typical SMC building block [124]. Output current as well as voltage from the EV are calculated and compared to standards before being sent to the PI controllers [125].

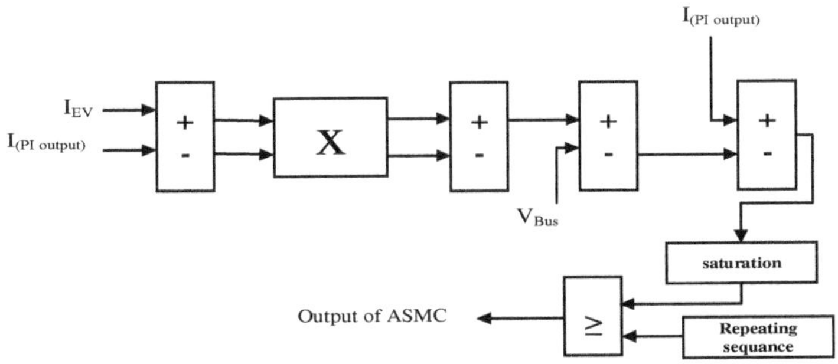

Figure 9. Average SMC for current control in its typical building shape [100].

4. Recent Studies on Various Topologies of Electric Vehicle/Photovoltaic Systems

Table 6 discuses the recent studies of the EV-PV systems.

Table 6. Summary of Topologies of EV/PV Systems based on the recent studies and their methodologies and outcomes.

Reference	Methodology	Outcomes
[126]	a high gain, fast charging DC–DC converter and a control algorithm for grid integrated Solar PV (SPV) based EVCS with battery backup	The proposed converter and its control algorithm's performance are investigated in three different modes using MATLAB/Simulink tool and the simulated results are validated with Real-Time Digital Simulation (RTDS) in OPAL-RT. The observed results meet the Power Quality limits of an IEC61000-3-2 [52] standard and met the level-4 dc charging standard IEC61851 [32]

Table 6. *Cont.*

Reference	Methodology	Outcomes
[127]	minimizing the cost incurred due to energy losses in an IEEE-37 bus system integrated with a commercial EVCS located in Qatar. The Particle Swarm Optimization (PSO) algorithm is used for the efficient location allocation of EVCS.	The system is analytically examined through the Thukaram Load Flow Algorithm and investigations are conducted to observe the beneficial impacts of load balancing between RES and utility
[128]	conceptualizes the interconnection of these components through a 750 V DC nanogrid as against a conventional three-phase 400 V AC system. The factors influencing the performance of a DC-based nanogrid are identified and a comparative analysis with respect to a conventional AC nanogrid is presented in terms of efficiency, stability, and protection.	It is proved how the minimization of grid energy exchange through power management is a vital system design choice. Secondly, the trade-off between stability, protection, and cost for sizing of the DC buffer capacitors is explored. The transient system response to different fault conditions for both AC and DC nanogrid is investigated.
[129]	, an application mode of EV charging network and distributed photovoltaic power generation local consumption is studied. The management idea of two-layer and four model has been established, including the regional distributed photovoltaic output model, electricity consumption model, EV consumption model, and regional grid load dispatching model, which can realize the scheduling of the energy flow formed by photovoltaic, induce the charging of EVs, and make the photovoltaic consumption in office building areas and residential building areas complementary	This mode is intended to guide the consumption of new energy through economic leverage, which can realize the unified regulation of distributed energy convergence, consumption and storage.
[130]	mainly focuses on the various standards for EV, PV systems and their interconnection with grid-connected systems.	For a better operation of EV for domestic and commercial use, it is necessary to follow certain international standards that are published and proved to be efficient. Some of the important standards to be followed in the design and manufacturing of EVs.
[131]	presents the control of a single phase EVCS powered by a wind energy generation system (WEGS) and a PV array that are installed in houses for EV charging and household supply.	Test results have proven the satisfactory operation of the CS for supplying the domestic loads while accomplishing the primary task of charging the EV.
[132]	r reviews the state-of-the-art literature on power electronics converter systems, which interface with the utility grid, PV systems, and EVs. Comparisons are made in terms of their topologies, isolation, power and voltage ranges, efficiency, and bi-directional power capability for V2G operation.	A brief description of EV charger types, their power levels, and standards is provided. It is anticipated that the studies and comparisons in this paper would be advantageous as an all-in-one source of information for researchers seeking information related to EV charging infrastructures.

5. Discussion and Conclusions

This paper provides an overview of PV-EV charging system technology, operation, and status. In addition, it provides information on the principles of electric cars, batteries, and a description of PV. To prove the technological and economic feasibility of PV-grid as well as PV solo charging, a case study is carried out by contrasting them with grid-only charging. It has been determined that PV-grid charging has the ability to create a profit. However, due to the limited capacity of the PV as well as the batteries, the Power system may not be cost effective. Furthermore, since PV is intermittent, it is probable that it will not be able to generate enough electricity to meet consumer demand.

6. Challenges and Future Work Recommendations

6.1. Modeling, Optimization and Control

In spite of the various efforts that have been done on PV–grid charge techniques, it is essential to the alternating nature of PV be recognized. Because of the unpredictability of the sun irradiation, there is no way to know how long the charcoal will keep its consistency. As a result, it is of the utmost importance to design the system to have an energy management function that is optimum. This development of reliable forecast model for power production of PV systems is one topic that is of interest [133]. This may be linked with an accurate model of the cost of grid energy (which is basing on dynamic tariff structures), which will guarantee that the owner of the charging station obtains the highest possible return on their investment [134]. The following works are evidence of the growing interest in the various energy management systems.

Static charging requires electric car owners to notify the station owner in advance of the anticipated demand for charging their vehicles. In more realistic dynamic charging situation, an electric vehicle (EV) may enter and depart the station at any moment, without the owner of the station having previous knowledge of this [135]. In order to accomplish this goal, optimum solutions basing linear programming & heuristic algorithms used, one for static issues & the other for the dynamic ones, respectively [136]. Despite the fact that the dynamic scenario presents a more accurate representation of reality, the answers provided by the static issues may serve as a standard against which performance can be measured. The optimization of electric vehicle charging has a number of possibilities for the use of approaches from soft computing [137].

The optimization of the photovoltaic charging system utilizing the soft computing approaches has a number of difficulties, despite the fact that these methods are successful in themselves. For instance, the fuzzy logic controller is simply appropriate for use in a system that has a limited number of charging outlets and electric vehicles that need to be charged [138]. The rule table will get more complex as well as the algorithm will take longer time to execute if there are a large number of energy sources and EVs. In addition, achieving exact tuning of the fuzzy variable quantity is quite challenging, yet with the assistance of methods such as the genetic algorithm [139].

6.2. Issue on the Integration with Smart Grid System

It has been determined that smart grid technology, which has renewable energy sources & electric vehicles play a vital part, will be that dominant trend in the fields in future power systems. For instance, electricity that is produced by PV may alleviate some of the strain placed on the grid, especially during peak hours [140]. However, utilities' worries about potential unintended consequences, such as impacts on system reliability, power quality, protection, as well as grid synchronisation, have increased in tandem with the widespread use of renewable energy sources. This is all tied up with PV's status as an intermittent energy source. It is regarded to be a big problem for the smart grid system to handle these qualms in PV–EV charging, which gives intriguing opportunities for additional exploration [139].

6.3. Challenges and Suggestions for Electric Vehicle Charging

- Range, refill time, and cost all play a role in determining how popular electric cars are. The availability of charging stations for EVs is crucial to these considerations. Below, we highlight some of the difficulties and recommendations associated with EV charging [141].
- There are not universal manufacturing requirements for charging devices. For instance, there are different standards for charging connectors in Japan, the US, and Europe. Homogeneity in charging standards and equipment may save costs and increase market acceptance of EVs.

- Currently, not each EV models can handle every charging levels, & Warning: not all public charging stations can handle high-power devices. Because of this, those who own electric vehicles have a hard time finding enough charging infrastructure [142].
- Currently, fast charging station users must pay a set monthly demand fee, which discourages EV owners from utilizing their vehicles since they are unable to charge them on demand based on a variable power tariff. Modifying the fixed demand fee policy may be able to alleviate EV owners' complaints [143].
- The charging facility layouts of the charging stations vary since they remain installed by various businesses in various locations. The users find it difficult to adjust to different charging facility layouts. The popularity of EVs will rise with a uniform charging facility structure comparable to ICEV refuelling stations [144].
- Private fast-charging facilities, such those in homes, are still difficult to set up and often need permission from local service companies & govt. This drawn-out procedure discourages EV owners from building their own private fast charging infrastructure for satisfying their needs.
- It's important to strategically deploy EV charging stations along major thoroughfares and in urban areas. Owners of EVs are concerned about the lack of planning for charging stations outside of major cities [145].
- Renewable energy sources, including solar or wind energy, may be used in charging stations. Such charging stations demand a lot of room and expensive design and execution. Vacant lots near roadways are excellent locations for renewable energy-powered EV charging stations [146].

Author Contributions: Conceptualization, A.J.A., M.S. and M.K.; methodology, K.Y. and M.M.; writing—original draft preparation, A.J., M.S. and M.K.; investigation, M.M. and M.K.; resources, A.J.A. and M.S.; writing—review and editing, M.S., M.M. and K.Y.; visualization, M.K. and A.J.A.; supervision, M.S. and M.K.; funding acquisition, M.M. All authors have read and agreed to the published version of the manuscript.

Funding: This research received no external funding.

Institutional Review Board Statement: Not applicable.

Informed Consent Statement: Not applicable.

Data Availability Statement: Data sharing not applicable.

Conflicts of Interest: The authors declare no conflict of interest.

References

1. Diniz, C.G.; de Almeida Souza, A.A.; Santos, D.C.; Dias, M.C.; Da Luz, N.C.; De Moraes, D.R.V.; Maia, J.S.A.; Gomes, A.R.; da Silva Narvaes, I.; Valeriano, D.M.; et al. DETER-B: The new Amazon near real-time deforestation detection system. *IEEE J. Sel. Top. Appl. Earth Obs. Remote Sens.* **2015**, *8*, 3619–3628. [CrossRef]
2. Wu, G.; Sun, H.; Zhang, X.; Egea-Àlvarez, A.; Zhao, B.; Xu, S.; Wang, S.; Zhou, X. Parameter design oriented analysis of the current control stability of the weak-grid-tied VSC. *IEEE Trans. Power Deliv.* **2020**, *36*, 1458–1470. [CrossRef]
3. Ratnam, K.S.; Palanisamy, K.; Yang, G. Future low-inertia power systems: Requirements, issues, and solutions—A review. *Renew. Sustain. Energy Rev.* **2020**, *124*, 109773. [CrossRef]
4. Johnson, S.C.; Rhodes, J.D.; Webber, M.E. Understanding the impact of non-synchronous wind and solar generation on grid stability and identifying mitigation pathways. *Appl. Energy* **2020**, *262*, 114492. [CrossRef]
5. Cheng, Y.; Azizipanah-Abarghooee, R.; Azizi, S.; Ding, L.; Terzija, V. Smart frequency control in low inertia energy systems based on frequency response techniques: A review. *Appl. Energy* **2020**, *279*, 115798. [CrossRef]
6. Johnson, S.C.; Papageorgiou, D.J.; Mallapragada, D.S.; Deetjen, T.A.; Rhodes, J.D.; Webber, M.E. Evaluating rotational inertia as a component of grid reliability with high penetrations of variable renewable energy. *Energy* **2019**, *180*, 258–271. [CrossRef]
7. Khan, S.; Ahmad, A.; Ahmad, F.; Shemami, M.S.; Alam, M.S.; Khateeb, S. A comprehensive review on solar powered electric vehicle charging system. *Smart Sci.* **2018**, *6*, 54–79. [CrossRef]
8. Eldeeb, H.H.; Faddel, S.; Mohammed, O.A. Multi-objective optimization technique for the operation of grid tied PV powered EV charging station. *Electr. Power Syst. Res.* **2018**, *164*, 201–211. [CrossRef]

9. Saleh, B.; Yousef, A.M.; Ebeed, M.; Abo-Elyousr, F.K.; Elnozahy, A.; Mohamed, M.; Abdelwahab, S.A.M. Design of PID controller with grid connected hybrid renewable energy system using optimization algorithms. *J. Electr. Eng. Technol.* **2021**, *16*, 3219–3233. [CrossRef]
10. Venkatasamy, B.; Kalaivani, L. Modeling and Power Quality Analysis of Grid-Connected PV Inverter with Active and Reactive Power Injection Mode. *J. Electr. Eng. Technol.* **2021**, *16*, 1375–1387. [CrossRef]
11. Chen, Y. Research on collaborative innovation of key common technologies in new energy vehicle industry based on digital twin technology. *Energy Rep.* **2022**, *8*, 15399–15407. [CrossRef]
12. Rajamand, S. Synchronous generator control concept and modified droop for frequency and voltage stability of microgrid including inverter-based DGs. *J. Electr. Eng. Technol.* **2020**, *15*, 1035–1044. [CrossRef]
13. Liu, Z.; Wang, Y.; Feng, J. Vehicle-type strategies for manufacturer's car sharing. *Kybernetes* **2022**, *ahead of print*. [CrossRef]
14. Mohamed, A.A.S.; El-Sayed, A.; Metwally, H.; Selem, S.I. Grid integration of a PV system supporting an EV charging station using Salp Swarm Optimization. *Sol. Energy* **2020**, *205*, 170–182. [CrossRef]
15. Zhang, W.; Zheng, Z.; Liu, H. Droop control method to achieve maximum power output of photovoltaic for parallel inverter system. *CSEE J. Power Energy Syst.* **2021**, *8*, 1636–1645.
16. Qiao, M.; Jiang, C.; Zhu, Y.; Li, G. Research on design method and electromagnetic vibration of six-phase fractional-slot concentrated-winding PM motor suitable for ship propulsion. *IEEE Access* **2016**, *4*, 8535–8543. [CrossRef]
17. Jang, Y.; Sun, Z.; Ji, S.; Lee, C.; Jeong, D.; Choung, S.; Bae, S. Grid-Connected Inverter for a PV-Powered Electric Vehicle Charging Station to Enhance the Stability of a Microgrid. *Sustainability* **2021**, *13*, 14022. [CrossRef]
18. Kannan, G.P.; Usha, S. Critical Review on and Analysis of Solar Powered Electric Vehicle Charging Station. *Int. J. Renew. Energy Res.* **2022**, *12*, 581–600.
19. Mastoi, M.S.; Zhuang, S.; Munir, H.M.; Haris, M.; Hassan, M.; Usman, M.; Bukhari, S.S.H.; Ro, J.S. An in-depth analysis of electric vehicle charging station infrastructure, policy implications, and future trends. *Energy Rep.* **2022**, *8*, 11504–11529. [CrossRef]
20. Rajendran, G.; Vaithilingam, C.A.; Misron, N.; Naidu, K.; Ahmed, M.R. A comprehensive review on system architecture and international standards for electric vehicle charging stations. *J. Energy Storage* **2021**, *42*, 103099. [CrossRef]
21. Sachan, S.; Deb, S.; Singh, P.P.; Alam, M.S.; Shariff, S.M. A comprehensive review of standards and best practices for utility grid integration with electric vehicle charging stations. *Wiley Interdiscip. Rev. Energy Environ.* **2022**, *11*, e424. [CrossRef]
22. Khalid, M.R.; Khan, I.A.; Hameed, S.; Asghar, M.S.J.; Ro, J.-S. A comprehensive review on structural topologies, power levels, energy storage systems, and standards for electric vehicle charging stations and their impacts on grid. *IEEE Access* **2021**, *9*, 128069–128094. [CrossRef]
23. Niccolai, A.; Bettini, L.; Zich, R. Optimization of electric vehicles charging station deployment by means of evolutionary algorithms. *Int. J. Intell. Syst.* **2021**, *36*, 5359–5383. [CrossRef]
24. Jin, H.; Lee, S.; Nengroo, S.H.; Har, D. Development of charging/discharging scheduling algorithm for economical and energy-efficient operation of multi-EV charging station. *Appl. Sci.* **2022**, *12*, 4786. [CrossRef]
25. Karolemeas, C.; Tsigdinos, S.; Tzouras, P.G.; Nikitas, A.; Bakogiannis, E. Determining electric vehicle charging station location suitability: A qualitative study of greek stakeholders employing thematic analysis and analytical hierarchy process. *Sustainability* **2021**, *13*, 2298. [CrossRef]
26. Metais, M.-O.; Jouini, O.; Perez, Y.; Berrada, J.; Suomalainen, E. Too much or not enough? Planning electric vehicle charging infrastructure: A review of modeling options. *Renew. Sustain. Energy Rev.* **2022**, *153*, 111719. [CrossRef]
27. Hall, D.; Lutsey, N. *Electric Vehicle Charging Guide for Cities*; ICCT: Washington, DC, USA, 2020.
28. Shariff, S.M.; Alam, M.S.; Ahmad, F.; Rafat, Y.; Asghar, M.S.J.; Khan, S. System design and realization of a solar-powered electric vehicle charging station. *IEEE Syst. J.* **2019**, *14*, 2748–2758. [CrossRef]
29. Laporte, S.; Coquery, G.; Deniau, V.; De Bernardinis, A.; Hautiere, N. Dynamic wireless power transfer charging infrastructure for future evs: From experimental track to real circulated roads demonstrations. *World Electr. Veh. J.* **2019**, *10*, 84. [CrossRef]
30. Guo, L.; Ye, C.; Ding, Y.; Wang, P. Allocation of centrally switched fault current limiters enabled by 5G in transmission system. *IEEE Trans. Power Deliv.* **2020**, *36*, 3231–3241. [CrossRef]
31. Pandey, S. PV Array Based EV Charging Station with Efficient Charging and Storage System. *i-Manag. J. Electron. Eng.* **2022**, *12*, 15.
32. Das, H.S.; Rahman, M.M.; Li, S.; Tan, C.W. Electric vehicles standards, charging infrastructure, and impact on grid integration: A technological review. *Renew. Sustain. Energy Rev.* **2020**, *120*, 109618. [CrossRef]
33. Lu, C.; Zhou, H.; Li, L.; Yang, A.; Xu, C.; Ou, Z.; Wang, J.; Wang, X.; Tian, F. Split-core magnetoelectric current sensor and wireless current measurement application. *Measurement* **2022**, *188*, 110527. [CrossRef]
34. Nour, M.; Chaves-Ávila, J.P.; Magdy, G.; Sánchez-Miralles, Á. Review of positive and negative impacts of electric vehicles charging on electric power systems. *Energies* **2020**, *13*, 4675. [CrossRef]
35. Acharige, S.S.G.; Haque, M.E.; Arif, M.T.; Hosseinzadeh, N. Review of Electric Vehicle Charging Technologies, Configurations, and Architectures. *arXiv* **2022**, arXiv:2209.15242.
36. Habib, S.; Khan, M.M.; Abbas, F.; Ali, A.; Faiz, M.T.; Ehsan, F.; Tang, H. Contemporary trends in power electronics converters for charging solutions of electric vehicles. *CSEE J. Power Energy Syst.* **2020**, *6*, 911–929.
37. Pandey, S. A Review on EV Charging Station using PV Source. *Int. J. Res. Eng. Sci.* **2022**, *10*, 28–35.

38. Khaligh, A.; D'Antonio, M. Global trends in high-power on-board chargers for electric vehicles. *IEEE Trans. Veh. Technol.* **2019**, *68*, 3306–3324. [CrossRef]
39. Rastogi, S.K.; Sankar, A.; Manglik, K.; Mishra, S.K.; Mohanty, S.P. Toward the Vision of All-Electric Vehicles in a Decade [Energy and Security]. *IEEE Consum. Electron. Mag.* **2019**, *8*, 103–107. [CrossRef]
40. Sanguesa, J.A.; Torres-Sanz, V.; Garrido, P.; Martinez, F.J.; Marquez-Barja, J.M. A review on electric vehicles: Technologies and challenges. *Smart Cities* **2021**, *4*, 372–404. [CrossRef]
41. Fachrizal, R.; Shepero, M.; Åberg, M.; Munkhammar, J. Optimal PV-EV sizing at solar powered workplace charging stations with smart charging schemes considering self-consumption and self-sufficiency balance. *Appl. Energy* **2022**, *307*, 118139. [CrossRef]
42. Mohamed, A.A.S.; Shaier, A.A.; Metwally, H.; Selem, S.I. An overview of dynamic inductive charging for electric vehicles. *Energies* **2022**, *15*, 5613. [CrossRef]
43. Huang, N.; Wang, S.; Wang, R.; Cai, G.; Liu, Y.; Dai, Q. Gated spatial-temporal graph neural network based short-term load forecasting for wide-area multiple buses. *Int. J. Electr. Power Energy Syst.* **2023**, *145*, 108651. [CrossRef]
44. Liu, S.; Song, Z.; Liu, Y.; Chen, Y.; Liu, C. Flux-Weakening Controller Design of Dual Three-Phase PMSM Drive System With Copper Loss Minimization. *IEEE Trans. Power Electron.* **2022**, *38*, 2351–2363. [CrossRef]
45. Tang, H.; Di, J.; Wu, Z.; Li, W. Temperature analysis for the asymmetric six-phase permanent magnet synchronous motor in healthy and fault-tolerant modes. *IEEE Trans. Ind. Electron.* **2022**, *70*, 6482–6493. [CrossRef]
46. Fachrizal, R.; Ramadhani, U.H.; Munkhammar, J.; Widén, J. Combined PV–EV hosting capacity assessment for a residential LV distribution grid with smart EV charging and PV curtailment. *Sustain. Energy Grids Netw.* **2021**, *26*, 100445. [CrossRef]
47. Han, Y.; Wang, B.; Guan, T.; Tian, D.; Yang, G.; Wei, W.; Tang, H.; Chuah, J.H. Research on road environmental sense method of intelligent vehicle based on tracking check. *IEEE Trans. Intell. Transp. Syst.* **2022**, *24*, 1261–1275. [CrossRef]
48. Yuan, J.; Dorn-Gomba, L.; Callegaro, A.D.; Reimers, J.; Emadi, A. A review of bidirectional on-board chargers for electric vehicles. *IEEE Access* **2021**, *9*, 51501–51518. [CrossRef]
49. Pan, L.; Yao, E.; MacKenzie, D. Modeling EV charging choice considering risk attitudes and attribute non-attendance. *Transp. Res. Part C Emerg. Technol.* **2019**, *102*, 60–72. [CrossRef]
50. Abdullah, H.M.; Gastli, A.; Ben-Brahim, L.; Mohammed, S.O. Planning and optimizing electric-vehicle charging infrastructure through system dynamics. *IEEE Access* **2022**, *10*, 17495–17514. [CrossRef]
51. Amjad, M.; Farooq-i-Azam, M.; Ni, Q.; Dong, M.; Ansari, E.A. Wireless charging systems for electric vehicles. *Renew. Sustain. Energy Rev.* **2022**, *167*, 112730. [CrossRef]
52. Suarez, C.; Martinez, W. Fast and ultra-fast charging for battery electric vehicles—A review. In Proceedings of the 2019 IEEE Energy Conversion Congress and Exposition (ECCE), Baltimore, MA, USA, 29 September–3 October 2019; pp. 569–575.
53. Fang, Y.; Min, H.; Wu, X.; Wang, W.; Zhao, X.; Mao, G. On-ramp merging strategies of connected and automated vehicles considering communication delay. *IEEE Trans. Intell. Transp. Syst.* **2022**, *23*, 15298–15312. [CrossRef]
54. Chen, P.; Wu, J.; Li, N. A Personalized Navigation Route Recommendation Strategy Based on Differential Perceptron Tracking User's Driving Preference. *Comput. Intell. Neurosci.* **2023**, *2023*, 8978398. [CrossRef]
55. Saldaña, G.; Martin, J.I.S.; Zamora, I.; Asensio, F.J.; Oñederra, O. Electric vehicle into the grid: Charging methodologies aimed at providing ancillary services considering battery degradation. *Energies* **2019**, *12*, 2443. [CrossRef]
56. Swain, D.R.; Biswal, S.S.; Rout, P.K.; Ray, P.K.; Jena, R.K. Optimal fractional sliding mode control for the frequency stability of a hybrid industrial microgrid. *AIMS Electron. Electr. Eng.* **2023**, *7*, 14–37. [CrossRef]
57. Ishak, M.F.; Yahaya, J.A.-F. Harmonic Analysis and Mitigation of Mass Adoption for Electric Vehicle Charging Station (100kW) in Distribution System. *Malays. J. Sci. Adv. Technol.* **2021**, *2021*, 57–61. [CrossRef]
58. Darbandsari, A.; Askarzadeh, A.R. Optimal Energy Management of a Smart Building Considering Rooftop Photovoltaic, Electric Vehicles and Schedulable Loads. In Proceedings of the 2022 9th Iranian Conference on Renewable Energy & Distributed Generation (ICREDG), Mashhad, Iran, 23–24 February 2022; pp. 1–8.
59. Hariri, A.O.; El Hariri, M.; Youssef, T.; Mohammed, O.A. A bilateral decision support platform for public charging of connected electric vehicles. *IEEE Trans. Veh. Technol.* **2018**, *68*, 129–140. [CrossRef]
60. Von Bonin, M.; Dörre, E.; Al-Khzouz, H.; Braun, M.; Zhou, X. Impact of dynamic electricity tariff and home PV system incentives on Electric Vehicle Charging Behavior: Study on potential grid implications and economic effects for households. *Energies* **2022**, *15*, 1079. [CrossRef]
61. Srivastava, M.; Tomar, P.S.; Verma, A.K. Emphasis on switch selection and its switching loss comparison for on-board electric vehicle charger. *IET Power Electron.* **2019**, *12*, 1385–1392. [CrossRef]
62. Taghizadeh, S.; Hossain, M.J.; Lu, J.; Water, W. A unified multi-functional on-board EV charger for power-quality control in household networks. *Appl. Energy* **2018**, *215*, 186–201. [CrossRef]
63. Mou, X.; Gladwin, D.T.; Zhao, R.; Sun, H. Survey on magnetic resonant coupling wireless power transfer technology for electric vehicle charging. *IET Power Electron.* **2019**, *12*, 3005–3020. [CrossRef]
64. Al-Thani, H.; Koç, M.; Isaifan, R.J.; Bicer, Y. A Review of the Integrated Renewable Energy Systems for Sustainable Urban Mobility. *Sustainability* **2022**, *14*, 10517. [CrossRef]
65. Parchomiuk, M.; Moradewicz, A.; Gawiński, H. An overview of electric vehicles fast charging infrastructure. In Proceedings of the 2019 Progress in Applied Electrical Engineering (PAEE), Koscielisko, Poland, 17–21 June 2019; pp. 1–5.

66. Kumar, K.J.; Kumar, S.; Nandakumar, V.S. Standards for electric vehicle charging stations in India: A review. *Energy Storage* **2022**, *4*, e261. [CrossRef]
67. Savari, G.F.; Sathik, M.J.; Raman, L.A.; El-Shahat, A.; Hasanien, H.M.; Almakhles, D.; Aleem, S.H.A.; Omar, A.I. Assessment of charging technologies, infrastructure and charging station recommendation schemes of electric vehicles: A review. *Ain Shams Eng. J.* **2022**, *2022*, 101938. [CrossRef]
68. Ali, S.S.; Rawdah, R.; Hasan, K.N. An Overview of Electric Vehicle Charging Data Acquisition and Grid Connection Standards for Power System Studies and EV-Grid Integration. In Proceedings of the 2021 31st Australasian Universities Power Engineering Conference (AUPEC), Virtual, 26–30 September 2021; pp. 1–6.
69. Arif, S.M.; Lie, T.T.; Seet, B.C.; Ayyadi, S.; Jensen, K. Review of electric vehicle technologies, charging methods, standards and optimization techniques. *Electronics* **2021**, *10*, 1910. [CrossRef]
70. Tavakoli, A.; Saha, S.; Arif, M.T.; Haque, M.E.; Mendis, N.; Oo, A.M.T. Impacts of grid integration of solar PV and electric vehicle on grid stability, power quality and energy economics: A review. *IET Energy Syst. Integr.* **2020**, *2*, 243–260. [CrossRef]
71. Rizvi, S.A.A.; Xin, A.; Masood, A.; Iqbal, S.; Jan, M.U.; Rehman, H. Electric vehicles and their impacts on integration into power grid: A review. In Proceedings of the 2018 2nd IEEE Conference on Energy Internet and Energy System Integration (EI2), Beijing, China, 20–22 October 2018; pp. 1–6.
72. Shongwe, S.; Hanif, M. Comparative analysis of different single-diode PV modeling methods. *IEEE J. Photovolt.* **2015**, *5*, 938–946. [CrossRef]
73. Shannan, N.M.A.A.; Yahaya, N.Z.; Singh, B. Single-diode model and two-diode model of PV modules: A comparison. In Proceedings of the 2013 IEEE International Conference on Control System, Computing and Engineering, Penang, Malaysia, 29 November–1 December 2013; pp. 210–214.
74. Kandpal, B.; Pareek, P.; Verma, A. A robust day-ahead scheduling strategy for EV charging stations in unbalanced distribution grid. *Energy* **2022**, *249*, 123737. [CrossRef]
75. Mohammad, A.; Zamora, R.; Lie, T.T. Integration of electric vehicles in the distribution network: A review of PV based electric vehicle modelling. *Energies* **2020**, *13*, 4541. [CrossRef]
76. Azam, F.; Priyadarshi, N.; Nagar, H.; Kumar, S.; Bhoi, A.K. An Overview of Solar-Powered Electric Vehicle Charging in Vehicular Adhoc Network. In *Electric Vehicles. Green Energy and Technology*; Springer: Singapore, 2021; pp. 95–102.
77. Ahmed, S.I.; Salehfar, H.; Selveraj, D.F. Grid Integration of PV Based Electric Vehicle Charging Stations: A Brief Review. In Proceedings of the 2022 North American Power Symposium (NAPS), Salt Lake City, UT, USA, 9–11 October 2022; pp. 1–6.
78. Wu, Y.; Aziz, S.M.; Haque, M.H. Techno-economic modelling for energy cost optimisation of households with electric vehicles and renewable sources under export limits. *Renew. Energy* **2022**, *198*, 1254–1266. [CrossRef]
79. Ibrahim, M.M. Investigation of a grid-connected solar pv system for the electric-vehicle charging station of an office building using pvsol software. *Polityka Energy* **2022**, *25*, 4235. [CrossRef]
80. Hajar, K.; Guo, B.; Hably, A.; Bacha, S. Smart charging impact on electric vehicles in presence of photovoltaics. In Proceedings of the 2021 22nd IEEE International Conference on Industrial Technology (ICIT), Valencia, Spain, 10–12 March 2021; Volume 1, pp. 643–648.
81. Guidi, G.; D'Arco, S.; Suul, J.A.; Iso, R.; Itoh, J.-I. A modular multilevel interface for transformerless grid integration of large-scale infrastructure for wireless electric vehicle charging. In Proceedings of the 2019 10th International Conference on Power Electronics and ECCE Asia (ICPE 2019–ECCE Asia), Busan, Republic of Korea, 27 May 2019; pp. 2059–2066.
82. Datta, U.; Kalam, A.; Shi, J. Smart control of BESS in PV integrated EV charging station for reducing transformer overloading and providing battery-to-grid service. *J. Energy Storage* **2020**, *28*, 101224. [CrossRef]
83. Fachrizal, R.; Munkhammar, J. Improved photovoltaic self-consumption in residential buildings with distributed and centralized smart charging of electric vehicles. *Energies* **2020**, *13*, 1153. [CrossRef]
84. Li, S.; Ye, C.; Ding, Y.; Song, Y. Reliability Assessment of Renewable Power Systems Considering Thermally-Induced Incidents of Large-Scale Battery Energy Storage. *IEEE Trans. Power Syst.* **2022**, 1–15. [CrossRef]
85. Tran, V.T.; Islam, M.R.; Muttaqi, K.M.; Sutanto, D. An efficient energy management approach for a solar-powered EV battery charging facility to support distribution grids. *IEEE Trans. Ind. Appl.* **2019**, *55*, 6517–6526. [CrossRef]
86. Zhao, P.; Ma, K.; Yang, J.; Yang, B.; Guerrero, J.M.; Dou, C.; Guan, X. Distributed Power Sharing Control Based on Adaptive Virtual Impedance in Seaport Microgrids with Cold Ironing. *IEEE Trans. Transp. Electrif.* **2022**. [CrossRef]
87. Abdalla, M.A.A.; Min, W.; Mohammed, O.A.A. Two-stage energy management strategy of EV and PV integrated smart home to minimize electricity cost and flatten power load profile. *Energies* **2020**, *13*, 6387. [CrossRef]
88. Joseph, P.K.; Devaraj, E.; Gopal, A. Overview of wireless charging and vehicle-to-grid integration of electric vehicles using renewable energy for sustainable transportation. *IET Power Electron.* **2019**, *12*, 627–638. [CrossRef]
89. Du, Y.; Qin, B.; Zhao, C.; Zhu, Y.; Cao, J.; Ji, Y. A novel spatio-temporal synchronization method of roadside asynchronous MMW radar-camera for sensor fusion. *IEEE Trans. Intell. Transp. Syst.* **2021**, *23*, 22278–22289. [CrossRef]
90. Mateen, S.; Haque, A.; Kurukuru, V.S.B.; Khan, M.A. Discrete Stochastic Control for Energy Management with Photovoltaic Electric Vehicle Charging Station. *CPSS Trans. Power Electron. Appl.* **2022**, *7*, 216–225. [CrossRef]
91. Patil, H.; Kalkhambkar, V.N. Grid integration of electric vehicles for economic benefits: A review. *J. Mod. Power Syst. Clean Energy* **2020**, *9*, 13–26. [CrossRef]

92. Liu, Y.; Zhang, D.; Gooi, H.B. Optimization strategy based on deep reinforcement learning for home energy management. *CSEE J. Power Energy Syst.* **2020**, *6*, 572–582.
93. Hu, J.; Liu, D.; Du, C.; Yan, F.; Lv, C. Intelligent energy management strategy of hybrid energy storage system for electric vehicle based on driving pattern recognition. *Energy* **2020**, *198*, 117298. [CrossRef]
94. Ge, L.; Du, T.; Li, C.; Li, Y.; Yan, J.; Rafiq, M.U. Virtual collection for distributed photovoltaic data: Challenges, methodologies, and applications. *Energies* **2022**, *15*, 8783. [CrossRef]
95. Rivera, S.; Kouro, S.; Vazquez, S.; Goetz, S.M.; Lizana, R.; Romero-Cadaval, E. Electric vehicle charging infrastructure: From grid to battery. *IEEE Ind. Electron. Mag.* **2021**, *15*, 37–51. [CrossRef]
96. Han, Y.; Wang, L.; Kang, R. Influence of consumer preference and government subsidy on prefabricated building developer's decision-making: A three-stage game model. *J. Civ. Eng. Manag.* **2023**, *29*, 35–49. [CrossRef]
97. Manickavasagam, K.; Thotakanama, N.K.; Puttaraj, V. Intelligent energy management system for renewable energy driven ship. *IET Electr. Syst. Transp.* **2019**, *9*, 24–34. [CrossRef]
98. Chen, J.; Sun, B.; Li, Y.; Jing, R.; Zeng, Y.; Li, M. Credible capacity calculation method of distributed generation based on equal power supply reliability criterion. *Renew. Energy* **2022**, *201*, 534–547. [CrossRef]
99. Jha, M.; Blaabjerg, F.; Khan, M.A.; Kurukuru, V.S.B.; Haque, A. Intelligent control of converter for electric vehicles charging station. *Energies* **2019**, *12*, 2334. [CrossRef]
100. De los Mozos, A.B.; Mouli, G.R.C.; Bauer, P. Evaluation of topologies for a solar powered bidirectional electric vehicle charger. *IET Power Electron.* **2019**, *12*, 3675–3687. [CrossRef]
101. Sami, B.S. Intelligent energy management for off-grid renewable hybrid system using multi-agent approach. *IEEE Access* **2020**, *8*, 8681–8696. [CrossRef]
102. Ugwu, J.; Odo, K.C.; Ohanu, C.P.; García, J.; Georgious, R. Comprehensive Review of Renewable Energy Communication Modeling for Smart Systems. *Energies* **2022**, *16*, 409. [CrossRef]
103. Ma, C.-T. System planning of grid-connected electric vehicle charging stations and key technologies: A review. *Energies* **2019**, *12*, 4201. [CrossRef]
104. Papas, I.; Estibals, B.; Ecrepont, C.; Alonso, C. Energy consumption optimization through dynamic simulations for an intelligent energy management of a BIPV building. In Proceedings of the 2018 7th International Conference on Renewable Energy Research and Applications (ICRERA), Paris, France, 14–17 October 2018; pp. 853–857.
105. Liu, T. Reinforcement learning-enabled intelligent energy management for hybrid electric vehicles. In *Synthesis Lectures on Advances in Automotive Technology (SLAAT)*; Springer: Berlin/Heidelberg, Germany, 2019; Volume 3, pp. 1–99.
106. Han, Y.; Yan, X.; Piroozfar, P. An overall review of research on prefabricated construction supply chain management. *Eng. Constr. Archit. Manag.* **2022**. *ahead of print*. [CrossRef]
107. Hannan, M.A.; Mollik, M.S.; Al-Shetwi, A.Q.; Rahman, S.A.; Mansor, M.; Begum, R.A.; Muttaqi, K.M.; Dong, Z.Y. Vehicle to grid connected technologies and charging strategies: Operation, control, issues and recommendations. *J. Clean. Prod.* **2022**, *339*, 130587. [CrossRef]
108. Trinh, H.A.; Nguyen, D.G.; Phan, V.D.; Duong, T.Q.; Truong, H.V.A.; Choi, S.J.; Ahn, K.K. Robust Adaptive Control Strategy for a Bidirectional DC-DC Converter Based on Extremum Seeking and Sliding Mode Control. *Sensors* **2023**, *23*, 457. [CrossRef]
109. Zhang, Z.; Li, W.; Yang, J. Analysis of stochastic process to model safety risk in construction industry. *J. Civ. Eng. Manag.* **2021**, *27*, 87–99. [CrossRef]
110. Aydin, E.; Aydemir, M.T.; Aksoz, A.; El Baghdadi, M.; Hegazy, O. Inductive Power Transfer for Electric Vehicle Charging Applications: A Comprehensive Review. *Energies* **2022**, *15*, 4962. [CrossRef]
111. Lim, K.L.; Speidel, S.; Bräunl, T. A comparative study of AC and DC public electric vehicle charging station usage in Western Australia. *Renew. Sustain. Energy Transit.* **2022**, *2*, 100021. [CrossRef]
112. Han, Y.; Xu, X.; Zhao, Y.; Wang, X.; Chen, Z.; Liu, J. Impact of consumer preference on the decision-making of prefabricated building developers. *J. Civ. Eng. Manag.* **2022**, *28*, 166–176. [CrossRef]
113. Das, H.S.; Salem, M.; Zainuri, M.A.A.M.; Dobi, A.M.; Li, S.; Ullah, H. A comprehensive review on power conditioning units and control techniques in fuel cell hybrid systems. *Energy Rep.* **2022**, *8*, 14236–14258. [CrossRef]
114. Qin, Y. Numerical modeling of energy storage unit during freezing of paraffin utilizing Al_2O_3 nanoparticles and Y-shape fin. *J. Energy Storage* **2021**, *44*, 103452. [CrossRef]
115. Zhang, M.; Yang, D.; Du, J.; Sun, H.; Li, L.; Wang, L.; Wang, K. A Review of SOH Prediction of Li-Ion Batteries Based on Data-Driven Algorithms. *Energies* **2023**, *16*, 3167. [CrossRef]
116. Zhang, M.; Liu, Y.; Li, D.; Cui, X.; Wang, L.; Li, L.; Wang, K. Electrochemical Impedance Spectroscopy: A New Chapter in the Fast and Accurate Estimation of the State of Health for Lithium-Ion Batteries. *Energies* **2023**, *16*, 1599. [CrossRef]
117. Jia, L.; Hu, Z.; Song, Y.; Luo, Z. Optimal siting and sizing of electric vehicle charging stations. In Proceedings of the 2012 IEEE International Electric Vehicle Conference, Greenville, SC, USA, 4–8 March 2012; pp. 1–6.
118. Bagheri, V.; Ehyaei, A.F.; Haeri, M. A Decentralized Energy Management Method for Load Curve Smoothing Considering Demand and Profit of Electric Vehicle Owners with Different Capacity of Batteries. *J. Oper. Autom. Power Eng.* **2023**, *11*, 223–229.
119. Guo, C.; Ye, C.; Ding, Y.; Wang, P. A multi-state model for transmission system resilience enhancement against short-circuit faults caused by extreme weather events. *IEEE Trans. Power Deliv.* **2020**, *36*, 2374–2385. [CrossRef]

120. Wang, B.; Dehghanian, P.; Wang, S.; Mitolo, M. Electrical safety considerations in large-scale electric vehicle charging stations. *IEEE Trans. Ind. Appl.* **2019**, *55*, 6603–6612. [CrossRef]
121. Shahapure, S.B.; Kulkarni, V.A.; Shinde, S.M. A Technology Review of Energy Storage Systems, Battery Charging Methods and Market Analysis of EV Based on Electric Drives. *Development* **2022**, *6*, 8. [CrossRef]
122. Zafar, S.; Blavette, A.; Camilleri, G.; Ahmed, H.B.; Agbodjan, J.-J.P. Decentralized optimal management of a large-scale EV fleet: Optimality and computational complexity comparison between an Adaptive MAS and MILP. *Int. J. Electr. Power Energy Syst.* **2023**, *147*, 108861. [CrossRef]
123. Liu, S.; Song, Z.; Dong, Z.; Liu, Y.; Liu, C. Generic carrier-based PWM solution for series-end winding PMSM traction system with adaptative overmodulation scheme. *IEEE Trans. Transp. Electrif.* **2022**, *9*, 712–726. [CrossRef]
124. Turan, M.T.; Gökalp, E. Integration Analysis of Electric Vehicle Charging Station Equipped with Solar Power Plant to Distribution Network and Protection System Design. *J. Electr. Eng. Technol.* **2022**, *17*, 903–912. [CrossRef]
125. Marinescu, C. Progress in the Development and Implementation of Residential EV Charging Stations Based on Renewable Energy Sources. *Energies* **2023**, *16*, 179. [CrossRef]
126. Prem, P.; Sivaraman, P.; Raj, J.S.S.S.; Sathik, M.J.; Almakhles, D. Fast charging converter and control algorithm for solar PV battery and electrical grid integrated electric vehicle charging station. *Automatika* **2020**, *61*, 614–625. [CrossRef]
127. Sultana, U.; Mujahid, A.; Jilani, H.A.; Perveen, U. Cost Minimization in Radial Distribution System Integrated with Commercial Electric Vehicle Charging Station. *Eng. Proc.* **2022**, *20*, 15. [CrossRef]
128. Sulaeman, I.; Mouli, G.R.C.; Shekhar, A.; Bauer, P. Comparison of ac and dc nanogrid for office buildings with ev charging, pv and battery storage. *Energies* **2021**, *14*, 5800. [CrossRef]
129. He, Q.; Wu, M.; Sun, P.; Guo, J.; Chen, L.; Jiang, L.; Zhang, Z. Research on a Charging Mechanism of Electric Vehicles for Photovoltaic Nearby Consumption Strategy. *Electronics* **2022**, *11*, 3407. [CrossRef]
130. Sunddararaj, S.P.; Rangarajan, S.S.; Nallusamy, S.; Collins, E.R.; Senjyu, T. A brief survey on important interconnection standards for photovoltaic systems and electric vehicles. *World Electr. Veh. J.* **2021**, *12*, 117. [CrossRef]
131. Singh, B.; Verma, A. Control of Solar PV and WEGS Powered EV Charging Station. In Proceedings of the 2020 IEEE Transportation Electrification Conference and Expo (ITEC), Virtual, 22 June 2020; pp. 609–614. [CrossRef]
132. Esfahani, F.N.; Darwish, A.; Williams, B.W. Power Converter Topologies for Grid-Tied Solar Photovoltaic (PV) Powered Electric Vehicles (EVs)—A Comprehensive Review. *Energies* **2022**, *15*, 4648. [CrossRef]
133. Huang, Z.; Xie, Z.; Zhang, C.; Chan, S.H.; Milewski, J.; Xie, Y.; Yang, Y.; Hu, X. Modeling and multi-objective optimization of a stand-alone PV-hydrogen-retired EV battery hybrid energy system. *Energy Convers. Manag.* **2019**, *181*, 80–92. [CrossRef]
134. Nutakki, M.; Mandava, S. Review on optimization techniques and role of Artificial Intelligence in home energy management systems. *Eng. Appl. Artif. Intell.* **2023**, *119*, 105721. [CrossRef]
135. Anaya-Ruiz, G.A.; Robles, D.R.; Caballero, L.E.U.; Moreno-Goytia, E.L. Design and prototyping of transformerless DC-DC converter with high voltage ratio for MVDC applications. *IEEE Lat. Am. Trans.* **2023**, *21*, 62–70. [CrossRef]
136. Gomes, I.S.F.; Perez, Y.; Suomalainen, E. Coupling small batteries and PV generation: A review. *Renew. Sustain. Energy Rev.* **2020**, *126*, 109835. [CrossRef]
137. Hemavathi, S.; Shinisha, A. A study on trends and developments in electric vehicle charging technologies. *J. Energy Storage* **2022**, *52*, 105013. [CrossRef]
138. Xie, W.; Luo, W.; Qin, Y. Integrated DC/DC Converter Topology Study for Fuel Cell Hybrid Vehicles with Two Energy Sources. *World Electr. Veh. J.* **2022**, *14*, 9. [CrossRef]
139. Hassoune, A.; Khafallah, M.; Mesbahi, A.; Bouragba, T. Power management strategies of electric vehicle charging station based grid tied PV-battery system. *Int. J. Renew. Energy Res.* **2018**, *8*, 851–860.
140. Taghavi, S.S.; Rezvanyvardom, M.; Mirzaei, A.; Gorji, S.A. High Step-Up Three-Level Soft Switching DC-DC Converter for Photovoltaic Generation Systems. *Energies* **2023**, *16*, 41. [CrossRef]
141. Mollah, M.B.; Zhao, J.; Niyato, D.; Lam, K.Y.; Zhang, X.; Ghias, A.M.; Koh, L.H.; Yang, L. Blockchain for future smart grid: A comprehensive survey. *IEEE Internet Things J.* **2020**, *8*, 18–43. [CrossRef]
142. Sadeghian, O.; Oshnoei, A.; Mohammadi-Ivatloo, B.; Vahidinasab, V.; Anvari-Moghaddam, A. A comprehensive review on electric vehicles smart charging: Solutions, strategies, technologies, and challenges. *J. Energy Storage* **2022**, *54*, 105241. [CrossRef]
143. Moon, H.E.; Ha, Y.H.; Kim, K.N. Comparative Economic Analysis of Solar PV and Reused EV Batteries in the Residential Sector of Three Emerging Countries—The Philippines, Indonesia, and Vietnam. *Energies* **2023**, *16*, 311. [CrossRef]
144. Rathor, S.K.; Saxena, D. Energy management system for smart grid: An overview and key issues. *Int. J. Energy Res.* **2020**, *44*, 4067–4109. [CrossRef]
145. Rancilio, G.; Bovera, F.; Delfanti, M. Tariff-based regulatory sandboxes for EV smart charging: Impacts on the tariff and the power system in a national framework. *Int. J. Energy Res.* **2022**, *46*, 14794–14813. [CrossRef]
146. Kakran, S.; Chanana, S. Smart operations of smart grids integrated with distributed generation: A review. *Renew. Sustain. Energy Rev.* **2018**, *81*, 524–535. [CrossRef]

MDPI AG
Grosspeteranlage 5
4052 Basel
Switzerland
Tel.: +41 61 683 77 34

Sustainability Editorial Office
E-mail: sustainability@mdpi.com
www.mdpi.com/journal/sustainability

www.ingramcontent.com/pod-product-compliance
Lightning Source LLC
Chambersburg PA
CBHW061450210926
51850CB00008B/1675
9783725854059